Hamiltonian Dynamics

Hamiltonian Dynamics

Gaetano Vilasi
University of Salerno

World Scientific
Singapore • New Jersey • London • Hong Kong

Published by

World Scientific Publishing Co. Pte. Ltd.
5 Toh Tuck Link, Singapore 596224
USA office: 27 Warren Street, Suite 401-402, Hackensack, NJ 07601
UK office: 57 Shelton Street, Covent Garden, London WC2H 9HE

British Library Cataloguing-in-Publication Data
A catalogue record for this book is available from the British Library.

HAMILTONIAN DYNAMICS

ISBN-13 978-981-02-3308-2
ISBN-10 981-02-3308-6

Preface

There are many books on classical mechanics. They can be roughly divided into two classes. One contains books which, in order to be more accessible, are sometimes less transparent with respect to the underlying theoretical structures; the other contains books giving the general, analytical and geometrical, structures of classical mechanics. In the latter, due to greater complexity of the mathematical tools involved, it is however difficult to find books suitable for teaching the subject to graduate students, often because they do not contain a teaching proposal but rather they appear to be written by authors for their colleagues. This book is intended to belong to the second class, but without the shortcoming that was just mentioned.

Part I, Part II and, partially, Part III are intended to be a teaching proposal suitable for graduate students. Thus, they are written from the point of view of a student but with the aim of giving a general understanding of the theory.

Part IV, instead, is concerned with the current research topic of completely integrable field theories and could be even used independently of the others. This part is not written with the same pedagogic spirit that animates the previous chapters and probably it would have required additional chapters concerning the Lagrangian and the Hamiltonian formulation of field theory. However, a pedagogic treatment of the last subject would have taken too much space–time.

I am grateful to my friends:

Giuseppe Marmo, for the invaluable help in reading the manuscript, criticism and important suggestions and for the very many years of common efforts toward an understanding of complete integrability in field theory.

Giovanni Sparano and Alexandre M. Vinogradov, for criticism and several remarks in special topics.

Vladimir S. Guerdjikov, Mario Rasetti and Geoffrey L. Sewell who strongly encouraged me to write this book.

I also wish to thank my friends:

Sergio De Filippo and Gianni Landi for many years of collaboration.

Finally, I wish to thank Roberto De Luca whose expertise, both in Physics and English, allows me to offer a readable final version. Of course, I am the only one responsible of remaining mistakes.

G.Vilasi
Salerno, April 1998

Introduction

A large amount of scientific activity has been devoted to the asymptotic and geometrical analysis of dynamical systems. This interest was born towards the close of the nineteenth century after the publication of *Les Méthodes Nouvelles de la Mécanique Céleste* in 1892, by Henri Poincaré. The proposed new methods insist on interpreting differential equations as integral curves of vector fields on manifolds, and to analyze the problems concerning long term stability of a dynamical system, for instance, the solar system, by studying their topological properties.

Henri Poincaré was the first to recognize the extraordinarily complicated behavior (today known as *chaos*) of orbits in the vicinity of a separatrix, whose analysis needed the introduction of entirely new mathematics.

Poincaré's suggestion lies in the origin of modern topology, with its powerful tools consisting of tangent and cotangent bundles, differential forms, exterior algebra and calculus, homology and cohomology. All such notions are usually associated with general relativity, string theories, or gauge theories, and not with their main source, *Classical Mechanics.*

On the other hand, in the last few decades there has been a renewed interest in completely integrable Hamiltonian systems, whose concept goes back to the last century, and which, loosely speaking, are dynamical systems admitting a Hamiltonian description and possessing sufficiently many constants of motion, so that they can be integrated by quadratures.

This interest, which previously had considerably weakened, for the really exiguous number of physically prominent examples of completely integrable dynamics with finitely many degrees of freedom, revived with the discovery of

Lax Representation and *Inverse Scattering Method.* The Lax Representation made possible the solution of many problems of remarkable physical interest as the ones described, for instance, by *Korteweg-de Vries*, *sine-Gordon*, *nonlinear Schrödinger* equations and *Toda's lattice.* All such dynamics are Hamiltonian dynamics on infinite dimensional *weakly-symplectic* manifolds, on which the classical Liouville criterion of integrability can be extended in terms of mixed tensor fields with vanishing Nijenhuis torsion.

Peculiar, in the approach to integrability in terms of invariant mixed tensor fields, is the direct construction of abelian maximal algebras of symmetries, leaving out the associated groups, so that only the algebraic aspect of the traditional methodology is reproduced. An exemplary case is given by the Kepler dynamics, in which both the integrability and the degeneration, classical and quantum, are inferred by identifying the corresponding invariance groups ($SO(3)$ and $SO(4)$).

On the other hand, it is just by means of the modern theory of Hamiltonian systems, based on the analysis of symmetries, that an algebraic group approach arises from the analysis of Lax dynamics. This approach arises from the observation that Hamiltonian dynamics, on the orbits of the coadjoint representation of a Lie group endowed with their natural symplectic structure, are Lax dynamics, provided that an internal product, invariant under the adjoint action, exists in the Lie algebra. The group approach analysis, even if from one side has the merit to be constructive, on the other, is not fit to investigate, *a priori*, the possible integrability of a given dynamics.

In these lectures we shall look at this geometric approach to the study of Hamiltonian dynamical systems, specially in connection with the kinds of problems which arise in completely integrable 2-dimensional field theories.

It would have been interesting to include a chapter concerning nonintegrable dynamics, an essential topic for the theory of particles dynamics in accelerators.

However, this last one is a vast subject and goes beyond the purposes of this book. We will spend however a few words to delineate the idea of invariant tori in phase space, to define and illustrate the structures for organizing dynamics and the origin of chaotic orbits in nonintegrable systems.

Finally, I also hope to lessen the impression, sometimes due to a formal approach, that classical mechanics is a closed subject with no mysteries left to explore.

Contents

Part I
Analytical Mechanics

The aim of this part is a self-contained treatment of Classical Mechanics in an advanced formulation. Many topics relevant to applications will not be treated, since they can be found in excellent textbooks.[9,21,23,24,32,38,39,47,57] Our treatment is inspired by two important classical textbooks, *Lezioni di Meccanica Razionale* by T. Levi-Civita and U. Amaldi, and *The Analytical Foundations of Celestial Mechanics* by A. Wintner.[36,58]

Definitions will be given for a *particle*, i.e. for a body whose space dimensions can be neglected with respect to the dimensions of the space in which it moves, and naturally extended to systems of particles and to continuous systems (*fields*). The simplicity of the formal extension from systems of particles to fields, and the difficulties for a rigorous extension, will limit the treatment to systems of particles.

Chapter 1

The Lagrangian Coordinates

1.1 A Primer for Various Formulations of Dynamics

1.1.1 *The Newtonian formulation of dynamics*

The Newton formulation of classical mechanics is based on three *principles*:

The First Principle or *Galilei's* Principle of Relativity*

- There exist special observers with respect to which a particle not being acted upon by any force moves with a rectilinear motion.
- Such an observer will be called an *inertial observer* or an *inertial frame*. He can define the time in such a way that the motion appears to be also uniform.
- Any observer moving with rectilinear and uniform motion with respect to an inertial observer is an inertial observer too.

*Galileo Galilei was born in Pisa on February 15, 1564 and died in Arcetri (Florence), Italy on January 8, 1642. The author of *Dialogo dei Massimi Sistemi* (*Landini ed. Florence, 1632*), and *Discorsi e dimostrazioni matematiche intorno a due nuove scienze attenenti alla meccanica e i movimenti locali* (*Leida, 1638*), Galilei is considered as the inventor of the dynamics.

The Second Principle or *Newton's*[†] *Second Law*

- In an inertial frame, once the time has been chosen as specified before, the motion of a particle is governed by the differential equation:

$$m\vec{a} = \vec{F},$$

  where m is the mass of the particle, $\vec{a}$ its acceleration and $\vec{F}$ the force acting on the particle.
- It is an experimental observation that forces acting on a particle can change with time t or with the position $\vec{r}$ and the velocity $\vec{v}$ of the particle. Thus, the force is represented as a vector function of variables $(t, \vec{r}, \vec{v})$, and the second law is more explicitly written in the form

$$m\frac{d^2\vec{r}}{dt^2} = \vec{F}\left(t, \vec{r}, \frac{d\vec{r}}{dt}\right). \tag{1.1}$$

Third Principle

- The total momentum $\vec{P}$ and the total angular momentum $\vec{L}$ of an isolated system of particles do not change in time.

$$\frac{d\vec{P}}{dt} = 0, \quad \frac{d\vec{L}}{dt} = 0. \tag{1.2}$$

In many elementary textbooks a statement can be found, namely: that the first principle is a particular case of the second principle when the force vanishes. So expressed, the statement is wrong. Actually, it suggests that the distinction between *kinematics* and *dynamics* is artificial, and that inertial frames can only be defined dynamically, as the following discussion well shows.

1.1.2 *A discussion on space and time*

In Newton's principles, at least three concepts are given as natural and absolute, namely:

- we are able to state that no forces act on a body;

[†]Isaac Newton was born in the castle of Woolsthorpe, a little village to the south of Grantham in Lincolshire, England, on Christmas 1642, eleven months after the death of Galilei. He died in a suburb of London in 1727. The author of the celebrated *Philosophiae Naturalis Principia Mathematica (London, 1687)*, in which the foundations of mechanics and mathematical physics are exposed, Newton invented, by himself, the main tool of investigation; i.e. the differential calculus. On his grave, in Westminster Abbey, it is written: *Sibi gratulentur Mortales tale tantumque extitisse Humani Generis Decus.*

- we have a notion of an absolute straight line;
- we have a notion of an absolute time as "*flowing uniformly*," to quote Newton.

Concerning the absence of forces, it is evident that Newton's definition of a *free* body as "*a body far away from any other body in the Universe*," presumes that all forces decrease with distance. Thus, Newton was only thinking of gravitational forces.

It is clear that it was an attempt by Newton to abstractly generalize the definition of an inertial observer given by Galilei, who defined inertial as a frame which is stationary with respect to the "fixed stars". However, after Mach, we are aware that the *inertia* is related to the surrounding Universe, so that the more pragmatic definition given by Galilei is much more acceptable.

Galilei recognized, as a result of clock measurements, that approximately free bodies move in an approximately inertial frame, along approximately straight lines with approximately constant velocities. His tools were an inclined plane to slow the fall, a water clock to measure its duration, and a pendulum to avoid rolling friction.

> "*Inoltre, è lecito aspettarsi che, qualunque grado di velocità si trovi in un mobile, gli sia per sua natura indclcbilmcntc impresso, purchè siano tolte le cause esterne di accelerazione e di ritardamento; il che accade soltanto nel piano orizzontale; infatti nei piani declivi è di già presente una causa di accelerazione, mentre in quelli acclivi di ritardamento; infatti, se è equabile, non scema o diminuisce, ne tanto meno cessa.*" (G. Galilei, *Discorsi e dimostrazioni matematiche intorno a due nuove scienze*, Terzo giorno)

Newton was aware that Galilei's conclusion might be only approximately true, but he was very impressed by the existence of numerous coordinate transformations leading to coordinate systems, in which the Galilei description *cannot* be given. Then he elevated the Galilei approximate empirical discovery to the position of a rigorous principle, the *inertia principle*, and stated that absolutely free bodies move, in an ideal inertial frame, with absolutely constant velocities along perfectly straight lines.

> "*Absolute space, in its own nature and with regard to anything external, always remains similar and unmovable. Relative*

> *space is some movable dimension or measure of absolute space, which our senses determine by its position with respect to other bodies, and is commonly taken for absolute space."*

From that, Newton also defines an absolute time congruence.

As far as the notion of a straight line is concerned, we need a structure of vector space, and we know that, on the same space, we can give different vector space structures. Thus, the notion of straight line is observer-dependent.

The same can be said about Newton's allusion to time, for a rate of flow can be recognized as uniform only when measured against some other rate of flow taken as standard. In other words, we need a *comparison dynamics*.

Even if, from a theoretical point of view, the law of inertia should allow us to get an accurate determination of congruent intervals, the impossibility to observe freely moving bodies, due to the presence of frictional and gravitational forces, suggested to define a frame to be Galilean if a perfectly rigid sphere rotating without friction about an axis, fixed in the frame, has a uniform or constant rate of rotation. Here, constant is understood as measured in terms of the standards of time congruence, defined by a freely moving body under the ideal conditions required by the principle of inertia. The previous definition is still far from perfect, but at least, is coherent with a definition of time congruence based on the principle of causality which, following Weyl, can be given as follows:

> *"If an absolutely isolated physical system reverts once again to exactly the same state as that it was at some earlier instant, then the same sequence of states will be repeated in time, and the whole sequence of events will constitute a cycle. In general, such a system is called a clock. Each period of the cycle lasts equally long."*

We now come to Einstein's definition of Galilean frame, as implicitly given in special relativity: *The velocity of a light ray passing through an inertial frame will be the same regardless of the relative motion of the luminous source and frame, and regardless of the direction of the ray.*

Remark 1 *Actually, this property of the light defines the conformal group which contains the Lorentz group as a subgroup.*

The optical definition presents a marked superiority over those of the pre-relativistic physics. While, with earth's rotation, we had to assume the correctness of Newton's law to determine the corrections required by earth's breathing and by the friction of the tides, the new definition is just based on the most highly refined experiments known to physicists.

The relevance of Einstein's definition lies in the consequences which follow from the attempt to correlate space and time measurements between two inertial frames in relative motion. The concepts of space and time congruence lose the classical attributes of universality given to them by the Newtonian physics. It is then found that congruence can only be defined in a universal way (independent from the observer) when we consider the extension to the 4-dimensional space-time.

> "*And now, in our time, there has been unloosed a cataclysm which has swept away space, time and matter, hitherto regarded as the firmest pillars of natural science, but only to make place for a view of things of wider scope, and entailing a deeper vision.*" H. Weyl (*Space, Time and Matter*).

Wonderful as they may appear, Einstein's previsions have thus far been verified in every detail.

After our short discussion on

- *the Galilean and Newtonian principles of relativity,*
- *the Einstein special principles of relativity,*

it appears useful, after 115 years from the appearance of Mach's book,[40] and after 83 years from Einstein's article,[90] to also discuss

- *The Einstein general principles of relativity.*

The principle is assumed after the results of the mentioned Galilei's experiments on free falling bodies, later confirmed by Eötvos' (1889) and Dicke's (1967) measurements, which suggest that, at any point in space-time, a reference frame can be chosen, henceforth called *locally inertial frame*, such that, in a sufficiently small neighborhood of the point, the motion of a free falling particle is described by the equation

$$\frac{d^2\xi^\alpha}{d\tau^2} = 0\,,$$

where the ξ's are the coordinates in the locally inertial frame, and τ is any parametrization of the curves (*principle of equivalence*).

Thus, by assuming that a differentiable map exists between the coordinates x^α in the *laboratory frame* and the coordinates ξ in the locally inertial frame, by the above equation we obtain

$$
\begin{aligned}
0 = \frac{d}{d\tau}\left(\frac{d\xi^\alpha}{d\tau}\right) &= \frac{d}{d\tau}\sum_{\mu=0}^{3}\left(\frac{\partial\xi^\alpha}{\partial x^\mu}\frac{dx^\mu}{d\tau}\right) \\
&= \sum_{\mu=0}^{3}\left(\frac{d}{d\tau}\frac{\partial\xi^\alpha}{\partial x^\mu}\right)\frac{dx^\mu}{d\tau} + \sum_{\mu=0}^{3}\frac{\partial\xi^\alpha}{\partial x^\mu}\frac{d^2x^\mu}{d\tau^2} \\
&= \sum_{\mu=0}^{3}\frac{\partial^2\xi^\alpha}{\partial x^\mu\partial x^\nu}\frac{dx^\nu}{d\tau}\frac{dx^\mu}{d\tau} + \sum_{\mu=0}^{3}\frac{\partial\xi^\alpha}{\partial x^\mu}\frac{d^2x^\mu}{d\tau^2}\,,
\end{aligned}
$$

so that, multiplying by $\partial x^\lambda/\partial\xi^\alpha$ and summing over λ, we finally have

$$
\frac{d^2x^\lambda}{d\tau^2} + \sum_{\mu,\nu=0}^{3}\Gamma^\lambda_{\mu\nu}(x)\frac{dx^\mu}{d\tau}\frac{dx^\nu}{d\tau} = 0\,, \tag{1.3}
$$

where the functions

$$
\Gamma^\lambda_{\mu\nu}(x) = \frac{\partial x^\lambda}{\partial\xi^\alpha}\frac{\partial^2\xi^\alpha}{\partial x^\mu\partial x^\nu}
$$

are called the *affine connection coefficients.* Since Eq. (1.3) represents the equation of a particle moving in a gravitational field, we are forced to interpret the affine connection coefficients as representing the gravitational force in the laboratory frame. We notice that no assumptions have been done on background mathematical structures as a vector space structure or a metric structure.

An alternative version of the principle of equivalence is given by the so-called *principle of general covariance* which states that an equation, which is taken to describe a physical phenomenon, will be true if the equation holds in absence of gravitation, and moreover, it is *form invariant* for any coordinate transformation.

Thus, this principle states that the mathematical expressions of the laws of the nature must maintain the same form regardless of our choice of a reference frame. Moreover, by extending the invariance of the laws of the nature to all types of motions of the reference frame, this principle marks the starting point for the possible relativization of acceleration.

1.1.3 *Inertial frames revisited*

At this point, we are in a position to revisit elementary mechanics, avoiding a lot of assumptions on the *space of events* or *carrier space*, as follows.

We shall start with a 4-dimensional smooth manifold, for instance $\Re^4$, as space of events, for the description of the evolution of particles. By using, in our laboratory frame, a coordinate system, say (x^0, x^1, x^2, x^3), it is an experimental observation that the evolution of states is governed by a second order differential equation of the type

$$\frac{d^2x^\lambda}{d\tau^2} = F^\lambda\left(x^\mu, \frac{dx^\mu}{d\tau}\right).$$

We can now state what a *free particle*, and subsequently, a *comparison dynamics* are in this space.

A motion of a particle is said to be a *free motion*, if there exists a coordinate system, namely $(\xi^0, \xi^1, \xi^2, \xi^3)$, such that the equations of the motion can be written in the following form:

$$\frac{d^2\xi^\alpha}{d\tau^2} = 0.$$

Solutions of the above equation will define a vector space structure on the carrier space and will represent the world-line of a *physical system*; i.e. a really existing system, iff $d\xi^0/d\tau \neq 0$.

Thus, inertial frames are dynamically defined relatively to some chosen *comparison system*, avoiding any reference to dynamical systems arising in a specific gravitational theory.

In this sense, Einstein's general theory of relativity is not a theory of invariance or covariance, as special relativity, which gives prescriptions about the choice of the reference frame, by requiring the parameter characterizing the frame (the velocity) not to appear in the transformed dynamical equations.

Einstein's theory is a dynamical theory of the gravitational field, since it does not require the parameter characterizing the reference frame (the affine connection) to be absent from the transformed equations of the motion; it just prescribes *how* this parameter should appear in these equations.

We conclude our revisiting by noticing that a system has to be physical with respect to all locally inertial observer, so that we shall call *equivalent*, two inertial frames, namely $(\xi^0, \xi^1, \xi^2, \xi^3)$ and $(\xi'^0, \xi'^1, \xi'^2, \xi'^3)$, if for any world-line for which $d\xi^0/d\tau \neq 0$, we also have $d\xi'^0/d\tau \neq 0$.

We finally mention the

- *Mach–Einstein principles of relativity.*

This principle tries to bring about the complete relativization of all kinds of motion, rotational and accelerated, as well as uniform. This is achieved by ascribing all the dynamical effects related to the acceleration and rotation of particle and electromagnetic systems, to motion, with respect to the universe as a whole. According to this principle, there can exist no observable difference between the rotation of a body with respect to the universe of stars and the rotation of the stars around the body. Thus, Mach's principle constitutes an attempt to vindicate the kinematical principle in spite of the difficulties, of a dynamical nature, which had been the cause of its rejection. It was in part, with the intention to satisfy Mach's principle, that Einstein elaborated the hypothesis of the cylindrical universe. This issue is still open.

1.1.4 *The Lagrangian formulation of dynamics*

Let us assume that the carrier space $\Re^3$ is endowed with the Euclidean metrics, so that the components of a generic vector coincide with the ones of the unique covector, naturally associated *via* the metric tensor.

In terms of the components (q_1, q_2, q_3) of the position vector $\vec{r}$ and of the components $(v_1 = \dot{q}_1, v_2 = \dot{q}_2, v_3 = \dot{q}_3)$ of the velocity vector $\vec{v}$,

$$\vec{r} = (q_1, q_2, q_3)\,, \quad \vec{v} = (v_1, v_2, v_3)\,, \tag{1.4}$$

Eq. (1.1) takes the form

$$\begin{cases} \dfrac{d}{dt} m v_h = F_h(t, q_1, q_2, q_3, v_1, v_2, v_3)\,, \\ \dfrac{d}{dt} q_h = v_h\,, \end{cases} \qquad h = 1, 2, 3\,, \tag{1.5}$$

where F_h is the h-component of the force. The kinetic energy $\mathcal{T} = \frac{1}{2} m v^2$ reads

$$\mathcal{T} = \frac{1}{2} m(\dot{q}_1^2 + \dot{q}_2^2 + \dot{q}_3^2)\,. \tag{1.6}$$

Therefore, by observing that

$$m a_h = \frac{d}{dt} m v_h = \frac{d}{dt} m \dot{q}_h = \frac{d}{dt} \frac{\partial \mathcal{T}}{\partial \dot{q}_h}\,, \tag{1.7}$$

and that $\partial\mathcal{T}/\partial q_h$, the equations of motion (1.5) can be written in the following form:

$$\begin{cases} \dfrac{d}{dt}\dfrac{\partial\mathcal{T}}{\partial v_h} - \dfrac{\partial\mathcal{T}}{\partial q_h} = F_h \,, \\ \dfrac{d}{dt} q_h = v_h \,. \end{cases} \tag{1.8}$$

In the *conservative* case, there exists a function $U(q_1, q_2, q_3)$, *the potential energy*, such that

$$F_h = -\frac{\partial U}{\partial q_h} \,. \tag{1.9}$$

In such cases, by observing that $\partial U/\partial v_h = 0$, Eq. (1.8) or Eq. (1.5) can also be written in the *Lagrangian form*:

$$\begin{cases} \dfrac{d}{dt}\dfrac{\partial\mathcal{L}}{\partial v_h} - \dfrac{\partial\mathcal{L}}{\partial q_h} = 0 \,, \\ \dfrac{d}{dt} q_h = v_h \,, \end{cases} \tag{1.10}$$

where the *Lagrange*‡ *function* or simply the *Lagrangian* $\mathcal{L}(q, v, t)$ is defined as the difference between the kinetic energy and the potential energy:

$$\mathcal{L} = \mathcal{T} - U \,. \tag{1.11}$$

Remark 2 *In the case of a generalized potential; i.e. in the case in which the force* $\vec{F}$ *can be expressed in terms of a function* $U(q_1, q_2, q_3, v_1, v_2, v_3)$, *which beyond the coordinates* q, *also depends on the velocity* $\vec{v}$, *as*

$$F_h = \frac{d}{dt}\frac{\partial U}{\partial v_h} - \frac{\partial U}{\partial q_h} \,, \tag{1.12}$$

it is possible to write the equations of the motion in the same form as in Eq. (1.10).

‡Giuseppe Luigi Lagrangia was born in Torino in 1736 and died in Paris in 1813. At the age of 19, he already was a professor of mathematics at Artillery's School in Torino, and soon after, an associate founder of the Academy of Sciences of Torino. Author of the *Mécanique Analytique* (Paris, 1788), Lagrange is considered as one of the greatest mathematicians of the modern age. For a more extended biography, see Appendix A.

This is the case, for instance, of a charged massive particle acted upon by an electromagnetic field $(\vec{E}, \vec{B})$. *The force, in this case, is the Lorentz force:*

$$\vec{F}(\vec{r}, \vec{v}) = e(\vec{E} + \vec{v} \wedge \vec{B}), \tag{1.13}$$

where e *is the charge of the particle and the symbol* $\wedge$ *denotes the vector product.*

The Lorentz force can be derived from the following generalized potential:

$$U(\vec{r}, \vec{v}) = e(\varphi - \vec{v} \cdot \vec{A}), \tag{1.14}$$

where φ *and* $\vec{A}$ *denote the scalar and the vector potential, respectively. Thus, the Lagrangian function describing the motion of such a particle is given by*

$$\mathcal{L}(\vec{r}, \vec{v}) = \frac{1}{2}mv^2 - e(\varphi - \vec{v} \cdot \vec{A}). \tag{1.15}$$

1.1.5 *The Hamiltonian formulation of dynamics*

In terms of the momentum vector $\vec{p} = (p_1, p_2, p_3)$, Eq. (1.1) takes the form

$$\begin{cases} \dfrac{d}{dt}p_h = F_h\left(t, q_1, q_2, q_3, \dfrac{p_1}{m}, \dfrac{p_2}{m}, \dfrac{p_3}{m}\right), \\ \dfrac{d}{dt}q_h = \dfrac{p_h}{m}. \end{cases} \tag{1.16}$$

The kinetic energy, on the other hand, can be written as follows:

$$\mathcal{T}^* = \frac{1}{2m}(p_1^2 + p_2^2 + p_3^2), \tag{1.17}$$

where the symbol $*$ indicates that the velocity has been expressed in terms of the momentum by means of $\vec{v} = \vec{p}/m$.

In the *conservative* case, the energy $E = \frac{1}{2}mv^2 + U(q_1, q_2, q_3)$, expressed in terms of momenta, is usually denoted by $\mathcal{H}$ and called the *Hamilton*[§] *function* or simply the *Hamiltonian*,

$$\mathcal{H} = \frac{1}{2m}(p_1^2 + p_2^2 + p_3^2) + U(q_1, q_2, q_3). \tag{1.18}$$

[§] William Rowan Hamilton was born in Dublin, Ireland in 1805 and died in Dunsik in 1865. He was a professor of astronomy at Dublin University and President of the Ireland Academy of Sciences. He invented the theory of quaternions and gave remarkable contributions to the Analytical Mechanics, in which, it successfully incorporated the theory of the light propagation.[11]

Therefore, since $\partial U/\partial p_h = 0$, the equations of motion can be written in the *Hamiltonian form*:

$$\begin{cases} \dfrac{d}{dt}p_h = -\dfrac{\partial \mathcal{H}}{\partial q_h}, \\ \dfrac{d}{dt}q_h = \dfrac{\partial \mathcal{H}}{\partial p_h}. \end{cases} \tag{1.19}$$

Remark 3 *It could seem to the reader that simple formal manipulations of the Newton equations have been done to transform to the Lagrange or to the Hamilton equations. Actually, an additional structure, a scalar product, has been used.*

Indeed, when a vector space V (in our case $\Re^3$) is supposed to be endowed with a scalar product defined by a metrics g (in our case the Euclidean metrics),

$$g : (\vec{u}, \vec{v}) \in V \times V \to g(\vec{u}, \vec{v}) \equiv \vec{u} \cdot \vec{v} \in \Re,$$

with any function f, where

$$f : (\vec{u}) \to f(\vec{u}) \in \Re,$$

we can associate a vector field, denoted by ∇f or $\delta f/\delta \vec{u}$, and called the gradient of f, by

$$\nabla f \cdot \vec{v} \equiv \frac{\delta f}{\delta \vec{u}} \cdot \vec{v} = \frac{d}{d\varepsilon} f((\vec{u} + \varepsilon \vec{v}))\bigg|_{\varepsilon=0}.$$

Therefore, a metric structure allows us to define a force $\vec{F}$ to be conservative, if a function U exists, such that

$$\vec{F} = -\nabla U.$$

Under a change of coordinates, forces on RHS of Newton's equations transform as accelerations, so that they are vector fields and do not require the use of any metric structure. These forces are measured with a dynamometer.

On the contrary, forces appearing in Lagrange's equations are mathematically defined and physically measured in terms of the scalar function called work. As a consequence, these forces are coefficients of a differential form and cannot be identified with vector fields, unless a metric structure is at our disposal. Of course, this metric structure cannot be chosen arbitrarily but must be inferred from the results of experiments.

Remark 4 *It could seem that the Lagrangian and the Hamiltonian formulations of dynamics do not bring in special advantages. This is certainly true in problems concerning particles not associated with other particles, for instance, as in a solid body or a fluid. In more complicated cases, the Newtonian approach is still applicable, provided some proper precautions in the analysis of forces are observed. This force-analysis sometimes becomes cumbersome and it is difficult to give a unique answer to the problem. The analytical approach (Lagrangian or Hamiltonian) to the problem of the motion is much more powerful. According to it, for a general system of N particles, subject to k limitations,* $n = N - k$ *special parameters, namely* $q_1, \ldots, q_n$, *can be found such that the equations of dynamics assume the form given by Eq. (1.10) or Eq. (1.19). Moreover, it will be shown that there is a unifying principle, the least action principle, which gives a meaning to the entire set of the analytical equations of dynamics (Lagrange or Hamilton equations). The statement of this principle is independent of any choice of the coordinate system and this implies that the analytical equations of dynamics are invariant with respect to any coordinate transformation. Unlike the Cauchy approach, which is local in nature, the unifying principle allows a global approach to the problem of the existence and uniqueness of the solution of dynamical equations.*

1.2 Constraints

A particle is said to be *constrained* if it cannot take all possible positions in the space. In the following examples, the space is supposed to be endowed with the Euclidean metrics.

Example 1 *Let us consider a particle* P *with coordinates* (x, y, z) *linked to a fixed point* P_0 *with coordinates* (x_0, y_0, z_0) *by means of a rigid bar whose length is* l.

Since the bar is rigid, the point P *is free to move in the space taking only positions at distance* l *from the fixed point* P_0. *In other words, the point* P *is constrained to move along the surface of a sphere with center in* P_0 *and radius* l; *in this way, it can fill only the positions whose coordinates satisfy the equation*

$$(x - x_0)^2 + (y - y_0)^2 + (z - z_0)^2 = l^2 . \tag{1.20}$$

Example 2 *Let us consider a particle P with coordinates (x, y, z) linked to a fixed point P_0 with coordinates (x_0, y_0, z_0) by means of a flexible and inextensible wire whose length is l.*

Of course, the particle P is free to move in the space inside the sphere with center P_0 and radius l; in this way, it can only fill the positions whose coordinates satisfy the equation

$$(x - x_0)^2 + (y - y_0)^2 + (z - z_0)^2 \leq l^2 . \tag{1.21}$$

The restrictions which impose a limitation to the mobility of P are called *constraints* or *links.* Equations (1.20) and (1.21) are mathematical expressions of constraints.

A constraint is said to be *two-sided* if it is expressed by an equality, *one-sided* if it is expressed by an inequality. Thus, the constraint (1.20) is two-sided while the constraint (1.21) is one-sided.

More generally, a particle constrained to be bound to the surface, represented by the equation

$$f(x, y, z) = 0 , \tag{1.22}$$

or bound to the curve represented by the equations

$$\begin{cases} f_1(x, y, z) = 0 , \\ f_2(x, y, z) = 0 , \end{cases} \tag{1.23}$$

is said to be subjected to two-sided constraints.

Let us now consider the case in which the particle P is bound not to pass through the surface σ, represented by Eq. (1.22). If the surface σ is supposed to divide the space into two regions, the one in which the particle is free to move is said to be the *exterior region*, and the remaining one the *interior region.*

Of course, the left-hand side of Eq. (1.22), vanishing on σ, will be positive in one of the two regions and negative in the other one. Moreover, since it is possible to multiply the left hand side of Eq. (1.22) by a nowhere vanishing factor, we can arrange the equation in such a way that it will be positive in the external region; that is in the region in which the particle P is free to move. Therefore, the positions filled by P are all and the only ones satisfying the inequality

$$f(x, y, z) \geq 0 . \tag{1.24}$$

In this way, the constraint for P is one-sided.

In the presence of such constraints, the possible positions of P are further distinguished in

- *ordinary positions* — the ones satisfying the condition:

$$f(x,y,z) > 0\,; \tag{1.25}$$

- *border positions* — the ones satisfying the condition:

$$f(x,y,z) = 0\,. \tag{1.26}$$

  In the last case, it is also said that the point is *leaning on the surface* σ.

Let us now extend the *constraint concept* to the case of an arbitrary system of particles.

It is worth recalling that in considering the motion $\mathcal{M}$ of a system of particles $\mathcal{S}$, it is usual, to speak of *aptitude* to indicate the *ensemble* of positions at a given time t, and of distribution of velocities $\vec{v}_P$ of points P of $\mathcal{S}$ at the same time. The aptitude is usually denoted by $\{P, \vec{v}_P(t)\}$.

Thus, for a system of particles $\mathcal{S}$, any restriction on the positions or on the aptitude of $\mathcal{S}$ is called a constraint. The constraint will be called *inner* if such a restriction translates an intrinsic property of the natural body represented by $\mathcal{S}$, *outer* if it originates from the presence of obstacles (other bodies) external to $\mathcal{S}$. For instance, the *rigidity constraint* schematizes an intrinsic property of solids (indeformability), so that it is an inner constraint.

While a particle is said to be *free* when it is not subjected to any constraints, a system of particles is said to be *free* when it is not subjected to outer constraints.

For a system of particles $\mathcal{S}$, it is important for further distinction to use terms introduced by Hertz¶: *holonomic*‖ *constraints* and *anholonomic constraints.*

- A constraint which directly imposes restrictions on the position of $\mathcal{S}$, is called *holonomic.*
- A constraint which directly imposes restrictions on the aptitude of $\mathcal{S}$, is called *anholonomic.*

¶H. Hertz was born in Hamburg in 1857 and died in Bonn in 1894. A physicist and mathematician, he first detected the electromagnetic waves.

‖From Greek $\acute{o}\lambda o\sigma$ (integer) and $\nu\acute{o}\mu o\sigma$ (law).

In this way, the equality or the inequality representing a given constraint will contain only parameters corresponding to the position of $\mathcal{S}$, if the constraint is holonomic; it will contain also the time derivative of such parameters, if the constraint is anholonomic, with the time derivatives specifying the aptitude of $\mathcal{S}$. A classical example of anholonomic constraint can be given as follows:

Let us consider a rigid sphere $\mathcal{S}$ on a plane π, rolling *without slipping* (see Remark 5) along it. According to the remark at the end of the section, this means that at any time t, the relative velocity of $\mathcal{S}$ with respect to π, in all contact points, is vanishing. As the *characteristic velocity* of rigid motions is

$$\vec{v}_Q = \vec{v}_P + \vec{\omega} \wedge (P - Q)\,, \tag{1.27}$$

where Q is a generic point of $\mathcal{S}$, and ω the angular velocity, it follows that the velocity of any point Q of the sphere $\mathcal{S}$, at time in which $\vec{v}_P = 0$, is given by

$$\vec{v}_Q = \vec{\omega} \wedge (P - Q)\,, \tag{1.28}$$

that is, the motion is a pure rotation. We can conclude that in such conditions, the aptitude of the sphere is a rotation around the instantaneous axis passing through the contact point P between $\mathcal{S}$ and π. Therefore, the requirement that a rigid sphere on a plane π rolls on it without slipping, constitutes an anholonomic constraint, since the aptitude of the sphere can only be a rotation about an axis passing through the contact point.

Remark 5 *When the motion of a system of particles is observed from two different frames, each one moving with respect to the other, it is a convention to assume one of them steady T_Ω, and the other one mobile T_O, with Ω and O denoting the origins of frames. The Euclidean space framed with T_Ω is called steady space, where else the one framed with T_O which moves with respect to T_Ω, is called mobile. Let us consider then a surface σ, a border of a natural body, of the mobile space, and a surface σ', a border of another body, of the steady space. When during the time σ and σ' share points and tangent planes at those points, it is said that during the motion, σ rolls on σ'.*

Let us suppose that during the motion, σ rolls on σ', and let H be one of the contact points. The velocity of the point P of σ, which at time t is laid upon H, is called slipping velocity of σ with respect to σ' at point H at time t. Finally, the circumstance that for all time t, the creeping velocity of σ with respect to σ' at any contact point is vanishing, is referred to saying that during the motion, σ rolls on σ', without slipping.

1.3 Degrees of Freedom and Lagrangian Coordinates

Let $\mathcal{S}$ be a generic system of particles constrained. The positions are that at time t, the constraints permit to $\mathcal{S}$ are said, "possible positions of $\mathcal{S}$ at time t" or also "compatible positions of $\mathcal{S}$ at time t." Let E^n be the n-dimensional Euclidean space and let us suppose, at first, that $\mathcal{S}$ is composed of one particle P only. Let us assume that P is constrained to move on a regular curve γ at rest in a Cartesian frame T_0. The regular curve will be represented in the frame T_O by the parametric equations:

$$\begin{cases} x = \varphi(\lambda)\,, \\ y = \psi(\lambda)\,, \\ z = \chi(\lambda)\,, \end{cases} \tag{1.29}$$

where φ, ψ, χ are three functions defined in the closed interval $[a,b]$ of $\Re^1$, and $\lambda \in [a,b]$ is a real parameter.

The regularity of γ means that the functions φ, ψ, χ are supposed to be continuous together with their first derivatives in the interval $[a,b]$, and to satisfy the conditions below:

- The function $H(\lambda) = \sqrt{\varphi'^2 + \psi'^2 + \chi'^2}$ is positive $\forall\, \lambda \in [a,b]$.
- There is no pair (λ', λ'') of distinct values of λ, such that, simultaneously, $\varphi(\lambda') = \varphi(\lambda'')$, $\psi(\lambda') = \psi(\lambda'')$, and $\chi(\lambda') = \chi(\lambda'')$.

As to the regularity of γ, the possible positions of P are, at any time t, in a one-to-one correspondence with real numbers in the interval $[a,b] \subset \Re^1$.

If a particle P is constrained to move on a moving regular curve γ, represented in the frame T_O by the parametric equations (a family of regular curves):

$$\begin{cases} x = \varphi(\lambda, t)\,, \\ y = \psi(\lambda, t)\,, \\ z = \chi(\lambda, t)\,, \end{cases} \tag{1.30}$$

at each time t, as the constraint depends on the time, the positions of P compatible with the constraint are in a one-to-one correspondence with real numbers in the interval $[a,b] \subset \Re^1$, generally changing in time.

The fact that it is possible to establish a one-to-one map between the positions of P, compatible with constraints at a given time t, the set of values that a parameter takes on an interval of $\Re^1$ is usually expressed by saying that:

In order to specify the position of a particle on a curve, only one parameter is necessary, or also that, *the number of degrees of freedom of a point constrained on a regular curve is 1.*

Let us now assume that P is constrained to move on a regular surface σ at rest in a Cartesian frame T_O. The regular surface will be represented in the frame T_O by the parametric equations:

$$\begin{cases} x = \varphi(\lambda_1, \lambda_2)\,, \\ y = \psi(\lambda_1, \lambda_2)\,, \\ z = \chi(\lambda_1, \lambda_2)\,, \end{cases} \tag{1.31}$$

where λ_1 and λ_2 are real parameters and the three functions φ, ψ, χ are defined in a simply connected bounded domain of $\Re^2$ by

$$a_i \leq \lambda_i \leq b_i\,, \quad i = 1, 2\,. \tag{1.32}$$

The regularity of σ means that the functions φ, ψ, χ are supposed to be continuous, together with their first partial derivatives in $[a_1, b_1] \times [a_2, b_2]$ and to satisfy the following conditions:

- the determinants

$$A = \det \begin{pmatrix} \dfrac{\partial\psi}{\partial\lambda_1} & \dfrac{\partial\psi}{\partial\lambda_2} \\ \dfrac{\partial\chi}{\partial\lambda_1} & \dfrac{\partial\chi}{\partial\lambda_2} \end{pmatrix}, \quad B = \det \begin{pmatrix} \dfrac{\partial\chi}{\partial\lambda_1} & \dfrac{\partial\chi}{\partial\lambda_2} \\ \dfrac{\partial\varphi}{\partial\lambda_1} & \dfrac{\partial\varphi}{\partial\lambda_2} \end{pmatrix}, \quad C = \det \begin{pmatrix} \dfrac{\partial\varphi}{\partial\lambda_1} & \dfrac{\partial\varphi}{\partial\lambda_2} \\ \dfrac{\partial\psi}{\partial\lambda_1} & \dfrac{\partial\psi}{\partial\lambda_2} \end{pmatrix}$$

  are nowhere vanishing, so that the function $W(\lambda_1, \lambda_2) = \sqrt{A^2 + B^2 + C^2}$ is always positive $\forall\, \lambda_1, \lambda_2 \in [a, b]$;
- there is no pair $((\lambda_1', \lambda_2'), (\lambda_1'', \lambda_2''))$ of distinct sets of values (λ_1, λ_2), such that, simultaneously, $\varphi(\lambda_1', \lambda_2') = \varphi(\lambda_1'', \lambda_2'')$, $\psi(\lambda_1', \lambda_2') = \psi(\lambda_1'', \lambda_2'')$, and $\chi(\lambda_1', \lambda_2') = \chi(\lambda_1'', \lambda_2'')$.

As to the regularity of σ, the possible positions of P are in a one-to-one correspondence with pairs of real numbers in the rectangle $[a, b] \subset \Re^2$.

We come to the same conclusion when the particle P is constrained to move on a moving regular surface σ'. In this case, of course, the rectangle, as the constraint, will depend on time.

We shall say that only two independent parameters are needed to specify the positions of a particle on a surface, or that, *the number of degrees of freedom of a point constrained on a regular surface is 2.*

In general, a system $\mathcal{S}$ of particles, however constrained, is said to have n degrees of freedom, if it is possible to establish a one-to-one map between the possible positions of $\mathcal{S}$ at time t and the values that n real parameters $(q_1, \ldots, q_n)$ take in an open subset of $\Re^n$. The parameters $(q_1, \ldots, q_n)$ are called *Lagrangian coordinates* of $\mathcal{S}$. Of course, the choice of Lagrangian coordinates is not unique.

Examples

- A particle P, constrained to lie on a curve, has *1* degree of freedom. It is possible to choose, as Lagrangian coordinate of P, a curvilinear coordinate.
- A particle P, constrained to lie on a plane, has *2* degrees of freedom. It is possible to choose, as Lagrangian coordinates of P, the Cartesian coordinates or the polar ones.
- A free particle P has *3* degrees of freedom. It is possible to choose as Lagrangian coordinates of P, the Cartesian coordinates, the cylindric coordinates, the spheric-polar ones, etc.
- A free rigid body has *6* degrees of freedom. Indeed, in order to specify the position of a frame $T_\Omega(\xi, \eta, \zeta)$ framed with the body, it is necessary to give the three coordinates of Ω and three of components of unit vectors along the (ξ, η, ζ) axis. It is possible to choose, as Lagrangian coordinates of the rigid body, the three coordinates of Ω and a triplet of parameters in a one-to-one correspondence with three of the components of unit vectors along the (ξ, η, ζ) axis.
- A rigid body with a fixed axis r has *1* degree of freedom. It is possible to choose as Lagrangian coordinate, the angle ϑ between two planes having r as intersection, one of them steady, the other framed with the body.
- A rigid body with a fixed point has *3* degrees of freedom since only three parameters suffice to specify the position of a frame framed with the body and having the origin in the fixed point.
- A system consisting of two rigid bodies which participate in a common axis (e.g., a compass), has *7* degrees of freedom. Indeed, six parameters are needed to specify the position of the first body and only one to specify the position of the second body with respect to the first.

The following definition will conclude the section:

A system of particles is called holonomic if it has finitely many degrees of freedom and if it is submitted to holonomic constraints only.

1.4 The Calculus of Variations and the Lagrange Equations

By considering the evolution of a holonomic system with n degrees of freedom as a sequence of equilibrium states (*d' Alembert*) under the action of all the forces (*effective, from constraints and inertial*), and by applying the *principle of the virtual works* to *Cardinal equations of dynamics*, the equations of the motion can be written in the elegant and powerful Lagrangian form:

$$\frac{d}{dt}\frac{\partial \mathcal{T}}{\partial \dot{q}_h} - \frac{\partial \mathcal{T}}{\partial q_h} = \mathcal{Q}_h \quad h \in \{1, \ldots, n\},$$

where q's denote the Lagrangian coordinates, $\dot{q}$'s the Lagrangian velocities and Q's the Lagrangian components of the "force." In case the forces are conservative, a function $\mathcal{U}(q/\dot{q}/t)$ (the potential energy) exists, such that

$$\mathcal{Q}_h = \frac{d}{dt}\frac{\partial \mathcal{U}}{\partial \dot{q}_h} - \frac{\partial \mathcal{U}}{\partial q_h},$$

so that the Lagrangian equations can be written as follows:

$$\frac{d}{dt}\frac{\partial \mathcal{L}}{\partial \dot{q}_h} - \frac{\partial \mathcal{L}}{\partial q_h} = 0 \quad h \in \{1, \ldots, n\},$$

where $\mathcal{L} = \mathcal{T} - \mathcal{U}$ is called the *Lagrangian function*. We do not report the derivation, as it can be found in almost all textbooks in classical mechanics. We just adopt here an axiomatic point of view according to which:

- The state of a system is completely defined by specifying its coordinates and velocities $(q/\dot{q})$.
- The evolution; that is the sequence of states is completely determined by giving a function

  $$\mathcal{L}(q/\dot{q}/t),$$

  defined on the sets of states, and two different configurations q_A and q_B of the system at two different time instant, t_A and t_B, such that

  $$q_A = q(t_A), q_B = q(t_B).$$

- Among all *close* curves $q_h = q_h(t)$ joining A and B, the one for which the action integral,

$$S[q] = \int_{t_A}^{t_B} \mathcal{L}(q/\dot{q}/t)dt\,, \tag{1.33}$$

takes the least value, will represent the evolution of the system.

Remark 6 *It is worth to remark that this formulation holds for a small part of the trajectory. On the whole trajectory the integral $S[q]$ may have just an extremum, not necessarily a minimum. However, the equations of the motion can be written by using only the extremum condition.*

The axiomatic approach is called the *principle of least action* or *the Hamilton principle.*

Integrals like those in Eq. (1.33) are defined on a space of functions

$$S : \mathcal{F} \longrightarrow \Re\,,$$

and could be called functions, but for historical reasons, are called *functionals.* A few words on their use will be spent after a short historical comment.

1.4.1 *Historical notes*

The Newton problem

The calculus of variations was founded simultaneously to the differential calculus (1686). In his *Philosophiae Naturalis Principia Mathematica,* Newton was the first to propose the problem of the body with the least opposition. It is a problem involving different concrete cases, as it concerns the best form of a body (submarine or a missile, etc.) in order to suffer, from the medium, the least opposition to its motion. So formulated, the problem is too difficult. For the sake of simplicity, the question was first proposed for a body with a form invariant for rotation about an axis parallel to the direction of the motion (equal inertial moments in the plane orthogonal to the velocity), and only for its *head* to avoid the difficult problem of vortices surrounding its *tail.* By fixing the length and the height of the head the problem becomes:

Given two points P and Q, find the planar curve joining them and generating, by rotation around the planar normal $\vec{n}$ at P, a revolution surface while moving parallel to $\vec{n}$, that suffers the least opposition from the medium.

The solution clearly depends on the *opposition law.* Newton assumed that the opposition, on a generic element of the surface, was proportional to the square of the projection of the velocity along $\vec{n}$.

By taking an orthogonal Cartesian frame, with the x axis coincident with the rotation axis and by denoting with $y = y(x)$ the unknown curve, it turns out that the opposition R is given by

$$R[y] = k \int_p^q \frac{yy'^3}{1+y'^2} dx \,,$$

where p, q are the x-coordinates of the points P, Q and k is a constant which depends on the velocity of the surface. The problem is then to find, among all curves $y = y(x)$ joining P and Q, the one for which the previous integral takes the least value.

The brachistochrone problem

Ten years later, the following problem was proposed and solved by Johann Bernoulli:

Given two points P and Q in a vertical plane, find among all planar curves joining them, the one which the time required for a particle to descend, without friction, from the origin P to Q, would be the least possible.

By choosing a suitable orthogonal Cartesian frame with a vertical y axis, the time for a particle to descend from P to Q along the curve $y = y(x)$ will be given by

$$T[y] = \int_p^q \sqrt{\frac{1+y'^2}{2gy}} dx \,,$$

where p, q are the x-coordinates of the points P, Q and $g \sim 9.8$ m/s^2 is the modulus of the acceleration due to gravity. Thus, the searched curve is the one for which the above integral takes the least value.

Johann Bernoulli found, as a solution, the curve whose parametric equations are

$$\begin{cases} x = k(\vartheta - \sin\vartheta) \,, \\ y = k(\vartheta - \cos\vartheta) \,, \end{cases}$$

where k represents the ratio between the distances of points P and Q from the origin. His method of solution was strongly criticized, from a mathematical point of view, by Newton, Leibnitz, l' Hospital and Jacob Bernoulli** who was able to find the same solution by using different methods. In particular, Jacob Bernoulli solved the problem by using a geometrical method, of larger applicability, based on the principle according to which, if a curve is characterized by a property of maximum or minimum, any part of it, no matter how small, is characterized by the same property. According to this principle, for instance, if a curve is a brachistochrone, any part of it is again a brachistochrone. It is, therefore, possible to replace the curve with a broken line in such a way that the problem reduces to find its vertex by using ordinary differential calculus methods.

The approach used by Jacob Bernoulli, who must be considered the founder of the calculus of variations, were generalized and wisely extended by Euler†† to a large category of problems. He started with the classification of problems in two classes:

- find among all curves satisfying suitable boundary conditions, the one for which a given integral takes an extremum (a minimum or a maximum) value.
- find among all curves satisfying suitable boundary conditions for which given integrals take assigned values (*constraints*), the one for which another integral takes an extremum value (*isoperimetric problems*).

He invented the *isoperimetric rule* which allows to reduce, at least on principle, a given problem of the second class to a problem belonging to the first one.

A more rigorous treatment was given by Lagrange, who introduced the concept of variation, which allows this kind of problems to be treated with ordinary differential calculus methods.

**Jacob Bernoulli was born at Basilea (Bâle) on December 27, 1654. He has been, for many years, a professor of mathematics at Basilea University. Supporter of Leibnitz' scientific ideas, he died at Basilea in 1705.

Johann Bernoulli, the brother of Jacob, was born at Bâle on August 7, 1667, and died there on January 1, 1748. He has been, for many years, a professor of mathematics at Groningen from 1695 to 1705 and at Bâle University, where he succeeded his brother, from 1705 to 1748. As an illustration of his character, it may be mentioned that he expelled his son Daniel from his house for obtaining a prize from the *Academie de France* which he had expected to receive himself.

††Leonard Euler, born in Basilea in 1707 and died in St. Petersburg in 1783, was director of the Academy of Sciences in Berlin, and soon after, the Academy of Sciences in St. Petersburg. He was one of the most important and fecund mathematicians of all times, both in calculus and in its physical applications.

Let $u = u_0(x)$ be the curve which extremizes the integral $I[u]$, among all the curves in the plane joining two given points P and Q. In order to analyze the properties of $I[u]$, it is necessary to compare the value $I[u_0]$, taken by the integral on the given curve $u = u_0(x)$, with the one $I[u]$ corresponding to a different curve. The difference for $u(x) - u_0(x)$ is the variation, for fixed x, in passing from the curve $u = u_0(x)$ to the curve $u = u(x)$. If the two curves are "very close," this variation will be "very small," similarly to the differential du that a given function u undergoes for a little change dx of x, but very different in nature. In order to distinguish between them, and at the same time to underline their analogy, Lagrange introduced the symbol δu for the variation and invented a calculus completely analogous to the ordinary one.

In correspondence with the variation δu of the curve $u = u_0(x)$, the integral $I[u_0]$ undergoes the variation

$$I[u] - I[u_0]$$

which decomposes in different parts conformable to the decomposition given to a function $f(x)$ from an increment of x. The different parts are given, save numerical factors, by succeeding differentials df, $d^2 f, \ldots$.

Similarly, the different parts of $I[u] - I[u_0]$, save numerical factors, are called first variation, second variation, ... of the integral I and denoted by δI, $\delta^2 I, \ldots$.

Lagrange was able to show that, similarly to the case of a function, the extrema of I satisfy the relation $\delta I = 0$. In order to distinguish extrema, in minima and maxima, Legendre analyzed the second variation and introduced an elegant transformation of $\delta^2 I$ leading to necessary conditions characterizing maxima or minima. The Legendre condition survived a deep criticism by Lagrange, who observed, among other questions, that the Riccati[‡‡], equation leading to the Legendre transformation of the second variation $\delta^2 I$, does not always admit a bounded continuous solution in all the considered interval and that the transformation does not always exist. Finally, the difficult question was genially solved by Jacobi (1837) who gave a new necessary condition.

In 1870, Weierstrass observed that, contrary to the case of functions, the sign of the second variation does not ensure the existence of the maximum or minimum on the considered extrema. According to Weierstrass and Scheeffer,

[‡‡]Jacopo Riccati was born in Venice in 1676 and died in Treviso in 1754. He was an aristocrat who studied mathematics privately. His famous equation is contained in *Acta Eruditorum* (Lipsia, 1722).

the reason for the difference lies in the complexity of a curve with respect to a point. Two curves, although "very close," can strongly differ by their tangents at close points and the integral I also contains the derivative u' which characterizes such tangents. Weierstrass was able to give sufficient conditions for the existence of the minimum or the maximum in terms of the sign of a particular function which today is called the *Weierstrass function.*

Weierstrass also criticized the representation $y = u(x)$ for a curve, since this representation meets at most in a point, the parallels to the vertical axis, and then it greatly restricts the field of possible solutions. He developed its theory by using parametric representations

$$x = x(t)\,, \quad y = y(t)\,, \quad t \in (t_0, t_1)$$

and replacing the integral I by

$$\mathcal{J} = \int_{t_0}^{t_1} F[x(t), y(t), x'(t), y'(t)]dt\,.$$

At the same time, Darboux found sufficient conditions for the minimum in the *geodesic* problem by introducing curvilinear coordinates, which allows us to write the considered integrals in a particularly simple form, where the minimum properties turn out to be evident.

Very important contributions were then given by Kneser, Lindeberg, Gauss, Ostrogradsky, Delauney, Clebsch, Schwarz, Volterra and Hilbert. For more details, see for instance, Ref. 53.

1.4.2 *A digression on the variation methods in problems with fixed boundaries*

Continuous functionals

A functional $F[u]$ will be called *continuous*, if to a "small change" of $u(x)$, there corresponds a "small change" in the functional $F[u]$.

With previous definition, only a few functionals will be continuous, since in general the function in the integrals will depend on the derivatives of u. For instance, in the action integral, first derivatives are also included.

Thus, it is natural to give the following definitions of *close* curves:

Two curves $u(x)$ and $v(x)$ are of zero-order proximity close, if the absolute value of their difference $|u(x) - v(x)|$ is small.

Two curves $u(x)$ and $v(x)$ are of first-order proximity close, if the absolute values of the differences $|u(x) - v(x)|$ and $|u'(x) - v'(x)|$ are small.

Two curves $u(x)$ and $v(x)$ are of kth-order proximity close, if the absolute values of the differences, $|u(x) - v(x)|$, $|u'(x) - v'(x)|, \ldots, |u^{(k)}(x) - v^{(k)}(x)|$, are small.

It is then possible to define the notion of distance σ between two curves. By assuming that u and v have continuous derivatives up to order k, the *distance of order k* is defined as

$$\sigma_k(u, v) = \sum_{h=1}^{k} \max_{x_0 \leq x \leq x_1} |u^{(h)}(x) - v^{(h)}(x)|,$$

and then close-lying curves would be curves with a small distance.

Derivatives in a vector space

1. *Strong derivative or Frechet derivative.* Let U and V be two normed vector spaces and F a map from an open subset A of U into V.

$$F : A \subseteq U \longrightarrow V .$$

The map F will be called *differentiable at point* $u \in U$, if a linear-bounded operator F' exists, such that

$$F(u + h) - F(u) = F'h + \sigma(u, h),$$

where σ is infinitesimal with respect to the distance in V given by the norm:

$$\lim_{||h|| \to 0} \frac{||\sigma(u, h)||}{||h||} = 0 .$$

The reader will recognize, by identifying U with $\Re^n$ and V with $\Re$, the usual definition of the differentiability for a numerical function of n real variables.

The linear part of the increment $F'h$, is called the *strong differential* (or the *Frechet differential*) of the map F at point u and the operator F' is called the *strong derivative* (or the *Frechet derivative*) of the map F at point u. It is easy to see that, if the map F is differentiable, the corresponding derivative is uniquely defined. Furthermore, the theorem on the derivative of composed maps can be easily proven.[28]

2. *Weak derivative or Gateaux derivative.* If G denotes a map from an open subset of U into V, the limit

$$DG(u,h) = \frac{d}{dt}G(u+th)|_{t=0} = \lim_{t\to 0}\frac{G(u+th)-G(u)}{t},$$

where the convergence is considered only with respect to the norm of the vector space V, is called the *weak differential* or the *Gateaux differential* of G at point u.

The Gateaux differential can be nonlinear with respect to h. In the case it is linear, for example, in the case a bounded linear operator G_u exists, such that

$$DG(u,h) = G_u h,$$

the operator G_u is called the *weak derivative* (or the *Gateaux derivative*) of G at point u.

Let us observe that, in general, for the weak derivatives, the theorem on the derivative of a composed map is not true.

It is easy to see that, if the strong derivative of a map F exists, then the weak derivative also exists and the two derivatives coincide.

As a matter of fact, if F is strongly differentiable, then

$$F(u+th) - F(u) = F'(u)(th) + \sigma(u,th) = tF'(u)(h) + \sigma(u,t),$$
$$\frac{F(u+th)-F(u)}{t} = F'(u)h + \frac{\sigma(u,th)}{t},$$

so that

$$\lim_{t\to 0}\frac{F(u+th)-F(u)}{t} = F'(u)h = F_u h.$$

The converse is not true, and generally never true in the finite dimensional spaces, as the following example well shows.

Example 3 *The function*

$$f : (x,y) \in \Re^2 \longrightarrow f(x,y) = \begin{cases} \dfrac{x^3 y}{x^2+y^2}, & \textit{if } (x,y) \neq (0,0) \\ 0, & \textit{if } (x,y) = (0,0) \end{cases}$$

is continuous on $\Re^2$*. At the point* $(0,0)$*, its weak differential exists:*

$$f_{(x,y)}h = \lim_{t \to 0} \frac{f(0+th_1, 0+th_2) - f(0,0)}{t} = 0\,.$$

Nevertheless, the weak differential is not the linear part of the increment of the function at the point $(0,0)$*, since*

$$\lim_{\|h\| \to 0} \frac{f(h_1,h_2) - f(0,0)}{\|h\|} = \frac{1}{2}\,.$$

The result of the previous exercise is not surprising, just like in the finite dimensional case, the existence of partial derivatives of a given function does not ensure its differentiability, which can be ensured by the continuity of the partial derivatives.

In this context, it is important the following theorem's proof can be found, for instance, in Ref. 28.

Theorem 1.1 *If the weak derivative* F_u *of the map* F *exists in a neighborhood of a point* u_0 *and represents a continuous operator at* u_0*, then the strong derivative of* F *also exists at* u_0 *and coincides with the weak one.*

3. *The gradient of a functional.* Let U be a vector space of numerical functions of a real variable, namely $u(x)$, endowed with a scalar product $(\cdot\,,\cdot)$ and let F be a functional defined in U,

$$F : u \in U \longrightarrow F[u] \in \Re\,.$$

The gradient of F, or the *functional derivative* of F, with respect to the scalar product is the function of U, denoted by $G = \delta F/\delta u$, defined by the relation

$$\delta F \equiv \frac{d}{d\varepsilon} F[u + \varepsilon\varphi]|_{\varepsilon=0} = \left(\frac{\delta F}{\delta u}, \varphi\right)\,. \tag{1.34}$$

The second functional derivative of F is given by the weak derivative G_u of G, which in this case, can be also defined as

$$\frac{d^2}{d\varepsilon d\tau} F[u + \varepsilon\phi + \tau\psi]|_{\varepsilon=\tau=0} =: (G_u\psi, \phi)\,.$$

As usual,
$G_u\psi = \frac{d}{d\tau} G(u + \tau\psi)|_{\tau=0}$.

Of course, the operator G_u is symmetric with respect to $(\cdot\,,\cdot)$; i.e.

$$G_u = G_u^+ ,$$

since

$$\begin{aligned}(G_u\psi, \phi) &= \frac{d^2}{d\varepsilon d\tau} F[u + \varepsilon\phi + \tau\psi]|_{\varepsilon=\tau=0} \\ &= \frac{d^2}{d\tau d\varepsilon} F[u + \varepsilon\phi + \tau\psi]|_{\varepsilon=\tau=0} =: (G_u\phi, \psi) .\end{aligned}$$

Example 4 *Let U be the space of all C^∞ functions $u(x)$, defined on the interval $I =]a, b[$ of the real axis x and going to zero at a and b, together with all their x derivatives, and let $F[u]$ be a functional,*

$$F[u] = \int_a^b f(u, u_x, u_{xx}, \ldots, u_{nx}, \ldots) dx ,$$

where f is a given function of u and of its x derivatives, which we are denoting with $u_x, u_{xx}, \ldots, u_{nx}, \ldots$.

According to the definition of Eq. (1.34), we have

$$\begin{aligned}\delta F &\equiv \frac{d}{d\varepsilon} F[u + \varepsilon h]|_{\varepsilon=0} \\ &= \int_a^b \left[\frac{\partial f}{\partial u} h + \frac{\partial f}{\partial u_x} h_x + \frac{\partial f}{\partial u_{xx}} h_{xx} + \cdots + \frac{\partial f}{\partial u_{nx}} h_{nx} + \cdots \right] dx ,\end{aligned}$$

and integrating by parts all the terms, except the first,

$$\delta F = \int_a^b \left[h\frac{\partial f}{\partial u} - h\frac{d}{dx}\frac{\partial f}{\partial u_x} + h\frac{d^2}{dx^2}\frac{\partial f}{\partial u_{xx}} + \cdots + h(-1)^n \frac{d^n}{dx^n}\frac{\partial f}{\partial u_{nx}} + \cdots \right] dx .$$

Therefore,

$$\delta F = \int_a^b \left(\frac{\partial f}{\partial u} - \frac{d}{dx}\frac{\partial f}{\partial u_x} + \frac{d^2}{dx^2}\frac{\partial f}{\partial u_{xx}} + \cdots + (-1)^n \frac{d^n}{dx^n}\frac{\partial f}{\partial u_{nx}} + \cdots , h \right) dx ,$$

so that

$$\frac{\delta F}{\delta u} = \frac{\partial f}{\partial u} - \frac{d}{dx}\frac{\partial f}{\partial u_x} + \frac{d^2}{dx^2}\frac{\partial f}{\partial u_{xx}} + \cdots + (-1)^n \frac{d^n}{dx^n}\frac{\partial f}{\partial u_{nx}} + \cdots .$$

In this way, in the simple case of a functional depending only on u,

$$F[u] = \int_a^b f(u)dx\,,$$

the functional derivative is given by

$$\frac{\delta F}{\delta u} = \frac{\partial f}{\partial u}\,.$$

The functional derivative of a functional depending only on u and u_x,

$$F[u] = \int_a^b f(u, u_x)dx\,,$$

is given by

$$\frac{\delta F}{\delta u} = \frac{\partial f}{\partial u} - \frac{d}{dx}\frac{\partial f}{\partial u_x}\,.$$

For instance, for square integrable functions, we have

$$\frac{\delta}{\delta u}\frac{1}{2}\int_a^b u^2 dx = u\,,$$

$$\frac{\delta}{\delta u}\frac{1}{2}\int_a^b u_x^2 dx = -u_{xx}\,,$$

$$\frac{\delta}{\delta u}\frac{1}{2}\int_a^b \left(\frac{1}{3}u^3 - u_x^2\right) dx = u^2 + u_{xx}\,,$$

and the Gateaux derivatives of

$$u\,, \quad -u_{xx}\,, \quad u^2 + u_{xx}$$

are the following operators

$$1\,, \quad -\partial_{xx}\,, \quad 2u + \partial_{xx}\,, \quad \partial_x \equiv \frac{d}{dx}\,,$$

which are symmetric with respect to the L_2 scalar product.

Example 5 *The function u_x cannot be the gradient, with respect to the L_2 scalar product, of any functional, since its Gateaux derivative is the skewsymmetric operator ∂_x.*

Example 6 *By using the same procedure, it is easy to see that for the functional*

$$F[u_1,\dots,u_n] = \int_a^b f\left(u_1,\dots,u_n,\frac{du_1}{dx},\dots,\frac{du_n}{dx}\right)dx\,,$$

we have

$$\delta F = \int_a^b \sum_{i=1}^n \left[\frac{\partial f}{\partial u_i} - \frac{d}{dx}\frac{\partial f}{\partial u_i}\right] h_i\,,$$

so that

$$\frac{\delta F}{\delta u} \equiv \left(\frac{\delta F}{\delta u_1},\dots,\frac{\delta F}{\delta u_n}\right) = \left(\frac{\partial f}{\partial u_1} - \frac{d}{dx}\frac{\partial f}{\partial u_1},\dots,\frac{\partial f}{\partial u_n} - \frac{d}{dx}\frac{\partial f}{\partial u_n}\right).$$

The Hamilton principle

From the previous example, by identifying the coordinate x with the time t, the functions u's with the Lagrangian coordinates q's and the derivatives u_x's with the Lagrangian velocities $\dot{q}$'s, it follows that Lagrange's equations

$$\frac{d}{dt}\frac{\partial \mathcal{L}}{\partial \dot{q}_h} - \frac{\partial \mathcal{L}}{\partial q_h} = 0\,, \quad h \in \{1,\dots,n\}$$

are equivalent to the vanishing of the functional derivatives of the functional

$$S[q] = \int_{t_P}^{t_Q} \mathcal{L}(q/\dot{q}/t)dt\,,$$

for functions $q(t)$ vanishing at the instants t_P and t_Q.

Thus, Lagrange's equations are the equations for the *extrema* (or *critical points*) of $S[q]$. When some further hypothesis are assumed on the $q(t)$'s, the Lagrangian equations give the curve on which the action integral takes its minimum value. For this reason, the Hamilton principle is also called *the least action principle.*

1.5 Remarks on Lagrange's Equations

1.5.1 *Equivalent Lagrangians*

Looking at the action integral, it turns evident that the Lagrangian equations corresponding to Lagrangian functions $\mathcal{L}(q/\dot{q}/t)$ and $\mathcal{L}'(q/\dot{q}/t)$, which differ by a term given by the derivative of a function depending only by the coordinates q and the time t,

$$\mathcal{L}'(q/\dot{q}/t) = \mathcal{L}(q/\dot{q}/t) + \frac{d}{dt} f(q/t)$$

coincide. Indeed, the two action integrals,

$$S[q] = \int_{t_P}^{t_Q} \mathcal{L}(q/\dot{q}/t)dt\,, \quad S'[q] = \int_{t_P}^{t_Q} \mathcal{L}'(q/\dot{q}/t)dt$$

differ by a constant term

$$S'[q] - S[q] = f(Q/t_Q) - f(P/t_P)\,,$$

which does not contribute to the variation

$$\delta S'[q] = \delta S[q]\,.$$

Two such Lagrangians are called *equivalent.*

This is not, of course, the most general case and there exist examples of Lagrangian dynamics admitting more than one, not equivalent, Lagrangian. An exhaustive treatment can be found in Ref. 157.

1.5.2 *Dynamical similitude*

Let us start by observing that two Lagrangian functions differing by a constant factor give the same Lagrangian equations. Thanks to this circumstance, it is possible to infer some properties of the motion without integrating the corresponding equations. This happens, for instance, for simple systems with kinetic energy $\mathcal{T} = \frac{1}{2}mv^2 = \frac{1}{2}m\sum_h \dot{q}_h^2$ and a potential energy $\mathcal{U}(q)$, which is a homogeneous function of the coordinates; i.e. a function satisfying the condition

$$\mathcal{U}(\lambda q) = \lambda^\alpha \mathcal{U}(q)\,,$$

where λ is an arbitrary constant and α is the homogeneity degree of $\mathcal{U}$.

By performing the transformation

$$q_h \to q'_h = \lambda q_h \,, \quad t \to t' = \mu t \,,$$

- the velocities $\dot{q}_h$ will be multiplied by the factor λ/μ

$$\dot{q}_h \to \dot{q}'_h = \frac{\lambda}{\mu}\dot{q}_h \,;$$

- the kinetic energy $\mathcal{T}$ will be multiplied by the factor $(\lambda/\mu)^2$

$$\mathcal{T} \to \left(\frac{\lambda}{\mu}\right)^2 \mathcal{T} \,;$$

- the potential energy $\mathcal{U}$ will be multiplied by the factor λ^α

$$\mathcal{U} \to \lambda^\alpha \mathcal{U} \,;$$

- the Lagrangian function $\mathcal{L}$ will change according to

$$\mathcal{L} \to \left(\frac{\lambda}{\mu}\right)^2 \mathcal{T} - \lambda^\alpha \mathcal{U} \,.$$

As a consequence, the Lagrangian function will be multiplied by a constant factor λ^α only if

$$\mu = \lambda^{1-\frac{\alpha}{2}} \,.$$

To change the coordinates by a constant factor λ means to pass from some trajectories to other ones geometrically similar to the first, the only difference lying in the linear dimensions (homothety).

We finally arrive to the following conclusion:

If the potential energy of a simple system is a homogeneous function of coordinates, the equations of the motion admit geometrically similar trajectories for which the time interval t' and t, between corresponding points on trajectories, have the ratio

$$\frac{t'}{t} = \left(\frac{l'}{l}\right)^{1-\frac{\alpha}{2}} \,. \tag{1.35}$$

Example 7 *For small oscillations, the potential energy is a quadratic function of coordinates ($\alpha = 2$). Equation (1.35) shows that the period P does not depend on their amplitudes (Galilei's observation on the candelabrum at Duomo in Pisa).*

Example 8 *In a homogeneous field, the potential energy is a linear function of coordinates* ($\alpha = 1$). *Equation (1.35) shows that for a free falling body in the gravity field, the squares of time of falling are in the ratio of their initial heights.*

Example 9 *In the case of the Newtonian attraction between two masses, or the Coulomb attraction of two charges, the potential energy is in the inverse proportion with their distance* ($\alpha = -1$). *Equation (1.35) shows that the squares of the revolution periods of planets, in their orbits, are proportional to the third power of their linear dimensions (Kepler's third law).*

The previous analysis can also be carried out, of course, by means of Newton's equations. The reader is invited to do it by himself.

1.5.3 *Electrical circuit analysis*

The circuital relations, for a network of coupled reactive impedances in which a system of electrical currents i_h is flowing, generated by electromotive forces V_h, are

$$\sum_k L_{hk} \frac{d^2 i_k}{dt^2} = F_h - \frac{1}{C_h} i_h \, ,$$

where $L_{hk} = L_{kh}$ are the mutual inductances ($h \neq k$) and self-inductances ($h = k$), C_h the capacitances, R_h the resistive impedances and $F_h = dV_h/dt$. They are the Lagrange equations associated with the Lagrangian function,[6,35]

$$\mathcal{L} = \frac{1}{2} \sum_{hk} L_{hk} \frac{di_h}{dt} \frac{di_k}{dt} + \sum_h F_h i_h - \frac{1}{2} \sum_h \frac{1}{C_h} i_h^2 \, .$$

Chapter 2

Hamiltonian Systems

Lagrange's equations constitute a system of n second order differential equations in the unknown curves $q_h = q_h(t)$. By writing

$$\frac{d}{dt}\frac{\partial \mathcal{L}}{\partial \dot{q}_h} = \sum_{k=1}^{n} \frac{\partial^2 \mathcal{L}}{\partial \dot{q}_h \partial \dot{q}_k}\ddot{q}_k + \sum_{k=1}^{n} \frac{\partial^2 \mathcal{L}}{\partial \dot{q}_h \partial q_k}\dot{q}_k + \frac{\partial^2 \mathcal{L}}{\partial \dot{q}_h \partial t},$$

they can be explicitly written in the form

$$\sum_{k=1}^{n} \frac{\partial^2 \mathcal{L}}{\partial \dot{q}_h \partial \dot{q}_k}\ddot{q}_k = F_h(q/\dot{q}/t),$$

where the force F^h is defined by

$$F_h(q/\dot{q}/t) \equiv \frac{\partial \mathcal{L}}{\partial q_h} - \sum_{k=1}^{n} \frac{\partial^2 \mathcal{L}}{\partial \dot{q}_h \partial q_k}\dot{q}_k - \frac{\partial^2 \mathcal{L}}{\partial \dot{q}_h \partial t}.$$

Thus, if the Lagrangian function is regular; that is, the *Hessian determinant* $\mathcal{J}$ of the matrix

$$L \equiv \left(\frac{\partial^2 \mathcal{L}}{\partial \dot{q}_h \partial \dot{q}_k}\right)$$

is not vanishing, Lagrange's equations can be written in the following *normal form:*

$$\ddot{q}_k = \sum_{h=1}^{n} (L^{-1})_{kh} F_h \,.$$

Moreover, if the relations $dq_h/dt = v_h$ are not interpreted any more as constraints, Lagrange's equations can be written, much more naturally, as a system of $2n$ first order differential equations in the form

$$\begin{cases} \dfrac{d}{dt}\dfrac{\partial \mathcal{L}}{\partial v_h} - \dfrac{\partial \mathcal{L}}{\partial q_h} = 0\,, \\ \dfrac{d}{dt} q_h = v_h\,, \end{cases} \qquad \forall\, h \in \{1, 2, \ldots, n\}\,, \tag{2.1}$$

and for a regular Lagrangian, in the following normal form:

$$\begin{cases} \dfrac{d}{dt} v_h = \displaystyle\sum_{k=1}^{n} (L^{-1})_{hk} F_k\,, \\ \dfrac{d}{dt} q_h = v_h\,, \end{cases} \qquad \forall\, h \in \{1, 2, \ldots, n\}\,.$$

2.1 The Legendre Transformation

The system (2.1) is not form invariant and can be transformed to a new system of $2n$ first order differential equations in infinitely many ways. Among them there exists a remarkable transformation, the so-called *Legendre* transformation*, leading to a particular system of $2n$ first order differential equations, called a *Hamiltonian system*, possessing very interesting symmetry properties. The Legendre transformation is naturally suggested by the Lagrangian system and it consists in introducing n new auxiliary coordinates p, defined by

$$p_h = \frac{\partial \mathcal{L}}{\partial v_h}(q/v/t)\,, \quad \forall\, h \in \{1, 2, \ldots, n\}\,. \tag{2.2}$$

*Adrien Marie Legendre was born in Toulouse on September 18, 1752 and died in Paris on January 10, 1833. He was appointed professor at the military school in Paris in 1777, and at the *École Normale* in 1795. The influence of Laplace was steadily exerted against his obtaining office public recognition, and Legendre, who was a timid student, accepted the obscurity to which the hostility of his colleague condemned him. Legendre's analysis is of high order of excellence, and is second only to that produced by Lagrange and Laplace, though it is not so original.

The p's are called *conjugate variables of* q's, or for their dynamical interpretation in some typical cases, *conjugate momenta of* q's.

The above system, considered as an algebraic system of equations, can be solved, for a regular Lagrangian, with respect to the unknown v_h in the following form:

$$v_h = \alpha_h(q/p/t)\,. \tag{2.3}$$

By using Eqs. (2.2) and (2.3), the Lagrangian equations (2.1) become

$$\begin{cases} \left[\dfrac{d}{dt}\dfrac{\partial \mathcal{L}}{\partial v_h}\right]_* = \left[\dfrac{\partial \mathcal{L}}{\partial q_h}\right]_*\,, \\ \dfrac{d}{dt}q_h = \alpha_h(q/p/t)\,, \end{cases} \tag{2.4}$$

where the symbol $*$ indicates that the velocities v_h have been expressed, by means of Eq. (2.3), in terms of Lagrangian coordinates q, their conjugate momenta p and time t.

Thus, we obtain the *normal* system of $2n$ first order differential equations with the $2n$ unknown functions q's and p's,

$$\begin{cases} \dfrac{d}{dt}p_h = \left[\dfrac{\partial \mathcal{L}}{\partial q_h}\right]_*\,, \\ \dfrac{d}{dt}q_h = \alpha_h(q/p/t)\,. \end{cases} \tag{2.5}$$

Of course, the above system is equivalent to the original Lagrangian system (2.1), since from one side, it follows from Eq. (2.1) by means of the described procedure ($\mathcal{J} \neq 0$), and vice versa, it is possible to go back to Eq. (2.1), replacing the momenta p in Eq. (2.5) with their expression given by Eq. (2.2).

2.2 The Hamilton Equations

2.2.1 *From Lagrange to Hamilton equations*

It is remarkable that the RHS of Eqs. (2.5) can be expressed in terms of a unique function of q's, p's and t, called Hamilton function or characteristic function or simply Hamiltonian.

In this way, the first order system will appear as simple as the original Lagrangian system (2.1), depending on the unique function $\mathcal{L}$. The mentioned

Hamiltonian function is essentially the function

$$E(q/v/t) = \sum_h v_h \frac{\partial \mathcal{L}}{\partial v_h} - \mathcal{L}\,,$$

representing, in the dynamical case, the total energy of the system. The only distinguishing difference is that it must be expressed in terms of q's, p's and t by means of Eqs. (2.2) and (2.3). In other words, the Hamiltonian function is the function

$$\mathcal{H}(p/q/t) = E(q/v/t)_* = \sum_h p_h \alpha_h(q/p/t) - \mathcal{L}(q/\alpha(q/p/t)/t)\,. \tag{2.6}$$

In order to recognize that RHS of Eqs. (2.5) can be expressed in a very simple way, in terms of the function defined by Eq. (2.6), it is enough to apply the following classical procedure first introduced by Hamilton. By considering the q's, p's and t as independent variables and the v's as function of them, given by Eq. (2.3), let us add, for fixed t, the arbitrary increments δp and δq, to q's and p's. In this way, the function $\mathcal{H}$ will increase by the differential of $\mathcal{H}(p/q/t)$, given by

$$\delta\mathcal{H} = \sum_k \frac{\partial \mathcal{H}}{\partial p_k}\delta p_k + \frac{\partial \mathcal{H}}{\partial q_k}\delta q_k\,.$$

On the other side†, from Eq. (2.6),

$$\begin{aligned}\delta\mathcal{H}(p/q/t) &= \delta[E(q/v/t)]_{|v_h=\alpha_h(q/p/t)} = [\delta E(q/v/t)]_{|v_h=\alpha_h(q/p/t)} \\ &= \sum_h \left[v_h \delta\left(\frac{\partial \mathcal{L}}{\partial v_h}\right) + \frac{\partial \mathcal{L}}{\partial v_h}\delta v_h - \delta\mathcal{L}\right]_{|v_h=\alpha_h(q/p/t)} \\ &= \sum_h \left[v_h \delta\left(\frac{\partial \mathcal{L}}{\partial v_h}\right) + \frac{\partial \mathcal{L}}{\partial v_h}\delta v_h - \frac{\partial \mathcal{L}}{\partial v_h}\delta v_h - \frac{\partial \mathcal{L}}{\partial q_h}\delta q_h\right]_{|v_h=\alpha_h(q/p/t)}\end{aligned}$$

†As for Eq. (2.3), the velocities v's are not independent variables. However, as it is well known from differential calculus, the first differential of $[\mathcal{L}]_{v=\alpha}$ coincides with the differential of $\mathcal{L}$ transformed with $v_h = \alpha_h(p/q/t)$. This follows from the observation that the first order differential of a function f is form invariant; i.e. the transformed of its differential coincides with the differential of the transformed function: $(df)_* = d(f)_*$, where the symbol $*$ denotes the transformation. This property, which does not hold for the second order differentials, will be more transparent after the geometrical considerations of Part II.

$$= \sum_h \left[v_h \delta \left(\frac{\partial \mathcal{L}}{\partial v_h} \right) - \frac{\partial \mathcal{L}}{\partial q_h} \delta q_h \right]_{|v_h = \alpha_h(q/p/t)}$$

$$= \sum_h \left(\alpha_h \delta p_h - \left[\frac{\partial \mathcal{L}}{\partial q_h} \right]_{|v_h=\alpha_h(q/p/t)} \delta q_h \right) .$$

Therefore, by comparison, the following equalities hold:

$$\begin{cases} \alpha_h(q/p/t) = \dfrac{\partial \mathcal{H}}{\partial p_h} , \\ \left[\dfrac{\partial \mathcal{L}}{\partial q_h} \right]_* = -\dfrac{\partial \mathcal{H}}{\partial q_h} , \end{cases} \quad \forall\, h \in \{1, 2, \ldots, n\} .$$

By using Eqs. (2.2) and (2.2.1), Lagrange's equations (2.1) give

$$\begin{cases} \dfrac{d}{dt} \dfrac{\partial \mathcal{L}}{\partial v_h} = \dfrac{\partial \mathcal{L}}{\partial q_h} \\ \dfrac{d}{dt} q_h = v_h \end{cases} \Longrightarrow \begin{cases} \left[\dfrac{d}{dt} \dfrac{\partial \mathcal{L}}{\partial v_h} \right]_* = \left[\dfrac{\partial \mathcal{L}}{\partial q_h} \right]_* = -\dfrac{\partial \mathcal{H}}{\partial q_h} , \\ \dfrac{d}{dt} q_h = \alpha_h(q/p/t) = \dfrac{\partial \mathcal{H}}{\partial p_h} , \end{cases} \tag{2.7}$$

or definitively

$$\begin{cases} \dfrac{d}{dt} p_h = -\dfrac{\partial \mathcal{H}}{\partial q_h} \\ \dfrac{d}{dt} q_h = \dfrac{\partial \mathcal{H}}{\partial p_h} \end{cases} , \quad \forall\, h \in \{1, 2, \ldots, n\} . \tag{2.8}$$

The above equations are called Hamilton's equations. Any system of the form (2.8), irrespective of how the function $\mathcal{H}(q/p/t)$ has been chosen, is called a *canonical system* or a *Hamiltonian system.* The p's and q's are called *canonical coordinates.* No essential distinctions there exists between them, as the system (2.8) is invariant under the interchange: $p \longleftrightarrow q$, $\mathcal{H} \longleftrightarrow -\mathcal{H}$.

2.2.2 *From Hamilton to Lagrange equations*

It has been shown, in the previous section, that a given Lagrangian system reducible to the normal form; i.e. such that $\mathcal{J} \neq 0$, can be suitably transformed to a new system of $2n$ first order differential equations in the unknown p's and q's. Moreover, Eqs. (2.2),

$$p_h = \frac{\partial \mathcal{L}}{\partial v_h}(q/v/t) , \quad \forall\, h \in \{1, 2, \ldots, n\} , \tag{2.9}$$

once solved with respect to v's, have the form

$$v_h = \frac{\partial \mathcal{H}}{\partial p_h}, \quad \forall\, h \in \{1, 2, \ldots, n\}, \tag{2.10}$$

where $\mathcal{H}$ denotes the Hamiltonian function defined by Eq. (2.6). It is, of course, possible to go back. Suppose that we start from Eq. (2.8) and that the Hessian determinant,

$$\Gamma = \det\left(\frac{\partial^2 \mathcal{H}}{\partial p_h \partial p_k}\right),$$

of $\mathcal{H}$ is nonvanishing, so that the equations

$$v_h \equiv \frac{d}{dt} q_h = \frac{\partial \mathcal{H}}{\partial p_h}(p/q/t),$$

can be solved with respect to p's to give

$$p_h = \beta_h(q/v/t), \quad \forall\, h \in \{1, 2, \ldots, n\}.$$

Let us then define the Lagrangian function corresponding to $\mathcal{H}$ by

$$\mathcal{L} = \sum_h v_h \beta_h(q/v/t) - \mathcal{H}(q/\beta(q/v/t)/t),$$

or shortly

$$\mathcal{L} = \left[\sum_h p_h \frac{\partial \mathcal{H}}{\partial p_h} - \mathcal{H}\right]_{p_h = \beta_h(q/v/t)},$$

and consider its differential,

$$\delta \mathcal{L} = \sum_k \frac{\partial \mathcal{L}}{\partial v_k} \delta v_k + \frac{\partial \mathcal{L}}{\partial q_k} \delta q_k .$$

By using again the observation that $(\delta f)_* = \delta(f)_*$, we obtain

$$\begin{aligned}
\delta \mathcal{L} &= \left[\sum_h \left(p_h \delta\left(\frac{\partial \mathcal{H}}{\partial p_h}\right) + \frac{\partial \mathcal{H}}{\partial p_h} \delta p_h\right) - \delta \mathcal{H}\right]_{|p_h = \beta_h(q/p/t)} \\
&= \sum_h \left[p_h \delta\left(\frac{\partial \mathcal{H}}{\partial p_h}\right) + \frac{\partial \mathcal{H}}{\partial p_h} \delta p_h - \frac{\partial \mathcal{H}}{\partial p_h} \delta p_h - \frac{\partial \mathcal{H}}{\partial q_h} \delta q_h\right]_{|p_h = \beta_h(q/p/t)}
\end{aligned}$$

$$= \sum_h \left[p_h \delta \left(\frac{\partial \mathcal{H}}{\partial p_h} \right) - \frac{\partial \mathcal{H}}{\partial q_h} \delta q_h \right]_{|p_h = \beta_h(q/p/t)}$$

$$= \sum_h \left(\beta_h \delta v_h - \left(\frac{\partial \mathcal{H}}{\partial q_h} \right)_{|p_h = \beta_h(q/p/t)} \delta q_h \right).$$

Therefore, we have

$$\begin{cases} \beta_h(q/v/t) = \dfrac{\partial \mathcal{L}}{\partial v_h}, \\ \dfrac{\partial \mathcal{L}}{\partial q_h} = -\left[\dfrac{\partial \mathcal{H}}{\partial q_h} \right]_*, \end{cases} \quad \forall\, h \in \{1, 2, \ldots, n\}. \tag{2.11}$$

By using the above relations, Hamilton's equations (2.8) give

$$\begin{cases} \dfrac{d}{dt} p_h = -\dfrac{\partial \mathcal{H}}{\partial q_h} \\ \dfrac{d}{dt} q_h = -\dfrac{\partial \mathcal{H}}{\partial p_h} \end{cases} \Longrightarrow \begin{cases} \dfrac{d}{dt} \dfrac{\partial \mathcal{L}}{\partial v_h} = \dfrac{d}{dt} \beta_h = -\left[\dfrac{\partial \mathcal{H}}{\partial q_h} \right]_* = \dfrac{\partial \mathcal{L}}{\partial q_h}, \\ \dfrac{d}{dt} q_h = v_h, \end{cases}$$

which are just the Lagrange equations (2.1).

We finally observe that the Hessian matrices

$$\left(\frac{\partial^2 \mathcal{L}}{\partial v_h \partial v_k} \right), \quad \left(\frac{\partial^2 \mathcal{H}}{\partial p_h \partial p_k} \right)$$

of the Lagrangian and of the Hamiltonian, respectively, are inverse to each other, since

$$\delta_h^k = \frac{\partial p_h}{\partial p_k} = \sum_r \frac{\partial \beta_h}{\partial v_r} \frac{\partial \alpha_r}{\partial p_k} = \sum_r \frac{\partial^2 \mathcal{L}}{\partial v_r \partial v_h} \frac{\partial^2 \mathcal{H}}{\partial p_k \partial p_r},$$

so that $\Gamma = \mathcal{J}^{-1}$.

2.2.3 *Remarks on Hamilton's equations*

The virial theorem

The time average $\bar{f}$ of a function $f(t)$ is defined by the following limit:

$$\bar{f} = \lim_{\tau \to \infty} \frac{1}{\tau} \int_0^\tau f(t) dt.$$

From the definition, it turns out that the time average of a function, which is the time derivative $f = dF/dt$ of a bounded function $F(t)$, is vanishing. Indeed,

$$\bar{f} = \lim_{\mathcal{T}\to\infty} \frac{1}{\mathcal{T}} \int_0^{\mathcal{T}} f(t)dt = \lim_{\mathcal{T}\to\infty} \frac{1}{\mathcal{T}} \int_0^{\mathcal{T}} \frac{dF}{dt} dt = \lim_{\mathcal{T}\to\infty} \frac{F(t) - F(0)}{\mathcal{T}} = 0\,.$$

If the motion of a simple system, whose potential energy $\mathcal{U}(q)$ is a homogeneous function of coordinates, develops in a bounded region of the space, there exists a very simple relation among the time averages of the potential and kinetic energies. This relation is known under the name of *the virial theorem.*

Let us start from Eq. (2.6) written in the following form:

$$\mathcal{H} = \left(\sum_h p_h \dot{q}_h - \mathcal{L} \right)_{|v_h = \alpha_h(q/p/t)} .$$

The above relation can also be written as follows:

$$\mathcal{H} + \mathcal{L}^* = \frac{d}{dt} \sum_h p_h q_h + \sum_h q_h \frac{\partial \mathcal{H}}{\partial q_h} .$$

In the case the motion takes place with bounded "velocities" in a bounded region of the configuration space, by taking the time averages of both sides, we have

$$\overline{\mathcal{H} + \mathcal{L}^*} = \overline{\sum_h q_h \frac{\partial \mathcal{H}}{\partial q_h}} .$$

For a simple system, the quantity $\mathcal{H} + \mathcal{L}^*$ is twice the kinetic energy $\mathcal{T}^*$, so that we can write

$$2\overline{\mathcal{T}^*} = \overline{\sum_h q_h \frac{\partial \mathcal{U}}{\partial q_h}} .$$

We can finally argue that, for a simple system whose potential energy $\mathcal{U}(q)$ is a homogeneous function of coordinates of degree α, and for motion with bounded velocities in a bounded region of space, the interesting relation

$$2\overline{\mathcal{T}^*} = \alpha \overline{\mathcal{U}}$$

holds.

Given that, for a simple system, $\mathcal{H} = \mathcal{T}^* + \mathcal{U}$, the above relation can be also written in the following equivalent form:

$$\begin{cases} \overline{\mathcal{U}} = \dfrac{2}{2+\alpha} E\,, \\ \overline{\mathcal{T}^*} = \dfrac{\alpha}{2+\alpha} E\,, \end{cases}$$

where E is the total energy.

In particular,

- for small oscillations: $\overline{\mathcal{T}^*} = \overline{\mathcal{U}} = (1/2)E$,
- for the Newtonian interaction: $2\overline{\mathcal{T}^*} = -\overline{\mathcal{U}}$, or $E = -\overline{\mathcal{T}^*}$, which given that $\overline{\mathcal{T}^*} > 0$, says that the Newtonian motion will be bounded in space only if the total energy E is negative.

The above analysis can be also carried out, of course, by means of Lagrange's equations. The reader is invited to do it himself.

2.3 The Poisson Bracket and the Jacobi–Poisson Theorem

2.3.1 *The state space*

The solution of the Lagrangian equations, describing the motion of a given arbitrary holonomic system $\mathcal{S}$ with n degrees of freedom and Lagrangian coordinates $q_1, q_2, \dots, q_n$, is given by n functions of time

$$q_h = q_h(t)\,, \quad \forall\, h \in \{1, 2, \dots, n\}\,, \tag{2.12}$$

representing, at each time t, the position of $\mathcal{S}$. Their derivatives

$$v_h = \dot{q}_h(t)\,, \quad \forall\, h \in \{1, 2, \dots, n\}\,, \tag{2.13}$$

will give at each time t the corresponding velocities. From Cauchy's theorem view point, the state of $\mathcal{S}$ is characterized by the $2n$ parameters $(q/\dot{q})$. In this way, it appears convenient in the analysis of the motion, to use a hyperspace representation, considering the $2n$ parameters q and $\dot{q}$ as Cartesian coordinates of a $2n$ dimensional space E. Since each point of this space represents a state $\{P, v_p\}$ of $\mathcal{S}$, E is called the *state space*. The motion in E, or to be more precise, the sequence of states in E will be represented by the parametric equations (2.12) and (2.13).

2.3.2 The phase space

An analogue geometrical representation is introduced for the canonical coordinates p, q, by considering them as Cartesian coordinates of a $2n$ dimensional Euclidean space Φ called, after Gibbs, the *phase space.* In this space any solution,

$$\begin{cases} p_h = p_h(t)\,, \\ q_h = q_h(t)\,, \end{cases} \quad \forall\, h \in \{1, 2, \ldots, n\}\,,$$

of the canonical system (2.8) is represented by a (integral) curve often called, regarding t as a measure of time, a trajectory. In this way, we will have ∞^{2n} trajectories corresponding to possible choices of the $2n$ arbitrary constants, from which the general integral of the canonical system depends.

2.3.3 First integrals

Let us consider the following canonical system again:

$$\begin{cases} \dfrac{d}{dt} p_h = -\dfrac{\partial \mathcal{H}}{\partial q_h}\,, \\ \dfrac{d}{dt} q_h = \dfrac{\partial \mathcal{H}}{\partial p_h}\,, \end{cases} \quad \forall\, h \in \{1, 2, \ldots, n\}\,.$$

As for any first order differential system of equations, any function f, such that the relation

$$f(p/q/t) = \text{constant} \tag{2.14}$$

is identically satisfied for all solutions of the system is called a *first integral*, or shortly, an *integral* of the canonical system. In other words, if

$$\begin{cases} p_h = p_h(t)\,, \\ q_h = q_h(t)\,, \end{cases} \quad \forall\, h \in \{1, 2, \ldots, n\}\,,$$

denotes an arbitrary solution of the given canonical system, representing the parametric equations of an integral curve in the phase space, the function f in the left hand side is such that

$$f(p(t)/q(t)/t) = \text{constant}\,.$$

For this reason, the function f is also called an *invariant*.

Of course, the function f may take on different constant values for different trajectories in the phase space. More precisely, denoting with p_0, q_0, t_0, the corresponding initial values of p, q, t, the constant must be chosen to coincide with $f(p_0, q_0, t_0)$.

Let us recall that, if the Lagrangian does not depend explicitly on time t, the total energy is a first integral:

$$E(q/v/t) = \sum_h v_h \frac{\partial \mathcal{L}}{\partial v_h} - \mathcal{L} = \text{constant}\,.$$

As for the equivalence between any Lagrangian system and the corresponding canonical system, it is expected that the Hamiltonian function, if not depending explicitly on time t, is a first integral of the canonical system.

It is interesting to prove the above statement directly, since the result follows from a general identity which will turn out useful later.

The rate of change of any function f, depending on the $2n$ coordinates (q/p) and the time t, under the evolution described by Eq. (2.8), is given by

$$\begin{aligned}\frac{d}{dt}f &= \frac{\partial f}{\partial t} + \sum_h \left(\frac{\partial f}{\partial q_h}\frac{d}{dt}q_h + \frac{\partial f}{\partial p_h}\frac{d}{dt}p_h \right) \\ &= \frac{\partial f}{\partial t} + \sum_h \left(\frac{\partial f}{\partial q_h}\frac{\partial \mathcal{H}}{\partial p_h} - \frac{\partial f}{\partial p_h}\frac{\partial \mathcal{H}}{\partial q_h} \right)\,. \end{aligned} \tag{2.15}$$

Thus, for $f = \mathcal{H}$, we have

$$\frac{d\mathcal{H}}{dt} = \frac{\partial \mathcal{H}}{\partial t}\,.$$

Therefore, if the Hamiltonian does not depend on time t, the Hamiltonian function $\mathcal{H}$ defines, for the canonical system, a first integral which can be also called the generalized integral of the energy.

Simple first integrals exist when the characteristic function $\mathcal{H}$ does not depend on some q's. For instance, if $\partial\mathcal{H}/\partial q_r = 0$, from Eq. (2.8) it follows:

$$p_r = \text{constant}\,.$$

These types of integrals also are called *generalized momenta*, as they coincide with the ones given by Lagrangian systems for cyclic coordinates. In this case, in fact, if the Lagrangian function does not depend, for instance, on q_r, the same is true for

$$p_h = \frac{\partial \mathcal{L}}{\partial \dot{q}_h}, \quad \forall\, h \in \{1, 2, \ldots, n\},$$

and for their inverse

$$\dot{q}_h = \alpha_h(q/p/t), \quad \forall\, h \in \{1, 2, \ldots, n\}.$$

It follows also that the characteristic function $\mathcal{H}$ will not depend on q_r.

Vice versa, if q_r does not appear in the characteristic function $\mathcal{H}(p/q/t)$ of a canonical systems, it does not appear in

$$\dot{q}_h = \frac{\partial \mathcal{H}}{\partial p_h}, \quad \forall\, h \in \{1, 2, \ldots, n\},$$

as well, and in their inverse

$$p_h = \beta_h(q/\dot{q}/t), \quad \forall\, h \in \{1, 2, \ldots, n\}.$$

2.3.4 *The Poisson Bracket*

Equation (2.15) can be written in the following form:

$$\frac{df}{dt} = \frac{\partial f}{\partial t} + \{f, \mathcal{H}\},$$

where the bracket of any two functions, f and g, defined by

$$\{f, g\} = \sum_h \left(\frac{\partial f}{\partial q_h}\frac{\partial g}{\partial p_h} - \frac{\partial f}{\partial p_h}\frac{\partial g}{\partial q_h} \right), \tag{2.16}$$

is called the *Poisson*[‡] *Bracket* of f and g.

The Poisson Bracket satisfies the following identities:

- *antisymmetry*

$$\{f, g\} = -\{g, f\}, \tag{2.17}$$

- *Jacobi identity*

$$\{f, \{g, h\}\} + \{g, \{h, f\}\} + \{h, \{f, g\}\} = 0, \tag{2.18}$$

[‡]Siméon Denis Poisson, author of the *Traitè de mécanique* (Paris, 1831), was born in Pithiviers (Loiret) in 1781 and died in Paris in 1840. He was a professor of mechanics at the Sorbonne University.

• *derivation*

$$\begin{cases} \{f, g+h\} = \{f, g\} + \{f, h\}\,, \\ \{f, gh\} = \{f, g\}h + g\{f, h\}\,, \\ \{h, c\} = 0 \quad \forall c \in \Re\,. \end{cases} \tag{2.19}$$

Properties (2.17) and (2.19) follow easily from the definition, and their proof is left to the reader. Property (2.17) simply expresses the antisymmetry of the bracket, while properties (2.19) simply say that the Poisson bracket has a natural compatibility with the usual associative product of functions, on which it acts as a derivative.

The Jacobi[§] identity also follows directly from the definition and the reader can check it by "brute force."

An elegant proof can be given as follows:

Let us observe that the left hand side of Eq. (2.18) is a sum of terms, each one being a product of first partial derivatives of two of the three functions f, g, h with a second partial derivative of the remaining function like

$$\frac{\partial f}{\partial q}\frac{\partial h}{\partial p}\frac{\partial^2 g}{\partial q \partial p}\,.$$

Therefore, the Jacobi identity will be proven if we are able to show that the left-hand side of Eq. (2.18) does not contain any second partial derivative. For this purpose, let us introduce, for any function f, the first order differential operator X_f defined by

$$X_f g = \{f, g\}\,,$$

which will be called the *Hamiltonian vector field associated with* f. The explicit expression of X_f is given by

$$X_f = \sum_h \left(\frac{\partial f}{\partial q_h}\frac{\partial}{\partial p_h} - \frac{\partial f}{\partial p_h}\frac{\partial}{\partial q_h} \right)\,.$$

[§]Karl Gustav Jacobi was born in Postdam in 1804 and died in Berlin in 1851. He is universally known for the investigations on elliptic function, for his papers on determinants and particularly the *Jacobian determinant*, for the *Jacobi identity*, which is basic almost everywhere in physics and mathematics, and for the *Hamilton–Jacobi theory*, which was a starting point for quantum theory. Most of the results of the researches are included in his *Vorlesungen über Dynamik*.

In terms of these operators, the left hand side of Eq. (2.18) can be handled as follows:

$$\begin{aligned}
&\{f,\{g,h\}\}+\{g,\{h,f\}\}+\{h,\{f,g\}\} \\
&\quad = \{f,\{g,h\}\}-\{g,\{f,h\}\}-\{\{f,g\},h\} \\
&\quad = X_f\{g,h\}-\{g,X_fh\}-\{X_fg,h\} \\
&\quad = X_fX_gh-X_gX_fh-X_{\{f,g\}}\,.
\end{aligned}$$

Therefore, the following remarkable relation holds:

$$\{f,\{g,h\}\}+\{g,\{h,f\}\}+\{h,\{f,g\}\}=[X_f,X_g]h-X_{\{f,g\}}h\,, \tag{2.20}$$

where the bracket $[X,Y]$ denotes the *commutator* of the differential operators X and Y.

The final observation is that

(a) the last term $X_{\{f,g\}}h$ does not contain second derivatives of h.

(b) the commutator $[X,Y]$ of two first order differential operators is again a first order differential operator.

Indeed, if

$$X=\sum_i X^i(x)\frac{\partial}{\partial x^i} \text{ and } Y=\sum_i Y^i(x)\frac{\partial}{\partial x^i}$$

denote two first order differential operators, we have

$$\begin{aligned}
[X,Y]\varphi &= \left[\sum_i X^i(x)\frac{\partial}{\partial x^i},\sum_i Y^i(x)\frac{\partial}{\partial x^i}\right]\varphi \\
&= \sum_{ij}\left(X^i(x)\frac{\partial}{\partial x^i}\left(Y^j(x)\frac{\partial\varphi}{\partial x^j}\right)-Y^j(x)\frac{\partial}{\partial x^j}\left(X^i(x)\frac{\partial\varphi}{\partial x^i}\right)\right) \\
&= \sum_{ij}\left(X^i(x)\frac{\partial Y^j}{\partial x^i}\frac{\partial\varphi}{\partial x^j}+X^i(x)Y^j(x)\frac{\partial^2\varphi}{\partial x^i\partial x^j}\right) \\
&\quad -\sum_{ij}\left(Y^j(x)\frac{\partial X^i}{\partial x^j}\frac{\partial\varphi}{\partial x^i}+Y^j(x)X^i(x)\frac{\partial^2\varphi}{\partial x^i\partial x^j}\right)
\end{aligned}$$

$$= \sum_{ij} \left(X^i(x) \frac{\partial Y^j}{\partial x^i} \frac{\partial \varphi}{\partial x^j} - Y^j(x) \frac{\partial X^i}{\partial x^j} \frac{\partial \varphi}{\partial x^i} \right)$$

$$= \sum_{ij} \left(X^i(x) \frac{\partial Y^j}{\partial x^i} \frac{\partial}{\partial x^j} - Y^j(x) \frac{\partial X^i}{\partial x^j} \frac{\partial}{\partial x^i} \right) \varphi$$

$$= \sum_{ij} \left(X^i(x) \frac{\partial Y^j}{\partial x^i} - Y^i(x) \frac{\partial X^j}{\partial x^i} \right) \frac{\partial \varphi}{\partial x^j} .$$

As a consequence of properties (a) and (b), the left-hand side of Eq. (2.20) does not contain any partial second order derivatives of the function h. The same is obviously true for the functions f and g. Therefore, the left-hand side of Eq. (2.20) does not contain any second partial derivative. The conclusion is that the sum of all terms in the left-hand side of Eq. (2.20) is vanishing.

2.3.5 *The Jacobi–Poisson theorem*

Let us suppose now that, for the canonical system

$$\begin{cases} \dfrac{d}{dt} p_h = -\dfrac{\partial \mathcal{H}}{\partial q_h} , \\ \dfrac{d}{dt} q_h = \dfrac{\partial \mathcal{H}}{\partial p_h} , \end{cases} \qquad \forall\, h \in \{1, 2, \ldots, n\} ,$$

two first integrals are known, namely, f and g. These first integrals satisfy the following relations:

$$\frac{\partial f}{\partial t} + \{f, \mathcal{H}\} = 0 , \quad \frac{\partial g}{\partial t} + \{f, \mathcal{H}\} = 0 .$$

It is easy to prove that the Poisson bracket $\{f, g\}$ of f and g is also a first integral. As a matter of fact, let us first observe that

$$\frac{\partial \{f, g\}}{\partial t} = \left\{ \frac{\partial f}{\partial t}, g \right\} + \left\{ f, \frac{\partial g}{\partial t} \right\} .$$

Then, by using the previous relation and the Jacobi identity,

$$\{\{f, g\}, \mathcal{H}\} = \{\{f, \mathcal{H}\}, g\} + \{f, \{g, \mathcal{H}\}\} ,$$

we can write

$$\frac{d\{f,g\}}{dt} = \left\{\frac{df}{dt}, g\right\} + \left\{f, \frac{dg}{dt}\right\} .$$

Therefore,

$$\begin{cases} \dfrac{\partial f}{\partial t} + \{f, \mathcal{H}\} = 0 \\ \dfrac{\partial g}{\partial t} + \{g, \mathcal{H}\} = 0 \end{cases} \Longrightarrow \frac{\partial\{f,g\}}{\partial t} + \{\{f,g\}, \mathcal{H}\} = 0 .$$

Of course, the Jacobi–Poisson theorem will not always give new first integrals, since their number is bounded to be $2n - 1$, where n is the number of degrees of freedom. In some cases, in fact, the Poisson bracket of first integrals simply reduces to a function of them or to a numeric constant. Two functions, f and g, with vanishing Poisson bracket $\{f, g\} = 0$, are said to be in *involution.*

Problems

1. Show that, if one of the functions f and g coincides with a momentum or a coordinate, the Poisson bracket simply reduces to a partial derivative

$$\{p_k, f\} = -\frac{\partial f}{\partial q_k} , \quad \{q_k, f\} = \frac{\partial f}{\partial p_k} .$$

As a particular case, notice that

$$\{q_h, q_k\} = 0 , \quad \{p_h, p_k\} = 0 , \quad \{q_h, p_k\} = \delta_{hk} ,$$

where δ_{hk} is the *Kronecker delta.*

2. Evaluate the Poisson bracket of the Cartesian components of the momentum $\vec{p}$ and the angular momentum $\vec{l} = \vec{r} \wedge \vec{p}$ of a particle.

Solution

$$\{p_x, l_x\} = -\frac{\partial l_x}{\partial x} = -\frac{\partial}{\partial x}(yp_z - zp_y) = 0 ,$$

$$\{p_y, l_x\} = -\frac{\partial l_x}{\partial y} = -\frac{\partial}{\partial y}(yp_z - zp_y) = -p_z ,$$

$$\{p_z, l_x\} = -\frac{\partial l_x}{\partial z} = -\frac{\partial}{\partial z}(yp_z - zp_y) = p_y ,$$

so that

$$\{l_x, p_x\} = 0\,, \quad \{l_x, p_y\} = p_z\,, \quad \{l_x, p_z\} = -p_y\,.$$

The remaining brackets follow from a cyclic permutation of the indices x, y, z. Finally, denoting with (p_1, p_2, p_3) and (l_1, l_2, l_3) the components of $\vec{p}$ and $\vec{l}$, respectively, we write

$$\{l_i, p_j\} = \sum_{h=1}^{3} \varepsilon_{ijh} p_h\,, \quad \forall\, i, j \in \{1, 2, 3\}\,,$$

where ε_{ijh} is the *Levi-Civita*¶ *tensor density* defined by

$$\varepsilon_{ijh} = \begin{cases} 1 & \text{if } i, j, h \text{ is an even permutation of } 1, 2, 3 \\ -1 & \text{if } i, j, h \text{ is an odd permutation of } 1, 2, 3 \\ 0 & \text{in the other cases; i.e., if two indices coincide.} \end{cases}$$

3. Show that

$$\{l_i, q_j\} = \sum_{h=1}^{3} \varepsilon_{ijh} q_h\,, \quad \forall\, i, j \in \{1, 2, 3\}\,.$$

4. Evaluate the Poisson bracket of the components of the angular momentum of a particle between them by using only the algebraic properties (2.19).

Solution

$$\{l_i, l_j\} = \sum_{h=1}^{3} \varepsilon_{ijh} l_h\,, \quad \forall\, i, j \in \{1, 2, 3\}\,.$$

¶Tullio Levi–Civita was born in Padova in 1873 and died in Rome in 1941. He obtained his degree at Padova University in 1894. He was a professor of mathematical physics, at the age of 24 years, at Padova University, where he taught until 1919. In this same year he moved to Rome University. In 1938, with the introduction of racial fascist laws, he was removed from the chair and his now classical books, including the *Lezioni di Meccanica Razionale* (*1929*) and *Lezioni di Calcolo Differenziale Assoluto* (*1923*), was interdicted. Fortunately, thanks to Whittaker, the last had been translated in English and published on 1926 by Blakie & Son. He died just in time to avoid to be forced also to hide because of racial persecutions.

Since the momenta and the coordinates of different particles are independent quantities, it is easy to verify that the resulting formulae of the previous problems still hold for the total momentum $\vec{P}$ and for the total angular momentum $\vec{L}$ of an arbitrary system of particles.

5. Prove that for any *scalar*[||] function φ of the coordinates and momenta of a particle, the following relation holds:

$$\{L_z, \varphi\} = 0\,.$$

Solution

A scalar function depends on the vectors $\vec{r}$ and $\vec{p}$ only by means of the combinations $r^2 = \vec{r}\cdot\vec{r}$, $p^2 = \vec{p}\cdot\vec{p}$ and $\vec{r}\cdot p$. Therefore, we can write

$$\frac{\partial\varphi}{\partial\vec{r}} = \frac{\partial\varphi}{\partial r^2}2\vec{r} + \frac{\partial\varphi}{\partial(\vec{r}\cdot p)}\vec{p}\,,$$

$$\frac{\partial\varphi}{\partial\vec{p}} = \frac{\partial\varphi}{\partial p^2}2\vec{p} + \frac{\partial\varphi}{\partial(\vec{r}\cdot\vec{p})}\vec{r}\,.$$

The relation $\{L_z, \varphi\} = 0$, then follows by applying the definition of Poisson bracket (2.16). The same relations hold for the remaining components of $\vec{l}$, so that, for any scalar function φ of the coordinates and momenta of a particle, we can write

$$\{l_x, \varphi\} = \{l_y, \varphi\} = \{l_z, \varphi\} = 0\,.$$

6. Prove that for any *vector* function $\vec{f}$ of coordinates and momenta of a particle, the following relation holds:

$$\{l_z, \vec{f}\} = \vec{n}\wedge\vec{f}\,,$$

where $\vec{n}$ is the unit vector along the z axis. Analogous formulae hold for the remaining components of $\vec{l}$.

Solution

Any vector function $\vec{f}$ of $\vec{r}$ and $\vec{p}$ can be written in the form $\vec{f} = \varphi_1\vec{r} + \varphi_2\vec{p} + \varphi_3(\vec{r}\wedge\vec{p})$, where $\varphi_1, \varphi_2, \varphi_3$ are scalar functions. Thus, by using the algebraic properties of the Poisson bracket, we finally obtain the solution to the problem.

[||] Here, scalar and vector functions are understood with respect to the rotation group $SO(3)$.

2.4 A More Compact Form of the Hamiltonian Dynamics

Let us start by considering a Hamiltonian system with $n = 2$ degrees of freedom.

By introducing the column vectors

$$u = \mathrm{col}(p_1, p_2, q_1, q_2)\,, \quad \nabla H = \mathrm{col}\left(\frac{\partial \mathcal{H}}{\partial p_1}, \frac{\partial \mathcal{H}}{\partial p_2}, \frac{\partial \mathcal{H}}{\partial q_1}, \frac{\partial \mathcal{H}}{\partial q_2}\right)$$

and the skew-symmetric matrix

$$E = \begin{pmatrix} 0 & -1 \\ 1 & 0 \end{pmatrix}\,, \tag{2.21}$$

Hamilton's equations can be written in the form

$$\dot{u} = E \nabla \mathcal{H}\,,$$

or in components, as follows:

$$\frac{du^h}{dt} = E^{hk} \frac{\partial \mathcal{H}}{\partial u^k}\,, \tag{2.22}$$

where the sum over the index k is understood. The Poisson bracket can then be written as follows:

$$\{f, g\} = \frac{\partial f}{\partial u^h} E^{hk} \frac{\partial g}{\partial u^k} = (\nabla f, E \nabla g)\,,$$

where the bracket $(\cdot, \cdot)$ denotes the Euclidean scalar product.

Of course, the previous notation can be also used for a Hamiltonian system with n degrees of freedom. In this case, the matrix E is given by

$$E = \begin{pmatrix} 0 & -I \\ I & 0 \end{pmatrix}\,, \tag{2.23}$$

where 0 and I denote the $n \times n$ nul and identity matrices.

2.4.1 *General Hamiltonian dynamics*

Let us now consider a generic dynamics described by an equation similar to Eq. (2.22):

$$\frac{du^h}{dt} = \Lambda^{hk} \frac{\partial \mathcal{H}}{\partial u^k}\,, \tag{2.24}$$

where the matrix Λ may depend on the point u. The evolution of a generic function f, defined on the phase space, will be given by

$$\frac{df}{dt} = \frac{\partial f}{du^h}\frac{du^h}{dt} = \frac{\partial f}{du^h}\Lambda^{hk}\frac{\partial \mathcal{H}}{\partial u^k} = (\nabla f, \Lambda \nabla \mathcal{H}) .$$

In order to have a Jacobi–Poisson theorem for this type of dynamics, we must require

- *skew-symmetry*

$$(\nabla f, \Lambda \nabla g) = -(\nabla g, \Lambda \nabla g) ,$$

- *Jacobi identity*

$$(\nabla(\nabla f, \Lambda \nabla g), \Lambda \nabla h) + (\nabla(\nabla g, \Lambda \nabla h), \Lambda \nabla f) + (\nabla(\nabla h, \Lambda \nabla f), \Lambda \nabla g) = 0 .$$

In this case, the bracket $(\nabla f, \Lambda \nabla g)$ will be called the Poisson bracket of f and g, and will be denoted with $\{f, g\}_\Lambda$, or simply, if no ambiguity arises, with $\{f, g\}$. In terms of the matrix Λ, the previous requirements are expressed by the following:

- *skewsymmetry* : $\Lambda = -\Lambda^T$,
- *Jacobi identity* : $\Lambda^{ij}\dfrac{\partial \Lambda^{hk}}{\partial u^j} + \Lambda^{hj}\dfrac{\partial \Lambda^{ki}}{\partial u^j} + \Lambda^{kj}\dfrac{\partial \Lambda^{ih}}{\partial u^j} = 0.$

Of course, these properties are trivially satisfied by the matrix E.

Definition 11 We shall call Hamiltonian a dynamical system with n degrees of freedom, and then with a $2n$-dimensional phase space, if it is described by the equation

$$\frac{df}{dt} = \{f, \mathcal{H}\}_\Lambda ,$$

where the bracket, beyond the usual derivation properties, satisfies the properties

$$\{f, g\}_\Lambda = -\{g, f\}_\Lambda$$
$$\{\{f, g\}_\Lambda, h\}_\Lambda + \{\{g, h\}_\Lambda, f\}_\Lambda + \{\{h, f\}_\Lambda, g\}_\Lambda = 0 .$$

2.4.2 *Jacobi–Poisson dynamics*

Let us finally observe that in the previous definition, no role is played by the even dimensionality of the phase space. Thus, it is natural to define more general dynamics according to the following definition.

Definition 12 A dynamics, described by the equations

$$\frac{df}{dt} = \{f, \mathcal{H}\}_P ,$$

with the bracket satisfying the properties

$$\begin{aligned}
&\{f, c\}_P = 0 \quad \forall c \in R \\
&\{f, g + h\}_P = \{f, g\}_P + \{f, h\}_P \\
&\{f, g, h\}_P = \{f, g\}_P h + g\{f, h\}_P \\
&\{f, g\}_P = -\{g, f\}_P , \\
&\{\{f, g\}_P, h\}_P + \{\{g, h\}_P, f\}_P + \{\{h, f\}_P, g\}_P = 0 ,
\end{aligned}$$

is called a Jacobi–Poisson dynamics.

Of course, a Hamiltonian dynamics is also a Jacobi–Poisson dynamics.

2.4.3 *More on the Poisson bracket*

We notice that properties expressed by Eqs. (2.17) and (2.18) endow the set $\mathcal{F}$, of differentiable functions defined on Φ, with a *Lie algebra structure.*

Remark 6 *A Lie algebra $\mathcal{A}$ is a vector space endowed with an internal composition law, denoted by $[\cdot, \cdot]$ and called a Lie bracket, satisfying the properties:*

$$[X, Y] = -[Y, X] , \quad \forall X, Y \in \mathcal{A} ,$$

$$[X, [Y, Z]] + [Y, [Z, X]] + [Z, [X, Y]] = 0 , \quad \forall X, Y, Z \in \mathcal{A} .$$

Examples of Lie algebras are

◇ *the set of vectors in $\Re^3$ endowed with the Lie bracket $[\cdot, \cdot]$ given by the vector product*

$$[\vec{u}, \vec{v}] = \vec{u} \wedge \vec{v} ;$$

◇ *the vector space of $n \times n$ matrices endowed with the Lie bracket $[\cdot,\cdot]$ given by the commutator*

$$[M, N] = MN - NM .$$

A geometrical definition of Lie algebra will be given in Part II.

Since it can be written in the following equivalent alternative forms:

$$X_{\{f,g\}}h = [X_f, X_g]h , \tag{2.25}$$

$$X_f\{g,h\} = \{X_f g, h\} + \{g, X_f h\} , \tag{2.26}$$

the Jacobi identity is equivalent to the following alternative statements suggested by Eqs. (2.25) and (2.26), respectively:

- The map

  $$f \mapsto X_f = \{f,\cdot\}$$
  $$\{f,g\} \mapsto X_{\{f,g\}}$$

  is a Lie algebra morphism

  $$(\mathcal{F},\{\cdot,\cdot\}) \mapsto (\mathcal{X}_{\mathcal{F}},[\cdot,\cdot])$$

  between $(\mathcal{F},\{\cdot,\cdot\})$ and the set of Hamiltonian vector fields $\mathcal{X}_{\mathcal{F}}$ endowed with the Lie bracket given by the commutator $[\cdot,\cdot]$.
- The operator $X_f = \{f,\cdot\}$ is a derivation of the Poisson bracket.

Last statement, as it will be shown in the next subsection, suggests the introduction of more general structures named *n-Poisson brackets.* For instance, a 3-Poisson bracket on $\mathcal{F}$ is bracket $\{\cdot,\cdot,\cdot\}$ satisfying the following properties:

$$\{f,g,h\} = -\{f,h,g\} = -\{g,f,h\} ,$$
$$\{f,g,\{u,v,w\}\} + \{v,w,\{f,g,u\}\} - \{u,w,\{f,g,v\}\} + \{u,v,\{f,g,w\}\} = 0 ,$$
$$\{f,g,h_1h_2\} = \{f,g,h_1\}h_2 + h_1\{f,g,h_2\} ,$$
$$\{f,g,c\} = 0 \quad \forall c \in R .$$

An algebraic formulation

We observe that properties expressed by Eqs. (2.17), (2.18) and (2.19) are purely algebraic in nature, so that the following abstract formulation can be introduced.

Let M be a Poisson manifold and $\mathcal{F}$ the ring of functions defined on it. This means that on M a bracket $\{\cdot,\cdot\}$ is defined such that

(1) it yields the structure of a Lie algebra on $\mathcal{F}$; i.e.

$$\{f,g\} = -\{g,f\}\,,$$

$$\{f,\{g,h\}\} + \{g,\{h,f\}\} + \{h,\{f,g\}\} = 0\,,$$

(2) it has a natural compatibility with the usual associative product of functions, which is

$$\{h,fg\} = \{h,f\}g + f\{h,g\}\,.$$

Therefore, we can define an abstract *Poisson algebra* as an associative commutative algebra endowed with a Lie bracket satisfying Eqs. (2.17), (2.18) and (2.19).

It is natural to generalize the notion of a Poisson manifold by relaxing condition (2) and requiring only that $\{f,g\}$ be just a *local type* operation:

$$\text{support}\ \{f,g\} \subseteq (\text{support}\ f) \cap (\text{support}\ g)\,.$$

The bracket $\{f,g\}$ is then called a *Jacobi bracket* and thc corresponding manifold a *Jacobi manifold.*

2.4.4 *Further generalizations of the Jacobi–Poisson dynamics*

The possibility of further generalizations of Jacobi–Poisson dynamics rely on the possibility to generalize the Poisson bracket.

Let us consider a dynamical system described by the equations

$$\frac{d}{dt}f = \{f,\mathcal{H}_1,\mathcal{H}_2\}\,,$$

where the *ternary bracket* in the right-hand side is supposed to be skewsymmetric. This dynamics will be called a *ternary Jacobi–Poisson dynamics* if the ternary bracket allows for a Jacobi–Poisson theorem on first integrals. In such a case the ternary bracket will be called *ternary Jacobi–Poisson bracket.* We are thus looking for a property of the ternary bracket such that

$$\{f_h,H_1,H_2\} = 0,\ h = 1,2,3 \Rightarrow \{\{f_1,f_2,f_3\},H_1,H_2\} = 0\,. \tag{2.27}$$

For this purpose it is useful to recall the form of Jacobi identity, for *binary bracket*, given in Eq. (2.26):

$$X_f\{g,h\} = \{X_f g, h\} + \{g, X_f h\}\,.$$

This form can be immediately generalized to skewsymmetric brackets with an arbitrary number of entries.

Indeed, given the ternary bracket $\{f,g,h\}$, we require that the operator X_{fg} (vector field), defined by

$$X_{fg}h := \{f,g,h\}\,,$$

be a derivation of the bracket; that is

$$X_{fg}\{h_1,h_2,h_3\} = \{X_{fg}h_1,h_2,h_3\} + \{h_1,X_{fg}h_2,h_3\} + \{h_1,h_2,X_{fg}h_3\}\,. \tag{2.28}$$

The above formula can be explicitly written as follows:

$$\{f,g,\{h_1,h_2,h_3\}\} = \{\{f,g,h_1\},h_2,h_3\} + \{h_1,\{f,g,h_2\},h_3\} + \{h_1,h_2,\{f,g,h_3\}\}\,,$$

which would be difficult to invent without a deep understanding of the significance of the usual Jacobi identity.

It is not difficult to prove that Eq. (2.27) is equivalent to Eq. (2.28). We will not go on further on this subject. Much more details can be found in Ref. 152 and references therein, where examples of n-ary Jacobi–Poisson dynamics are explicitly given, and the following important property, here reported just for the case $n = 3$, is proven:

If $\{f,g,h\}$ is a ternary Jacobi–Poisson bracket, the binary bracket $\{f,g\}_h = \{f,g,h\}$, obtained by fixing one of the functions, is a binary Jacobi-Poisson bracket. Furthermore, a linear combination of two of them $c_1\{f,g\}_{h_1} + c_2\{f,g\}_{h_2}$ is again a binary Jacobi–Poisson bracket.

2.5 The Variational Principle for the Hamilton Equations

It has been shown that Lagrange's equations (2.1) are differential equations for the unknown functions $q_h(t)$, which are required to be an extremum of the action

$$S[q] = \int_{t_a}^{t_b} \mathcal{L}(q/\dot{q}/t)dt\,.$$

On the other hand, from Eq. (2.6), we have

$$(\mathcal{L})_* dt = \sum_h p_h dq_h - \mathcal{H}dt\,,$$

where $\mathcal{L}_*$ is the Lagrangian $\mathcal{L}$ in which the velocities $\dot{q}_h$ have been expressed in terms of momenta and coordinates by means of Eq. (2.3).

It is then natural to argue that Hamilton's equations can be obtained as the equations for the extrema of

$$S^*[q/p/t] = \int_{t_a}^{t_b} \mathcal{L}_*(q/p/t)dt\,,$$

the transformed functional of $S[q]$:

$$S^*[q/p/t] = \int_{t_a}^{t_b} \sum_h p_h dq_h - \mathcal{H}dt\,.$$

This is easily verified, since

$$\begin{aligned}
\delta S^* &= \int_{t_a}^{t_b} \sum_h (\delta p_h dq_h + p_h \delta dq_h) - \delta\mathcal{H}dt \\
&= [p_h \delta q_h]_{t_A}^{t_B} + \int_{t_a}^{t_b} \sum_h (\delta p_h dq_h - dp_h \delta q_h) - \delta\mathcal{H}dt \\
&= [p_h \delta q_h]_{t_A}^{t_B} + \int_{t_a}^{t_b} \sum_h \left(\delta p_h dq_h - dp_h \delta q_h - \frac{\partial\mathcal{H}}{\partial p_h}\delta p_h dt - \frac{\partial\mathcal{H}}{\partial q_h}\delta q_h dt\right) \\
&= [p_h \delta q_h]_{t_A}^{t_B} + \int_{t_a}^{t_b} \sum_h \left[\left(dq_h - \frac{\partial\mathcal{H}}{\partial p_h}dt\right)\delta p_h - \left(dp_h + \frac{\partial\mathcal{H}}{\partial q_h}dt\right)\delta q_h\right].
\end{aligned}$$

As before, by imposing $\delta q_h(t_A) = \delta q_h(t_B) = 0$, we obtain

$$\delta S^* = \int_{t_a}^{t_b} \sum_h \left[\left(dq_h - \frac{\partial\mathcal{H}}{\partial p_h}dt\right)\delta p_h - \left(dp_h + \frac{\partial\mathcal{H}}{\partial q_h}dt\right)\delta q_h\right].$$

In this way, Hamilton's equations of the motion follow from the vanishing of δS^* for any choice of δp_h's and δq_h's.

Chapter 3

Transformation Theory

3.1 Canonical, Completely Canonical and Symplectic Transformations

3.1.1 *Canonical transformations*

The differential equations of motion have been brought into a particularly desirable form, the *canonical form*:

$$\begin{cases} \dfrac{d}{dt}p_h = -\dfrac{\partial \mathcal{H}}{\partial q_h}, \\ \dfrac{d}{dt}q_h = \dfrac{\partial \mathcal{H}}{\partial p_h}, \end{cases} \quad \forall\, h \in \{1, 2, \ldots n\}.$$

However, no direct integration method of the canonical system is known. There exist indirect methods which allow to highly simplify the integration problem. One of them is the method of coordinate transformations, whose goal is to find new coordinates, namely (π, χ), in which the characteristic function $\mathcal{H}$ of the canonical system is "more simple." For a generic coordinate transformation, the canonical system is not *form invariant*; i.e., its form is not preserved. Therefore, the interesting preliminary problem is to characterize the set $\mathcal{C}$ of invertible differentiable transformations which preserve the canonical form of equations of motion. Any transformation satisfying this requirement, will be called a *canonical transformation.**

It was already clear from the Lagrangian form of dynamics that a proper choice of coordinates can greatly facilitate the search for the solutions of the differential equations of motion. For instance, since a first integral of the Lagrange equation is known whenever one of the Lagrangian coordinates is cyclic, it is of great interest to produce cyclic coordinates by transforming the original ones.

Let

$$\begin{cases} \pi_h = \pi_h(q/p/t) \\ \chi_h = \chi_h(q/p/t) \end{cases} \tag{3.1}$$

be an invertible differentiable transformation from the coordinates (q/p) to (χ, π), which may depend on time t. It was proven by Sophus Lie† that

A sufficient condition for Eq. (3.1) to define a canonical transformation is that there exist two functions $\mathcal{H}_0$ *and* $\mathcal{F}$ *of* $(q/p/\chi/\pi/t)$ *such that the relation*

$$\sum_h p_h dq_h = \sum_h \pi_h d\chi_h + \mathcal{H}_0 dt + d\mathcal{F} \tag{3.2}$$

identically‡ holds. The new characteristic function is $\mathcal{K} = (\mathcal{H} - \mathcal{H}_0)$,* *where the symbol * indicates that all coordinates* (p, q) *have been expressed in terms of* (π/χ).

*The theory of canonical transformations is essentially due to Jacobi whose efforts were too much bent on the integration problem to which Hamilton was only incidentally interested. The resulting integration theory played an important part in the modern development of atomic physics.

†Marius Sophus Lie was born at Nordfjordeide (Norway) on 1842 and died in Oslo on 1899. He was a professor at Oslo University from 1872 to 1885, at Lipsia University from 1866 to 1987 and again at Oslo University from 1898 to 1889. It is difficult to illustrate, in a short note, its enormous contribution to mathematics. He invented, in particular, the theory of contact transformations, the theory of (finite and infinite) Lie groups, the theory of minimal surfaces, the theory of translation surfaces, the theory of surfaces with geodesical groups, the theory of surfaces with constant curvature. Many results of M.S.Lie have been recovered, independently, by excellent modern mathematicians, after almost 100 years. We just mention here the Konstant–Kirillov–Souriau symplectic structure.

‡Here *identically* means that once the transformation (3.1) has been performed, the relation (3.2) reduces to an identity.

It is a trivial exercisc to vcrify that the transformation

$$\begin{cases} \pi_h = \alpha p_h \\ \chi_h = \beta q_h \end{cases}$$

is a canonical one; the new canonical system being

$$\begin{cases} \dfrac{d}{dt}\pi_h = -\dfrac{\partial \mathcal{K}}{\partial \chi_h} \\ \dfrac{d}{dt}\chi_h = \dfrac{\partial \mathcal{K}}{\partial \pi_h} \end{cases}$$

with $\mathcal{K} = \alpha\beta\mathcal{H}^*$, but it does not satisfy the condition (3.2).

We can argue that the set C of canonical transformations is larger than the one characterized by the Lie condition (3.2).

It was later proven by Lee Hwa-Chung[24] how the condition (3.2) can be generalized, in order to express a necessary and sufficient condition for Eq. (3.1) to be a canonical transformation.

A heuristic way to find a necessary and sufficient condition for a differentiable invertible transformation to be canonical is the following.

By using the variational principle, the Hamilton equations can be written as

$$\delta S = 0\,,$$

where

$$S[q/p/t] = \int_{t_a}^{t_b} \sum_h p_h dq_h - \mathcal{H}dt\,.$$

In this way, associated with the transformation

$$\begin{cases} \pi_h = \pi_h(q/p/t)\,, \\ \chi_h = \chi_h(q/p/t)\,, \end{cases}$$

we have the following picture:

$$\boxed{S[q/p/t] = \int_{t_a}^{t_b} \sum_h p_h dq_h - \mathcal{H}dt} \qquad \boxed{S^*[\chi/\pi/t] = \int_{t_a}^{t_b} \sum_h \pi_h d\chi_h - \mathcal{K}dt}$$

$$\downarrow \delta S = 0 \qquad\qquad \downarrow \delta S^* = 0$$

$$\boxed{\begin{cases} \dfrac{d}{dt} p_h = -\dfrac{\partial \mathcal{H}}{\partial q_h} \\ \dfrac{d}{dt} q_h = \dfrac{\partial \mathcal{H}}{\partial p_h} \end{cases}} \qquad \boxed{\begin{cases} \dfrac{d}{dt} \pi_h = -\dfrac{\partial \mathcal{K}}{\partial \chi_h} \\ \dfrac{d}{dt} \chi_h = \dfrac{\partial \mathcal{K}}{\partial \pi_h} \end{cases}}$$

Therefore, the necessary and sufficient condition for a differentiable invertible transformation to be canonical is that

$$\delta S^* = 0 \Leftrightarrow \delta S = 0\,,$$

where $S^*[\chi/\pi/t]$ is the transformed of $S[q/p/t]$.

The above equivalence will be certainly true if differential forms, up to a multiplicative constant c, differ by an exact differential form $d\mathcal{F}$:

$$\sum_h p_h dq_h - \mathcal{H}dt = c\left(\sum_h \pi_h d\chi_h - \mathcal{K}dt\right) + d\mathcal{F}\,.$$

It was shown by Lee Hua-Chung that the condition is also necessary.

We can conclude that

A necessary and sufficient condition for a differentiable invertible transformation (3.1) to be canonical is the existence of a constant c and of two functions, $\mathcal{H}_0$ and $\mathcal{F}$, of $(q/p/\chi/\pi/t)$, such that the relation

$$\sum_h p_h dq_h = c\sum_h \pi_h d\chi_h + \mathcal{H}_0 dt + d\mathcal{F} \tag{3.3}$$

*identically holds. The new characteristic function is $\mathcal{K} = 1/c(\mathcal{H} - \mathcal{H}_0)^*$, where the symbol * indicates that all coordinates (p, q) have been expressed in terms of (π/χ).*

A simple example of canonical transformation, with $\mathcal{H}_0 = 0$ and $\mathcal{F} = pq$, is given by

$$\begin{cases} \pi = \dfrac{1}{2\lambda}(p^2 + \lambda^2 q^2)\,, \\ \chi = \arctan(\lambda q/p)\,. \end{cases} \tag{3.4}$$

Indeed,

$$\begin{aligned} pdq - \pi d\chi &= pdq - \frac{1}{2\lambda}(p^2 + \lambda^2 q^2) d\left(\arctan\left(\frac{\lambda q}{p}\right)\right) \\ &= pdq - \frac{1}{2\lambda}(p^2 + \lambda^2 q^2)\frac{p^2}{(p^2 + \lambda^2 q^2)}\left(\frac{\lambda pdq - \lambda qdp}{p^2}\right) \\ &= pdq - \frac{1}{2}(pdq - qdp) \\ &= d(pq)\,. \end{aligned}$$

3.1.2 *A general class of canonical transformations*

A particular class of canonical transformations is generated by an arbitrary function V which depends on "one half" of original coordinates and on "one half" of the new ones, for instance on q's and π's.

The function is only required to satisfy the condition that the mixed functional determinant

$$\mathcal{J} = \det\left(\frac{\partial^2 V}{\partial q_h \partial \pi_k}\right)$$

does not identically vanish: $\mathcal{J} \neq 0$.

It is easy to see that the relations

$$p_h = \frac{\partial V}{\partial q_h}\,, \qquad \chi_h = \frac{\partial V}{\partial \pi_h}\,, \tag{3.5}$$

implicitly define an invertible differentiable coordinate transformation between the p, q's and π, χ's. In fact, since $\mathcal{J} \neq 0$, by the implicit function theorem, the second of relations (3.5) can be solved in the form

$$q_h = q_h(\pi/\chi/t)$$

and associated to the first one to give an explicit one-to-one transformation between the coordinates (p, q) and (π, χ).

As for the canonical character, it is enough to observe that Lie's condition is satisfied:

$$\sum_h p_h dq_h + \sum_h \chi_h d\pi_h = \sum_h \left(\frac{\partial V}{\partial q_h} dq_h + \frac{\partial V}{\partial \pi_h} d\pi_h\right) = dV - \frac{\partial V}{\partial t} dt\,,$$

from which, by using $\sum_h \chi_h d\pi_h = d(\sum_h \pi_h \chi_h) - \sum_h \pi_h d\chi_h$, we obtain

$$\sum_h p_h dq_h = \sum_h \pi_h d\chi_h - \frac{\partial V}{\partial t} dt + d\left(V - \sum_h \pi_h \chi_h\right) .$$

Then, the transformation defined by relations (3.5) with $\mathcal{J} \neq 0$, is canonical with

$$\mathcal{H}_0 = -\frac{\partial V}{\partial t}, \quad F = V - \sum_h \pi_h \chi_h ,$$

and leads to the new characteristic function

$$\mathcal{K} = \mathcal{H} + \frac{\partial V}{\partial t},$$

expressed, of course, in terms of (p/q) coordinates.

The function V is called the *generating function* of the canonical transformation.

A different choice could be to choose a function V' depending on q's and on χ's, and satisfying the condition that the mixed functional determinant

$$\mathcal{J}' = \det\left(\frac{\partial^2 V'}{\partial q_h \partial \chi_k}\right)$$

does not identically vanish.

The relations

$$p_h = \frac{\partial V'}{\partial q_h}, \quad \pi_h = -\frac{\partial V'}{\partial \chi_h}, \tag{3.6}$$

implicitly define an invertible differentiable coordinate transformation between the p, q's and π, χ's. In fact, since $\mathcal{J}' \neq 0$, by the implicit function theorem, the second of relations (3.6) can be solved in the form

$$q_h = q_h(\pi/\chi/t),$$

and associated to the first one to give an explicit one-to-one transformation between the coordinates (p, q) and (π, χ).

As for the canonical character, it is enough to observe that Lie's condition is satisfied:

$$\sum_h p_h dq_h - \sum_h \pi_h d\chi_h = \sum_h \left(\frac{\partial V'}{\partial q_h} dq_h + \frac{\partial V'}{\partial \chi_h} d\chi_h\right) = dV' - \frac{\partial V'}{\partial t} dt,$$

from which we obtain

$$\sum_h p_h dq_h = \sum_h \pi_h d\chi_h - \frac{\partial V'}{\partial t} dt + dV' .$$

Then, the transformation defined by relations (3.6) with $\mathcal{J}' \neq 0$, is canonical with

$$\mathcal{H}_0 = -\frac{\partial V'}{\partial t} , \quad \mathcal{F} = V' ,$$

and leads to the new characteristic function

$$\mathcal{K} = \mathcal{H} + \frac{\partial V'}{\partial t} ,$$

expressed, of course, in terms of (p/q) coordinates.

The function V' is also called the *generating function* of the canonical transformation.

3.1.3 *Completely canonical transformations*

It has been shown that a canonical transformation of a given canonical system with characteristic function $\mathcal{H}$ leads to a canonical system having $\mathcal{K} = (1/c)(\mathcal{H} - \mathcal{H}_0)^*$ as characteristic function. When the function $\mathcal{K} = (1/c)(\mathcal{H} - \mathcal{H}_0)^*$ reduces to $\mathcal{K} = (1/c)(\mathcal{H})^*$ the transformation is called *a completely canonical transformation.* Then, a necessary and sufficient condition for a canonical transformation to be a completely canonical one is that $\mathcal{H}_0 = 0$.

It is easy to see that a canonical transformation

$$\begin{cases} \pi_h = \varphi_h(q/p) \\ \chi_h = \psi_h(q/p) \end{cases} , \tag{3.7}$$

which does not explicitly depend on time t, is a completely canonical one.

The Lie condition (3.3) for the transformation (3.7) gives

$$\begin{aligned} \sum_k p_k dq_k &= c \sum_h \pi_h d\chi_h + \mathcal{H}_0 dt + d\mathcal{F} \\ &= c \sum_h \varphi_h \sum_k \left(\frac{\partial \psi_h}{\partial p_k} dp_k + \frac{\partial \psi_h}{\partial q_k} dq_k \right) + \mathcal{H}_0 dt \\ &\quad + \frac{\partial \mathcal{F}}{\partial t} dt + \sum_k \left(\frac{\partial \mathcal{F}}{\partial p_k} dp_k + \frac{\partial \mathcal{F}}{\partial q_k} dq_k \right) , \end{aligned}$$

or equivalently,

$$\sum_k \left[\left(\Phi_k + \frac{\partial F}{\partial p_k} \right) dp_k + \left(\Psi_k + \frac{\partial F}{\partial q_k} \right) dq_k \right] + \left(\mathcal{H}_0 + \frac{\partial F}{\partial t} \right) dt = 0 \,,$$

where the functions Φ_k and Ψ_k are defined by

$$\Phi_k(p/q) = c \sum_i \varphi_i \frac{\partial \psi_i}{\partial p_k} \,, \quad \Psi_k(p/q) = c \sum_i \varphi_i \frac{\partial \psi_i}{\partial q_k} - p_k \,.$$

It follows that F has to satisfy the following relations:

$$\Phi_k = -\frac{\partial F}{\partial p_k} \,, \quad \Psi_k = -\frac{\partial F}{\partial q_k} \,, \quad \mathcal{H}_0 + \frac{\partial F}{\partial t} = 0 \,.$$

Moreover, as Φ_k and Ψ_k do not explicitly depend on time t, we also have

$$\frac{\partial}{\partial t} \left(\frac{\partial F}{\partial p_k} \right) = \frac{\partial}{\partial t} \left(\frac{\partial F}{\partial q_k} \right) = 0 \,,$$

or equivalently,

$$\frac{\partial}{\partial p_k} \left(\frac{\partial F}{\partial t} \right) = \frac{\partial}{\partial q_k} \left(\frac{\partial F}{\partial t} \right) = 0 \,,$$

which implies that $\partial F / \partial t$ does not depend on the p's and the q's but only on t.

From

$$\frac{\partial F}{\partial t} = f(t) \,,$$

it follows that

$$F = f(t) + F_1(p/q)$$

and

$$\mathcal{H}_0 + dF = dF_1 \,.$$

Therefore, Lie's condition (3.3) becomes

$$\sum_h p_h dq_h = c \sum_h \pi_h d\chi_h + dF_1 \,,$$

so that the new characteristic function is simply $\mathcal{K} = (1/c)(\mathcal{H})^*$.

3.1.4 *Symplectic transformations*

A canonical transformation, which leads to a new characteristic function $\mathcal{K}$ of the form $\mathcal{K} = (\mathcal{H})^*$, is called a *symplectic transformation.* Therefore, a symplectic transformation is a completely canonical transformation with $c = 1$. The following picture summarizes all the cases well:

$$\begin{aligned} \mathcal{K} &= \frac{1}{c}(\mathcal{H} - \mathcal{H}_0)^* \quad && \text{(canonical)}\,, \\ \mathcal{K} &= \frac{1}{c}(\mathcal{H})^* && \text{(completely canonical)}\,, \\ \mathcal{K} &= (\mathcal{H})^* && \text{(symplectic)}\,. \end{aligned} \tag{3.8}$$

The transformation (3.4) of the previous example is a symplectic transformation with a generating function given by

$$V = \frac{q}{2}\sqrt{2\lambda\pi - \lambda^2 q^2} + \arcsin\frac{\lambda q}{\sqrt{2\lambda\pi}}.$$

3.1.5 *Area preserving transformations*

An invertible differentiable map from $\Re^2$ to itself,

$$(p, q) \in \Re^2 \longleftrightarrow (\pi, \chi) \in \Re^2\,, \tag{3.9}$$

will transform a given Lebesgue measurable region $S \subseteq \Re^2$ in a measurable region $\Sigma \subseteq \Re^2$,

$$S \longleftrightarrow \Sigma\,.$$

The map will be said an *area-preserving transformation*, or simply, an *equivalent transformation*, if the measures of S and Σ coincide: $m_2(S) = m_2(\Sigma)$.

Theorem 13 *A necessary and sufficient condition for an invertible transformation on* $\Re^2$,

$$(p, q) \in \Re^2 \longleftrightarrow (\pi, \chi) \in \Re^2\,;$$

to be symplectic is to be an area preserving transformation.

Proof

Let the transformation (3.9) be symplectic. Then there exists a function G, such that, identically,

$$pdq = \pi d\chi + dG\,,$$

i.e.

$$p\frac{\partial q}{\partial \pi}d\pi + \left(p\frac{\partial q}{\partial \chi} - \pi\right)d\chi = dG\,.$$

Since the RHS of the above equation is an exact differential, the equality of crossed derivatives

$$\frac{\partial}{\partial \chi}\left(p\frac{\partial q}{\partial \pi}\right) = \frac{\partial}{\partial \pi}\left(p\frac{\partial q}{\partial \chi} - \pi\right),$$

gives

$$\frac{\partial p}{\partial \chi}\frac{\partial q}{\partial \pi} - \frac{\partial p}{\partial \pi}\frac{\partial q}{\partial \chi} = 1\,,$$

which can be, equivalently, expressed by

$$\frac{\partial(p,q)}{\partial(\pi,\chi)} = \det\begin{pmatrix} \dfrac{\partial p}{\partial \pi} & \dfrac{\partial p}{\partial \chi} \\ \dfrac{\partial q}{\partial \pi} & \dfrac{\partial q}{\partial \chi} \end{pmatrix} = 1\,.$$

It follows that

$$m_2(S) = \iint_S dpdq = \iint_\Sigma \left|\frac{\partial(p,q)}{\partial(\pi,\chi)}\right| d\pi d\chi = \iint_\Sigma d\pi d\chi = m_2(\Sigma)\,.$$

Conversely, if the transformation (3.9) is area preserving, then

$$0 = m_2(S) - m_2(\Sigma) = \iint_S dpdq - \iint_\Sigma d\pi d\chi = \iint_\Sigma \left(\left|\frac{\partial(p,q)}{\partial(\pi,\chi)}\right| - 1\right) d\pi d\chi\,.$$

The arbitrariness of Σ and the continuity of the Jacobian determinant imply

$$\frac{\partial(p,q)}{\partial(\pi,\chi)} = \pm 1\,.$$

If the Jacobian is 1, the transformation (3.9) is[§] *symplectic. The same is true if the Jacobian is -1; it is sufficient to interchange the names of the variables π's and χ's to go back to the first case.*

3.1.6 *Volume preserving transformation*

An invertible differentiable map from $\Re^n$ to itself,

$$(p, q) \in \Re^n \longleftrightarrow (\pi, \chi) \in \Re^n \,, \tag{3.10}$$

will transform a given Lebesgue measurable region $S \subseteq \Re^n$ in a measurable region $\Sigma \subseteq \Re^n$,

$$S \longleftrightarrow \Sigma \,.$$

The map will be said a *volume preserving transformation*, or simply, an *equivalent transformation*, if the measures of S and Σ coincide: $m_n(\Sigma) = m_n(\Sigma)$.

It will be shown, in the next section, that a symplectic transformation on $\Re^n$ is also equivalent. *It is worth mentioning that the converse is true only in the case $n = 2$.*

3.2 A New Characterization of Completely Canonical Transformations

The Lie condition is not easily handled for checking the canonical nature of a differentiable invertible transformation. Thus, we are looking for conditions to be directly required to the functions φ_h, ψ_h, in the invertible differentiable transformation

$$\begin{cases} \pi_h = \varphi_h(q/p)\,, \\ \chi_h = \psi_h(q/p)\,, \end{cases} \tag{3.11}$$

to define a completely canonical transformations.

The condition from which we start is the usual one, namely

$$\sum_h p_h dq_h = c \sum_h \pi_h d\chi_h + dG \,,$$

[§]With the usual assumption that the Σ and S be linearly simple-connected.

in which the transformation (3.11) must be performed. In this way, we have

$$-dG(p/q) = \sum_h (\Phi_h dq_h + \Psi_h dp_h)\,, \tag{3.12}$$

where

$$\begin{cases} \Phi_h = c\displaystyle\sum_i \varphi_i \frac{\partial \psi_i}{\partial q_h} - p_h\,, \\ \Psi_h = c\displaystyle\sum_i \varphi_i \frac{\partial \psi_i}{\partial p_h}\,. \end{cases} \tag{3.13}$$

If the right-hand side of Eq. (3.12) has to be an exact differential, the following conditions

$$\begin{cases} \dfrac{\partial \Phi_k}{\partial q_h} = \dfrac{\partial \Phi_h}{\partial q_k}\,, \\ \dfrac{\partial \Psi_k}{\partial p_h} = \dfrac{\partial \Psi_h}{\partial p_k}\,, \\ \dfrac{\partial \Phi_k}{\partial p_h} = \dfrac{\partial \Psi_h}{\partial q_k}\,, \end{cases}$$

must be satisfied.

Replacing Eq. (3.13) in the above conditions, we find

$$\begin{cases} \displaystyle\sum_k \left(\frac{\partial \psi_k}{\partial q_i}\frac{\partial \varphi_k}{\partial p_j} - \frac{\partial \psi_k}{\partial p_j}\frac{\partial \varphi_k}{\partial q_i} \right) = \frac{1}{c}\delta_{ij}\,, \\ \displaystyle\sum_k \left(\frac{\partial \psi_k}{\partial q_i}\frac{\partial \varphi_k}{\partial q_j} - \frac{\partial \psi_k}{\partial q_j}\frac{\partial \varphi_k}{\partial q_i} \right) = 0\,, \\ \displaystyle\sum_k \left(\frac{\partial \psi_k}{\partial p_i}\frac{\partial \varphi_k}{\partial p_j} - \frac{\partial \psi_k}{\partial p_j}\frac{\partial \varphi_k}{\partial p_i} \right) = 0\,. \end{cases}$$

The above relations can be written in a very compact form by introducing the *Lagrange bracket*, which, for given $2n$ functions φ_h, ψ_h of variables u, v, are defined by[¶]

$$[u, v] = \sum_{k=1}^{n} \left(\frac{\partial \varphi_k}{\partial u}\frac{\partial \psi_k}{\partial v} - \frac{\partial \varphi_k}{\partial v}\frac{\partial \psi_k}{\partial u} \right) = \sum_{k=1}^{n} \frac{\partial(\varphi_k, \psi_k)}{\partial(u, v)}\,. \tag{3.14}$$

[¶]Many authors define as Lagrange bracket the analogue, but associated with the inverse transformation.

By using this bracket, the necessary and sufficient conditions for a transformation to be completely canonical can be expressed as follows:

$$[q_i, q_j] = 0\,, \quad [p_i, p_j] = 0\,, \quad [q_i, p_j] = \frac{1}{c}\delta_{ij}\,, \quad \forall\, (i,j) \in \{1, 2, \ldots, n\}\,. \qquad (3.15)$$

It has been already shown that, in the plane $\Re^2$, the symplectic transformations coincide with area preserving transformations. Thus, it appears interesting to analyze, from this point of view, the properties of a completely canonical transformation more closely.

Let then

$$\begin{cases} \pi_h = \varphi_h(q/p) \\ \chi_h = \psi_h(q/p) \end{cases}$$

be a completely canonical transformation and let

$$J = \det\left(\frac{\partial(\varphi, \psi)}{\partial(p, q)}\right)$$

be its Jacobian determinant, which, more explicitly, can be written in a *block form* as

$$J = \det\begin{pmatrix} \dfrac{\partial\varphi_h}{\partial p_i} & \dfrac{\partial\varphi_h}{\partial q_i} \\ \dfrac{\partial\psi_h}{\partial p_i} & \dfrac{\partial\psi_h}{\partial q_i} \end{pmatrix}, \qquad i, h \in \{1, \ldots, n\}\,.$$

By performing, on the Jacobian matrix

$$M = \begin{pmatrix} \dfrac{\partial\varphi_h}{\partial p_i} & \dfrac{\partial\varphi_h}{\partial q_i} \\ \dfrac{\partial\psi_h}{\partial p_i} & \dfrac{\partial\psi_h}{\partial q_i} \end{pmatrix}, \qquad i, h \in \{1, \ldots, n\}\,,$$

the following interchanges:

- row number j with row number $n + j$, $\forall\, j \in \{1, \ldots, n\}$,
- column number j with column number $n + j$, $\forall\, j \in \{1, \ldots, n\}$,
- inversion of the sign in first n rows,
- inversion of the sign in first n columns,

we obtain the matrix

$$M' = \begin{pmatrix} \dfrac{\partial \psi_h}{\partial q_i} & -\dfrac{\partial \psi_h}{\partial p_i} \\ -\dfrac{\partial \varphi_h}{\partial q_i} & \dfrac{\partial \varphi_h}{\partial p_i} \end{pmatrix}, \quad i, h \in \{1, \dots, n\}.$$

Of course, since an even number ($4n$) of interchanges, each one introducing a factor -1, has been performed, it turns out that

$$J = \det M = \det M' .$$

It follows that the product of $M^{\intercal}$, the transposed of M, with M', has the following determinant

$$\det(M^{\intercal} \cdot M') = (\det M)(\det M') = J^2 .$$

More directly, the matrix M' can be obtained from M, by using the matrix

$$E = \begin{pmatrix} 0 & -I \\ I & 0 \end{pmatrix},$$

introduced in Eq. (2.23) of the previous chapter. Accordingly,

$$M' = E \cdot M \cdot E^{\intercal} ,$$

with $E^{\intercal}$ being the transposed of E. Trivially, $E^{\intercal} = -E = E^{-1}$.

The elements c_{ij} of the product $M^{\intercal} \cdot M'$ can be divided into the following four groups: $(i \le n, j \le n)$, $(i \le n, j > n)$, $(i > n, j \le n)$, $(i > n, j > n)$. In other words, similarly to M and M', also the product $M^{\intercal} \cdot M'$ can be divided into four sectors, which separately evaluated, give

- for $(i \le n, j \le n)$, $c_{ij} = \sum_h \left(\dfrac{\partial \varphi_h}{\partial p_i} \dfrac{\partial \psi_h}{\partial q_j} - \dfrac{\partial \psi_h}{\partial p_i} \dfrac{\partial \varphi_h}{\partial q_j} \right) = [p_i, q_j]$,
- if $(i \le n, j > n)$, $c_{ij} = \sum_h \left(\dfrac{\partial \varphi_h}{\partial p_j} \dfrac{\partial \psi_h}{\partial p_i} - \dfrac{\partial \psi_h}{\partial p_j} \dfrac{\partial \varphi_h}{\partial p_i} \right) = [p_j, p_i]$,
- for $(i > n, j \le n)$, $c_{ij} = \sum_h \left(\dfrac{\partial \varphi_h}{\partial q_i} \dfrac{\partial \psi_h}{\partial q_j} - \dfrac{\partial \psi_h}{\partial q_i} \dfrac{\partial \varphi_h}{\partial q_j} \right) = [q_i, q_j]$,
- if $(i > n, j > n)$, $c_{ij} = \sum_h \left(\dfrac{\partial \varphi_h}{\partial p_j} \dfrac{\partial \psi_h}{\partial q_i} - \dfrac{\partial \psi_h}{\partial p_j} \dfrac{\partial \varphi_h}{\partial q_i} \right) = [p_j, q_i]$.

Therefore, we have

$$M^{\intercal} \cdot M' = \begin{pmatrix} [p_i, q_j] & [p_j, p_i] \\ [q_i, q_j] & [p_j, q_i] \end{pmatrix} = \begin{pmatrix} \frac{1}{c}\delta_{ij} & 0 \\ 0 & \frac{1}{c}\delta_{ij} \end{pmatrix} = \frac{1}{c} I ,$$

the last equality following from Eq. (3.15). Thus, the Jacobian determinant J of a completely canonical transformation in $\Re^{2n}$ satisfies the relation

$$J^2 = c^{-2n} . \tag{3.16}$$

From the expression

$$M^{\intercal} \cdot M' = \frac{1}{c} I ,$$

we have

$$M^{\intercal} = \frac{1}{c} (M')^{-1} ,$$

so that

$$M \cdot (M')^{\intercal} = \frac{1}{c} ((M')^{-1})^{\intercal} \cdot (M')^{\intercal} = \frac{1}{c} (M' \cdot (M')^{-1})^{\intercal} = \frac{1}{c} I .$$

By performing the product $M \cdot (M')^{\intercal}$, we finally obtain

$$\begin{pmatrix} \frac{1}{c}\delta_{ij} & 0 \\ 0 & \frac{1}{c}\delta_{ij} \end{pmatrix} = M \cdot (M')^{\intercal} = \begin{pmatrix} \{\varphi_h, \psi_k\} & \{\psi_h, \psi_k\} \\ \{\varphi_h, \varphi_k\} & \{\varphi_h, \varphi_k\} \end{pmatrix} .$$

Therefore, it follows that

The necessary and sufficient conditions for a transformation to be completely canonical can be expressed as:

$$\{q_i, q_j\} = 0 , \quad \{p_i, p_j\} = 0 , \quad \{q_i, p_j\} = \frac{1}{c}\delta_{ij} , \quad \forall\ (i, j) \in \{1, 2, \ldots, n\} . \tag{3.17}$$

3.3 New Characterization of Symplectic Transformations

From relations (3.15), it turns out that

The necessary and sufficient conditions for a transformation to be symplectic is given by

$$[q_i, q_j] = 0\,, \quad [p_i, p_j] = 0\,, \quad [q_i, p_j] = \delta_{ij}\,, \quad \forall\ (i,j) \in \{1, 2, \ldots, n\}\,,$$

or

$$\{q_i, q_j\} = 0\,, \quad \{p_i, p_j\} = 0\,, \quad \{q_i, p_j\} = \delta_{ij}\,, \quad \forall\ (i,j) \in \{1, 2, \ldots, n\}\,.$$

Moreover, from Eq. (3.16), it follows that

A symplectic transformation preserves the volume of any given region of the phase space.

Chapter 4

The Integration Methods

4.1 Integrals Invariants of a Differential System

Let us consider the first order differential system

$$\frac{dx^i}{dt} = X^i(x/t), \quad \forall i \in (1, 2, \ldots n), \tag{4.1}$$

and denote with

$$x^i = x^i(t, x_0)$$

the solution, which at time t_0, takes the value $x_0 : x_0^i = x^i(t_0, x_0)$.

The above relations can be equivalently expressed by

$$Q = Q(t, Q_0), \tag{4.2}$$

where Q and Q_0 denote the points whose coordinates are $(x^1, \ldots, x^n)$ and $(x_0^1, \ldots, x_0^n)$, respectively. Given any submanifold $U_0 \subseteq \Re^n$, whose points will be denoted by Q_0, let U be the submanifold, depending on t, of points Q given by Eq. (4.2). In other words, U represents the evolution at time t, according to Eq. (4.1), of U_0.

Equation (4.2) thus define a map between U_0 and U. The map is one-to-one by the existence and uniqueness theorem, which is taken to hold for the system (4.1).

Let us consider an arbitrary function ρ of x and t and the integral

$$I = \int_U \rho(x,t)dU\,, \tag{4.3}$$

which, of course, will generally depend on time t. In the case I does not depend on time, no matter how U is chosen, we shall say that *I is an integral invariant for the system* (4.1).

In order for I to be an integral invariant, ρ is required to satisfy suitable conditions. Let us start from the natural characterization of an integral invariant, which is given by

$$\frac{d}{dt}\int_U \rho(x,t)dU = 0\,.$$

In the above expression, the transfer of time derivative under the integral sign is not permitted, since the integration region depends on time. The difficulty is easily overcome by using the change of variables given by Eq. (4.2). In this way, we obtain

$$\int_U \rho(x,t)dU = \int_{U_0} \tilde{\rho}(x_0,t)JdU_0\,,$$

where $\tilde{\rho} = \rho \circ Q$ is the composed function between ρ and Eq. (4.2); i.e. $\tilde{\rho}(x_0,t) = \rho(x(x_0,t),t)$ and J is the Jacobian determinant

$$J = \frac{\partial(x^1, x^2, \ldots, x^n)}{\partial(x_0^1, x_0^2, \ldots, x_0^n)}\,, \tag{4.4}$$

of the transformation Eq. (4.2).

Remark 7 *The theorem on the change of variables in the integrals requires, really, the absolute value of the determinant. In our case, however, the Jacobian matrix at initial time t_0 coincides with the unit matrix $\mathcal{I}$. Then, by the continuity, it exists a neighborhood of t_0 (an interval of time) in which the Jacobian determinant is always positive.*

It follows that

$$\frac{d}{dt}\int_U \rho dU = \frac{d}{dt}\int_{U_0} \tilde{\rho}\, JdU_0 = \int_{U_0} \frac{d}{dt}(\tilde{\rho}\, J)dU_0 = \int_U \frac{1}{J}\frac{d}{dt}(\rho\, J)dU\,,$$

where we have used the property that the Jacobian of the inverse transformation is the inverse of the Jacobian.

Therefore,

$$\frac{dI}{dt} = \int_U \left(\frac{d\rho}{dt} + \rho\, J^{-1} \frac{dJ}{dt} \right) dU \,.$$

In order to explicitly calculate the derivative dJ/dt, let us observe that J depends on t via the elements $g_{ij} = \partial x^i / \partial x_0^j$ of Jacobian matrix. Then

$$\frac{dJ}{dt} = \sum_{ij} \frac{\partial J}{\partial g_{ij}} \frac{dg_{ij}}{dt} \,.$$

By using the Laplace* expansion, the Jacobian determinant can be written as follows:

$$J = g_{i1} G_{i1} + \cdots + g_{in} G_{in} = \sum_{j=1}^{n} g_{ij} G_{ij} \,, \quad \forall\, i \in \{1, 2, \ldots, n\} \,.$$

It is important to note that in the above expression the algebraic complement G_{ij} of the element g_{ij} does not contain the elements $g_{i1}, \ldots, g_{in}$. In this way

$$\begin{aligned}
\frac{dJ}{dt} &= \sum_{ij} \frac{\partial J}{\partial g_{ij}} \frac{dg_{ij}}{dt} = \sum_{ij} G_{ij} \frac{dg_{ij}}{dt} \\
&= \sum_{ij} G_{ij} \frac{d}{dt} \frac{\partial x^i}{\partial x_0^j} = \sum_{ij} G_{ij} \frac{\partial}{\partial x_0^j} \frac{dx^i}{dt} \\
&= \sum_{ij} G_{ij} \frac{\partial X^i}{\partial x_0^j} = \sum_{ij} G_{ij} \sum_k \frac{\partial X^i}{\partial x^k} \frac{\partial x^k}{\partial x_0^j} \\
&= \sum_{ij} G_{ij} \sum_k \frac{\partial X^i}{\partial x^k} g_{kj} = \sum_{ijk} G_{ij} \frac{\partial X^i}{\partial x^k} g_{kj}
\end{aligned}$$

---

*Pierre Simon Laplace was born in 1749, in a little village of Calvados, a region of France, and died in Paris in 1827. He covered public charges and was a member of the Science Academy of the *Institute de France* and a professor at the *Ècole Normale*. He was appointed Earl, Marquis and Peer of France by Napoleon. His results on celestial mechanics, acoustics and electromagnetism are very important; the treatises on the celestial mechanics (five volumes) and on the calculus of probability as well as the divulgation works *Exposition du Système du monde* (two volumes) and *Essai philosophique sur le probabilités* are now considered as classical works. His complete production fill up 14 volumes.

$$= \sum_{ik} \frac{\partial X^i}{\partial x^k} \sum_j G_{ij}\, g_{kj} = \sum_{ik} \frac{\partial X^i}{\partial x^k} J \delta_{ik}$$

$$= J \sum_i \frac{\partial X^i}{\partial x^i} = J \text{ div } \vec{X} \,,$$

and

$$J^{-1} \frac{dJ}{dt} = \text{div } \vec{X} \,.$$

It thus follows that

$$\frac{dI}{dt} = \int_U \left(\frac{d\rho}{dt} + \rho \text{ div } \vec{X} \right) dU \,. \tag{4.5}$$

Since the function in the integral is continuous and the region U is arbitrary, we can conclude that

The necessary and sufficient condition for $I = \int_U \rho(x,t) dU$ to be an integral invariant is that

$$\frac{d\rho}{dt} + \rho \text{ div } \vec{X} = 0 \,. \tag{4.6}$$

A function ρ satisfying the previous equation is called a *Jacobi multiplier.*

Finally, let us observe that, by using the identity

$$\frac{d\rho}{dt} = \frac{\partial \rho}{\partial t} + \sum_i \frac{\partial \rho}{\partial x^i} \frac{dx^i}{dt} = \frac{\partial \rho}{\partial t} + \sum_i \frac{\partial \rho}{\partial x^i} X^i \,,$$

Eq. (4.6) can also be written in the more familiar form

$$\frac{\partial \rho}{\partial t} + \text{div} \, (\rho \vec{X}) = 0 \,,$$

which the reader has already met, for instance, in electrodynamics, where ρ has to be identified with the *electric charge density* and $\rho \vec{X}$ with the *current density.*

Finally, let us address that in the case of *divergenceless* vector field $\vec{X}$, any constant is a Jacobi multiplier. For these types of dynamics we get the important result that the *measure* $\mu(U)$ of any region U does not change in time. This is certainly the case for Hamiltonian systems

$$\begin{cases} \dfrac{d}{dt}p_h = -\dfrac{\partial \mathcal{H}}{\partial q_h}\,, \\ \dfrac{d}{dt}q_h = \dfrac{\partial \mathcal{H}}{\partial p_h}\,, \end{cases}$$

for which we have

$$\sum_{i=1}^{2n} \frac{\partial X^i}{\partial x^i} = \sum_{h=1}^{n} \left(-\frac{\partial^2 \mathcal{H}}{\partial q_h \partial p_h} + \frac{\partial^2 \mathcal{H}}{\partial p_h \partial q_h} \right) = 0\,.$$

It has already been remarked that the Jacobian determinant of a completely canonical transformation is 1, and that this type of transformation are volume preserving.

It has also already been remarked that the Hamiltonian evolution is itself a completely canonical transformation between a submanifold U_0 of the phase space Φ and the submanifold U of the same space whose points are the evolutes, at time t, of points of U_0, having the Hamiltonian function as a generator.

Thus, for canonical systems, a double conservation of the measure holds (*Liouville remark*).

4.2 A Primer on the Lie Derivative

Let us consider the differential system

$$\frac{dx^i}{dt} = X^i(x)\,, \quad \forall i \in (1, 2, \ldots, n)\,, \tag{4.7}$$

where the X's do not depend explicitly on time, and evaluate the rate of change of any function of the coordinates x along the solutions of the system, shortly the "*time*" *derivative.* We have

$$\frac{df}{dt} = \sum_{i=1}^{n} \frac{\partial f}{\partial x^i}\frac{dx^i}{dt} = \sum_{i=1}^{n} X^i \frac{\partial f}{\partial x^i}\,.$$

Therefore, one can naturally associate, with a differential system, the first order differential operator

$$X = \sum_{i=1}^{n} X^i \frac{\partial}{\partial x^i}\,,$$

whose action on an arbitrary function f gives its "time" derivative. Since with each point x of the space we can associate a vector having the real numbers $X^i(x)$ as components, the previous operator X is also called a *vector field.*

Vice versa, with any vector field we can associate a system of differential equations

$$\frac{dx^i}{dt} = X^i(x)\,, \quad \forall i \in (1, 2, \ldots, n)\,,$$

defining the curves $x^i = x^i(t)$, such that the tangent vector $\vec{v} \equiv (dx^1/dt, \ldots, dx^n/dt)$, in a generic point x, is just given by the components of X^i at point x. Such curves are called the *integral curves of the vector field* X.

By denoting, as before, with Q and Q_0 the points whose coordinates are $(x^1, x^2, \ldots, x^n)$ and $(x_0^1, x_0^2, \ldots, x_0^n)$, respectively, the equations

$$Q = Q(t, Q_0) \tag{4.8}$$

represent the solutions of the differential system which, at time t_0, takes the value Q_0. They locally define a one-to-one global map, which depends on a parameter t,

$$\varphi_t : Q_0 \to Q = \varphi_t(Q_0)\,,$$

called *the flow generated by the vector field* $(X^1, X^2, \ldots, X^n)$, satisfying the *group properties*

$$\begin{cases} \varphi_0 = \text{identity map}\,, \\ (\varphi_t)^{-1} = \varphi_{-t}\,, \\ \varphi_{t_1} \circ \varphi_{t_2} = \varphi_{t_1+t_2}\,. \end{cases}$$

If the differential system is canonical with characteristic function $\mathcal{H}$, the map φ_t is also called the *Hamiltonian flow generated* by $\mathcal{H}$.

The derivative of a function f along the solutions of the differential system (4.7) is denoted by

$$L_X f = \sum_{i=1}^{n} X^i \frac{\partial f}{\partial x^i}\,,$$

and is known as the *Lie derivative of f with respect to X*.

From equations $x^i = x^i(t, x_0)$, it follows

$$dx^i = \sum_{j=1}^{n} \frac{\partial x^i}{\partial x_0^j} dx_0^j .$$

It is then easy to evaluate the "time" derivative of the differentials dx^i. We obtain

$$\begin{aligned}\frac{d(dx^i)}{dt} &= \frac{d}{dt}\sum_{j=1}^{n}\left(\frac{\partial x^i}{\partial x_0^j}\right) dx_0^j = \sum_{j=1}^{n} \frac{\partial}{\partial x_0^j}\frac{dx^i}{dt} dx_0^j \\ &= \sum_{j=1}^{n} dx_0^j \frac{\partial}{\partial x_0^j} X^i = \sum_{j=1}^{n} dx_0^j \left(\sum_{k=1}^{n} \frac{\partial x^k}{\partial x_0^j}\frac{\partial}{\partial x^k} X^i\right) \\ &= \sum_{k=1}^{n}\left(\sum_{j=1}^{n}\frac{\partial x^k}{\partial x_0^j} dx_0^j\right)\frac{\partial X^i}{\partial x^k} = \sum_{k=1}^{n}\frac{\partial X^i}{\partial x^k} dx^k . \end{aligned} \tag{4.9}$$

The above "time" derivative is known as the *Lie derivative of dx^i with respect to X* and is denoted with $L_X dx^i$, so that the above equation can be written in the following form:

$$L_X dx^i = dX^i ,$$

without any reference to the parameter t.

By now using

$$\frac{\partial}{\partial x^i} = \sum_{j=1}^{n} \frac{\partial x_0^j}{\partial x^i}\frac{\partial}{\partial x_0^j} ,$$

which also follows from $x^i = x^i(t, x_0)$, it is possible to evaluate the "time" derivative of partial derivatives $\partial/\partial x^i$. We thus get

$$L_X \frac{\partial}{\partial x^i} = \frac{d}{dt}\frac{\partial}{\partial x^i} = \frac{d}{dt}\sum_{j=1}^{n}\frac{\partial x_0^j}{\partial x^i}\frac{\partial}{\partial x_0^j} = \sum_{j=1}^{n}\left(\frac{d}{dt}\frac{\partial x_0^j}{\partial x^i}\right)\frac{\partial}{\partial x_0^j} = -\sum_{k=1}^{n}\frac{\partial X^k}{\partial x^i}\frac{\partial}{\partial x^k} . \tag{4.10}$$

The last step in the above formula is explicitly performed in the Appendix B.

By observing that

$$-\sum_{k=1}^{n} \frac{\partial X^k}{\partial x^i}\frac{\partial}{\partial x^k} = \left[X, \frac{\partial}{\partial x^i}\right], \tag{4.11}$$

Eq. (4.10) can also be written in the following form:

$$L_X \frac{\partial}{\partial x^i} = \left[X, \frac{\partial}{\partial x^i}\right]. \tag{4.12}$$

The Lie derivative with respect to a vector field X has been defined on functions f, on differentials dx^i and on vector fields $\partial/\partial x^i$, with the transparent physical significance to be a "time" derivative; i.e. a derivative along the solutions of the evolutive first order differential system. Since, by definition, L_X satisfies the Leibnitz rule, the Lie derivative, with respect to a vector field X, is defined for generic differential forms $\alpha = \alpha_i(x)dx^i$ and vector fields $Y = Y^i(x)\partial/\partial x^i$, as follows:

$$L_X f = \sum_{i=1}^{n} X^i \frac{\partial f}{\partial x^i} \tag{4.13}$$

$$L_X\alpha = \sum_{i=1}^{n}(L_X\alpha_i dx^i + \alpha_i dX^i) = \sum_{i=1}^{n}\left(\frac{\partial \alpha_i}{\partial x^j} + \alpha_i \frac{\partial X^i}{\partial x^j}\right) dx^j \tag{4.14}$$

$$\begin{aligned} L_X Y &= \sum_{j=1}^{n} L_X\left(Y^j \frac{\partial}{\partial x^i}\right) = \sum_{j=1}^{n}\left(\sum_{i=1}^{n}\left(X^i\frac{\partial Y^j}{\partial x^i} - Y^j \frac{\partial X^i}{\partial x^j}\frac{\partial}{\partial x^i}\right)\right) \\ &= \sum_{i,j=1}^{n}\left[X^i\frac{\partial}{\partial x^i}, Y^j\frac{\partial}{\partial x^j}\right] \\ &= [X, Y]. \end{aligned} \tag{4.15}$$

Remark 8 *Alternatively, once Lie's derivative has been defined on functions, the Leibnitz rule suggests to define*

$$\left(L_X\frac{\partial}{\partial x^i}\right) f = L_X\left(\frac{\partial f}{\partial x^i}\right) - \frac{\partial}{\partial x^i}(L_X f).$$

The interested reader will prove that the two definitions are equivalent.

4.3 The Kepler Dynamics

The gravitational potential energy of two bodies with mass m_1 and m_2 located, with respect to a chosen frame, at $\vec{r}_1$ and $\vec{r}_2$ is given by:

$$\mathcal{U}(\vec{r}_1, \vec{r}_2) = -G\frac{m_1 m_2}{|\vec{r}_1 - \vec{r}_2|},$$

where $G = 6.6 \cdot 10^{-11}\ \mathrm{Nm}^2/\mathrm{kg}^2$ is the *gravitational universal constant.*

The Lagrangian function $\mathcal{L}$, obtained subtracting $\mathcal{U}$ from the kinetic energy $\mathcal{T}$, is

$$\mathcal{L}(\vec{r}_1, \vec{r}_2/\vec{v}_1, \vec{v}_2) = \frac{1}{2}m_1 v_1^2 + \frac{1}{2}m_2 v_2^2 + \frac{k}{|\vec{r}_1 - \vec{r}_2|}.$$

The coordinates $\vec{r}_1, \vec{r}_2$ can be expressed in terms of *center of mass coordinate* $\vec{R}$ and *relative coordinate* $\vec{r}$ defined by

$$\vec{R} = \frac{m_1\vec{r}_1 + m_2\vec{r}_2}{m_1 + m_2}, \qquad \vec{r} = \vec{r}_1 - \vec{r}_2 .$$

We have

$$\vec{r}_1 = \frac{m_2\vec{r}}{m_1 + m_2} + \vec{R}, \quad \vec{r}_2 = -\frac{m_1\vec{r}}{m_1 + m_2} + \vec{R}.$$

The velocities $\vec{v}_1$ and $\vec{v}_2$ can also be expressed in terms of the center mass velocity $\vec{V}$ and relative velocity $\vec{v}$ as follows:

$$\vec{v}_1 = \vec{V} + \frac{m_2\vec{v}}{m_1 + m_2}, \qquad \vec{v}_2 = \vec{V} - \frac{m_1\vec{v}}{m_1 + m_2}.$$

The Lagrangian $\mathcal{L}$ expressed in terms $\vec{r}, \vec{R}, \vec{v}$ and $\vec{V}$ becomes

$$\mathcal{L} = \frac{1}{2}(m_1 + m_2)V^2 + \frac{1}{2}\mu v^2 + \frac{k}{r},$$

where

$$\mu = \frac{m_1 m_2}{m_1 + m_2}$$

is the *reduced mass.*

Thus, the Lagrangian $\mathcal{L}$ is the sum of a *free* Lagrangian $\mathcal{L}_R = [(m_1 + m_2)V^2]/2$ and a Lagrangian $\mathcal{L}_r = (1/2)\mu v^2 - k/r$ of a system with *1* degree of freedom. The first Lagrangian describes the motion of the center mass which turns out, of course, to be uniform. The second describes the motion of a particle with the reduced mass μ in the gravitational field force located at center of mass coordinate. We may notice that if $m_2 \ll m_1$, then $\mu \sim m_2$ and $\vec{r}_1 \sim \vec{R}$. The corresponding Hamiltonian function, with $m_2 = m$ and $m_1 = M$, is given by

$$\mathcal{H} = \frac{1}{2m}\vec{p}\cdot\vec{p} - \frac{k}{r}\,. \tag{4.16}$$

It is worth observing, by using the results of problems in the section devoted to the Poisson bracket, that the angular momentum $\vec{L}$ is a constant of the motion, so that

$$\{\vec{L}, \mathcal{H}\} = 0\,. \tag{4.17}$$

Of course the last property is shared by all central potentials; that is, by all potentials $\mathcal{U}(r)$ depending only on the modulus r of the vector position $\vec{r}$. Therefore, the trajectory lies in the plane determined by the initial values of position $\vec{r}_0$ and velocity $\vec{v}_0$, which is the plane orthogonal to the angular momentum $\vec{L}$.

4.3.1 *The Laplace–Runge–Lenz vector*

For the Kepler potential, beyond the angular momentum $\vec{L}$, there exist specific constants of the motion expressed by the so-called *Laplace–Runge–Lenz vector* given by

$$\vec{B} = \frac{1}{m}\vec{p}\wedge\vec{L} - k\frac{\vec{r}}{r}\,, \tag{4.18}$$

whose components are not independent, since

$$\vec{B}\cdot\vec{L} = 0\,,$$

which simply says that the Laplace–Runge–Lenz vector lies on the plane of the motion.

It is interesting to evaluate the Poisson brackets involving $\vec{B}$. The reader is invited to verify that

$$\{\vec{B}, \mathcal{H}\} = 0\,.$$

Moreover, the Poisson bracket

$$\{\vec{L}, \vec{B}\} = \vec{n} \wedge \vec{B} \,,$$

which has already been proven to hold for any vector $\vec{B}$, can also be written in terms of components as follows:

$$\{L_i, B_j\} = \sum_{k=1}^{3} \varepsilon_{ijk} B_k \,.$$

More tedious, but important, is to verify that

$$\{B_i, B_j\} = -\frac{2\mathcal{H}}{m} \sum_{k=1}^{3} \varepsilon_{ijk} L_k \,.$$

It is well known that for a negative total energy, $E < 0$, the motion is bounded and the orbits are ellipses with the sun located in one of two focuses. In this case, we introduce the vector

$$\vec{A} = \left(-\frac{2\mathcal{H}}{m}\right)^{-\frac{1}{2}} \vec{B} \,,$$

so that the previous Poisson brackets can be written in the form

$$\{L_i, A_j\} = \sum_{k=1}^{3} \varepsilon_{ijk} A_k \,, \quad \{A_i, A_j\} = \sum_{k=1}^{3} \varepsilon_{ijk} L_k \,.$$

As a consequence, the vectors defined by

$$\vec{I} = \frac{1}{2}(\vec{L} + \vec{A}) \,,$$

$$\vec{J} = \frac{1}{2}(\vec{L} - \vec{A}) \,,$$

will have the following Poisson brackets:

$$\{I_h, I_k\} = \sum_{l=1}^{3} \varepsilon_{hkl} I_l \,, \quad \{J_h, J_k\} = \sum_{l=1}^{3} \varepsilon_{hkl} J_l \,, \quad \{I_h, J_k\} = 0 \,,$$

in close analogy with the ones of the angular momentum.

For readers familiar with Lie algebras, this shows that the Lie algebra of symmetries for the Kepler dynamics is *twice* $so\,(3)$, or better, $su\,(2) \otimes su\,(2)$, which is locally isomorphic with $so\,(4)$.

More precisely, we observe that the Hamiltonian $\mathcal{H}$ can be written as

$$\mathcal{H} = -\frac{mk^2}{2(L^2 + A^2)} = -\frac{mk^2}{4(I^2 + J^2)}\,.$$

In terms of the *generators* of $SO(4)$, $L_{\alpha\beta} = -L_{\beta\alpha}(\alpha, \beta = 1, 2, 3, 4)$, defined by

$$L_{hk} = \sum_{i=1}^{3} \varepsilon_{hki} L_i\,, \quad h, k = 1, 2, 3\,,$$

$$L_{h4} = -L_{4h} = A_h\,, \quad h = 1, 2, 3\,,$$

the Hamiltonian $\mathcal{H}$ becomes

$$\mathcal{H} = -\frac{mk^2}{C_1}\,,$$

where $C_1 = \sum_{\alpha,\beta} L_{\alpha\beta} L_{\alpha\beta}$ is the first Casimir of $SO\,(4)$.

4.3.2 *The hydrogen atom*

The $SO\,(4)$ invariance explains why the degeneracy of the quantum energy levels of the hydrogen atom is greater than what is naturally expected from the central symmetry ($SO\,(3)$ invariance).

Quantization rules roughly consist in replacing classical dynamical variables with self-adjoint operators in the Hilbert space of complex squared integrable functions, according to what follows:

$$\begin{cases} p_h \to \hat{p}_h = \dfrac{\hbar}{i} \dfrac{\partial}{\partial x^h}\,, \\ x^h \to \hat{x}^h = x^h\,, \end{cases}$$

where $\hbar = 1.052 \cdot 10^{-27}$ erg · s is the Planck† constant (divided by 2π) and i the imaginary unity. Thus, the quantum angular momentum and the Hamiltonian operator corresponding to the classical Hamiltonian function $\mathcal{H} = (1/2m)$

†Max Planck was born in Kiel in 1858, and died in Gottingen in 1947. He was appointed to a theoretical physics chair in 1880 at Kiel University and in 1884 at Berlin University. Revolutionary against his will, at the beginning Planck was persuaded that the discontinuity concept, characterized by the so-called *quantum of action* $\hbar$, was a "purely mathematical lucky violence against the laws of classical physics." It was really just the first example of the renormalization procedure, after systematically introduced in field theory to cancel the infinities. He was appointed to a Nobel Prize in 1918.

$[\vec{p}\cdot\vec{p}-e^2/r]$ will be given by

$$\hat{L}_i = \frac{\hbar}{i}\sum_{h,k}\varepsilon_{ihk}x^h\frac{\partial}{\partial x^k}\,,$$

$$\hat{H} = -\frac{\hbar^2}{2m}\nabla^2 - \frac{e^2}{r}\,,$$

where ∇^2 is the Laplace operator and e the electric charge of the proton.

The Laplace–Runge–Lenz vector has to be written in the form

$$\vec{B} = \frac{1}{2m}(\vec{p}\wedge\vec{L}-\vec{L}\wedge\vec{p}) - e^2\frac{\vec{r}}{r}\,,$$

in order for the corresponding vector-operator, which is called the *Pauli vector*,

$$\hat{B}_i = \frac{1}{2m}\sum_{h,k=1}^{3}\varepsilon_{ihk}(\hat{p}_h\hat{L}_k - \hat{L}_h\hat{p}_k) - e^2\frac{x^i}{r}\,,$$

to be a self-adjoint operator.

Classical formulae will be replaced by the corresponding quantum ones

$$\sum_{h=1}^{3}\hat{B}_h\hat{L}_h = 0\,,\quad [\hat{B}_h,\hat{H}] = 0\,,\quad h=1,2,3\,,$$

$$[\hat{L}_i,\hat{L}_j] = i\hbar\sum_{k=1}^{3}\varepsilon_{ijk}\hat{L}_k\,,\quad [\hat{L}_i,\hat{B}_j] = i\hbar\sum_{k=1}^{3}\varepsilon_{ijk}\hat{B}_k\,,$$

where the bracket $[\cdot,\cdot]$ denotes the commutator-operator. Moreover, we have

$$[\hat{B}_i,\hat{B}_j] = -i\hbar\frac{2\hat{H}}{m}\sum_{k=1}^{3}\varepsilon_{ijk}\hat{L}_k\,.$$

We can now restrict ourselves to the Hilbert subspace which corresponds to bound states; that is, to states with negative definite eigenvalues, $E<0$, of $\hat{H}$. In this subspace we may define another self-adjoint vector-operator $(\hat{A}_1,\hat{A}_2,\hat{A}_3)$:

$$\hat{A}_i = \left(-\frac{2\mathcal{H}}{m}\right)^{-\frac{1}{2}}\hat{B}_i\,,$$

so that the previous commutation relations can be written in the form

$$[\hat{L}_i, \hat{A}_j] = \sum_{k=1}^{3} \varepsilon_{ijk} \hat{A}_k \,, \quad [\hat{A}_i, \hat{A}_j] = \sum_{k=1}^{3} \varepsilon_{ijk} \hat{L}_k \,.$$

As a consequence, the operators defined by

$$\hat{I}_k = \frac{1}{2}(\hat{L}_k + \hat{A}_k) \,,$$
$$\hat{J}_k = \frac{1}{2}(\hat{L}_k - \hat{A}_k) \,,$$

will satisfy the following commutation relations:

$$[\hat{I}_h, \hat{I}_k] = i\hbar \sum_{l=1}^{3} \varepsilon_{hkl} \hat{I}_l \,, \quad [\hat{J}_h, \hat{J}_k] = i\hbar \sum_{l=1}^{3} \varepsilon_{hkl} \hat{J}_l \,, \quad [\hat{I}_h, \hat{J}_k] = 0 \,.$$

The Hamiltonian operator will thus be

$$\hat{H} = -\frac{me^4}{2(\hat{L}^2 + \hat{A}^2 + \hbar^2)} = -\frac{me^4}{2(2(\hat{I}^2 + \hat{J}^2) + \hbar^2)} \,,$$

and by observing that

$$\hat{I}^2 = \hat{J}^2 \,,$$

it can be finally written as

$$\hat{H} = -\frac{me^4}{2(4\hat{I}^2 + \hbar^2)} \,.$$

The previous formula allows us to find the (quantum) energy spectrum E_n, and the corresponding degeneracy, of the hydrogen atom, by using only the knowledge of the irreducible representations of $SU(2) \otimes SU(2)$, without any mention to the Schrödinger‡ equation.

‡Erwin Schrödinger was born in Vienna in 1887, and died there in 1961. After his degree, obtained at Vienna University in 1906, he moved to Stoccard, Zurich and Berlin. After Hitler's advent, Schrödinger moved to Oxford and Dublin. Finally, he returned to Vienna in 1956. According to Niels Bohr, he was a "universal man"; indeed he was a scientist with large cultural interests covering physics, philosophy, politics, biology and classical Greek culture. He established the basic equation of nonrelativistic quantum mechanics, and was appointed, together with Dirac, to the Nobel Prize in 1933.

In fact, since the operators $\hat{I}_h$ satisfy the commutation relations of an angular momentum, $\hat{I}^2$ can be quantized accordingly: $\hat{I}^2 = \hbar^2 l(l+1)$, with l integer or half integer, so that the energy levels of the hydrogen atom will be given by

$$E_n = -\frac{me^4}{2\hbar^2(2l+1)^2} = -\frac{me^4}{2\hbar^2 n^2},$$

which is, of course, Bohr's[§] formula, with $n = 2l+1$.

The degeneracy of the energy levels will be

$$\deg E_n = \sum_{l=1}^{n-1}(2l+1) = n^2 = D\left(\frac{n-1}{2}, \frac{n-1}{2}\right),$$

where $D(i,j)$ denotes the dimension of the (i,j)-irreducible representation of $SU(2) \otimes SU(2) \sim SO(4)$:

$$D(i,j) = (2i+1)(2j+1).$$

4.4 The Hamilton–Jacobi Integration Method

Important concepts, as *first integral* and *integral invariant*, concerning canonical systems have been discussed in the previous sections. It is now time to briefly discuss problems concerning the effective integration of canonical systems. Let us start with the classical Hamilton–Jacobi integration method.

This method brings the integration of any canonical systems of rank $2n$ to the determination of a so-called *complete integral* for a partial differential equation in $n+1$ independent variables.

Given then the canonical systems

$$\begin{cases} \dfrac{d}{dt}p_h = -\dfrac{\partial \mathcal{H}}{\partial q_h}, \\ \dfrac{d}{dt}q_h = \dfrac{\partial \mathcal{H}}{\partial p_h}, \end{cases} \quad \forall\, h \in \{1, \ldots, n\},$$

[§]Niels Henrik Bohr was born in Copenhagen in 1885, and died there in 1962. Soon after his degree, he moved to Cambridge and then to Rutherford's laboratory in Manchester. He solved the contradiction between the Rutherford's atomic model and the electrodynamics classical laws. Indeed he was able to agree on four physical theories: the classical electrodynamics, the quantum black-body radiation by M. Planck, the Rutherford atomic model and the atomic spectra observed by J. J. Balmer. He was appointed to the Nobel Prize in 1922, and was an associated founder of the CERN in Geneva.

let us try to find, if any, new canonical coordinates (π, χ), such that the new Hamiltonian function $\mathcal{K}$ is the simplest one; that is $\mathcal{K} = 0$. Then the integration of the transformed canonical system

$$\begin{cases} \dot{\pi}_h = 0\,, \\ \dot{\chi}_h = 0\,, \end{cases} \qquad \forall\, h \in \{1, \ldots, n\}\,,$$

becomes trivial

$$\begin{cases} \pi_h = \text{constant}\,, \\ \chi_h = \text{constant}\,, \end{cases} \qquad \forall\, h \in \{1, \ldots, n\}\,.$$

Let us take advantage of the general method, previously introduced, consisting in generating a canonical transformation

$$p_h = \frac{\partial V}{\partial q_h}\,, \qquad \chi_h = \frac{\partial V}{\partial \pi_h}\,,$$

by using an arbitrary function V depending on the q's and the π's and satisfying the condition

$$\mathcal{J} = \det\left(\frac{\partial^2 V}{\partial q_h \partial \pi_k}\right) \neq 0\,.$$

The new characteristic function will thus be

$$\mathcal{K} = \left[\mathcal{H}\left(\frac{\partial V}{\partial q} \Big/ q/t\right) + \frac{\partial V}{\partial t}\right]^*\,,$$

where the $*$ indicates that the transformation has to be completed expressing the q's in terms of the (π, χ)'s by using the relations

$$\chi_h = \frac{\partial V}{\partial \pi_h}(q/\pi/t)\,.$$

Our goal will be achieved if V is such that $\mathcal{K} = 0$. Therefore, V has to be a solution of the celebrated partial differential equation

$$\mathcal{H}\left(\frac{\partial V}{\partial q} \Big/ q/t\right) + \frac{\partial V}{\partial t} = 0\,, \tag{4.19}$$

known as *the Hamilton–Jacobi equation.*

The Hamilton–Jacobi integration method can be summarized as follows:

- Once given the canonical systems

$$\begin{cases} \dfrac{d}{dt} p_h = -\dfrac{\partial \mathcal{H}}{\partial q_h} \,, \\ \dfrac{d}{dt} q_h = \dfrac{\partial \mathcal{H}}{\partial p_h} \,, \end{cases} \qquad \forall\, h \in \{1, \ldots, n\} \,,$$

  replace the momenta p's in $\mathcal{H}(p/q/t)$ with the symbol

$$p_h = \frac{\partial V}{\partial q_h} \,,$$

  where V is an unknown function.
- Write down the Hamilton–Jacobi equation

$$\mathcal{H}\left(\frac{\partial V}{\partial q} \Big/ q/t\right) + \frac{\partial V}{\partial t} = 0 \,.$$

- Find a *complete integral* $V(q/\pi/t)$ of the Hamilton–Jacobi equation; that is, any solution V of the equation

$$\mathcal{H}\left(\frac{\partial V}{\partial q} \Big/ q/t\right) + \frac{\partial V}{\partial t} = 0 \,,$$

  depending, besides the q's and the time t, also on n arbitrary integration constants, namely $(\pi_1, \pi_2, \ldots, \pi_n)$ and satisfying the condition $\mathcal{J} \neq 0$.
- Write down the canonical transformation

$$\begin{cases} p_h = \dfrac{\partial V}{\partial q_h} \,, \\ \chi_h = \dfrac{\partial V}{\partial \pi_h} \,, \end{cases} \qquad \forall\, h \in \{1, \ldots, n\} \,, \tag{4.20}$$

  leading to the trivial solutions π_h = constant, χ_h = constant of the *new* canonical system.
- Explicitly write down the above transformation in the form

$$\begin{cases} p_h = p_h(\pi/\chi/t) \,, \\ q_h = q_h(\pi/\chi/t) \,, \end{cases} \qquad \forall\, h \in \{1, \ldots, n\} \,, \tag{4.21}$$

  representing the general integral of the canonical system.

- Fix up the values of constants π's and χ's according to initial data:

$$\begin{cases} p_h^0 = p_h(\pi/\chi/t_0)\,, \\ q_h^0 = p_h(\pi/\chi/t_0)\,, \end{cases} \quad \forall\, h \in \{1, \ldots, n\}\,. \tag{4.22}$$

- Compose the two mappings (4.21) and (4.22) to obtain

$$\begin{cases} p_h = p_h(p_0/q_0/t)\,, \\ q_h = q_h(p_0/q_0/t)\,, \end{cases} \quad \forall\, h \in \{1, \ldots, n\}\,, \tag{4.23}$$

  representing, finally, the integral of the canonical system in terms of the initial conditions.

Example 14 *Let us consider the harmonic oscillator with 1 degree of freedom whose Hamiltonian is given by*

$$H = \frac{1}{2m}(p^2 + m^2\omega^2 q^2)\,,$$

so that the corresponding Hamilton–Jacobi equation can be written as follows:

$$\frac{1}{2m}\left(\left(\frac{dV}{dq}\right)^2 + m^2\omega^2 q^2\right) + \frac{\partial V}{\partial t} = 0\,.$$

Let us try to find a solution of the form

$$V = -Et + W\,,$$

where E is an arbitrary constant.

Then, the Hamilton–Jacobi equation simplifies to

$$\frac{1}{2m}\left(\left(\frac{dW}{dq}\right)^2 + m^2\omega^2 q^2\right) = E\,,$$

for which the solution is easily found in the form

$$W = \frac{q}{2}\sqrt{2mE - m^2\omega^2 q^2} + \frac{E}{\omega}\arcsin\frac{m\omega q}{\sqrt{2mE}}\,,$$

so that

$$V = -Et + \frac{q}{2}\sqrt{2mE - m^2\omega^2 q^2} + \frac{E}{\omega}\arcsin\frac{m\omega q}{\sqrt{2mE}}\,. \tag{4.24}$$

The function V generates now the canonical map

$$\begin{cases} p = \dfrac{dV}{dq} = \sqrt{2mE - m^2\omega^2 q^2}\,, \\[2mm] \chi = \dfrac{dV}{dE} = -t + \dfrac{1}{\omega}\arcsin\dfrac{m\omega q}{\sqrt{2mE}}\,, \end{cases}$$

with

$$\mathcal{J} = \frac{d^2V}{dqdE} = \frac{m}{\sqrt{(2mE - m^2\omega^2 q^2)}}\,.$$

The explicit canonical transformation turns out to be

$$\begin{cases} \pi = \dfrac{1}{2m}(p^2 + m^2\omega^2 q^2)\,, \\[2mm] \chi = -t + \dfrac{1}{\omega}\arctan\dfrac{m\omega q}{p}\,, \end{cases}$$

and its inverse

$$\begin{cases} p = \sqrt{2m\pi}\cos(\omega(\chi + t)) \\[2mm] q = \dfrac{1}{\omega}\sqrt{\dfrac{2\pi}{m}}\sin(\omega(\chi + t)) \end{cases}$$

represents the general integral of the canonical system.

4.4.1 *Remarks on the Hamilton–Jacobi equation*

The Hamilton–Jacobi equation can be considered the most elegant form of dynamics and gives an important physical example of the deep connection between first-order partial differential equations and first-order ordinary differential systems.

It was first introduced in 1834 by W. R. Hamilton,[112] in his investigation on analytical dynamics, and it has been the starting point for Schrödinger to state the wave equation in quantum mechanics, of which is the approximate version in the cases in which the Planck constant can be neglected.

The proof that once a complete integral is found, then the dynamical problem can be completely solved by using Eq. (4.20), is instead due to Jacobi,[115] hence the name "Hamilton–Jacobi" for the equation and the ensuing method of solution.

Each complete integral of the Hamilton–Jacobi equation gives rise to a family of solutions of Hamilton's equations, and according to Dirac¶,[85] "*while the family does not have any importance from the point of view of Newtonian mechanics, ... it ... corresponds to one state of motion in the quantum theory, so presumably the family has some deep significance in nature, not yet properly understood.*"

Once the full dynamical problem has already been solved, an explicit solution of the Hamilton–Jacobi equation is given by

$$S = \int_{\bar{t}}^{t} \left[\sum_h p_h(\tau) q'_h - \mathcal{H}(p/q/\tau) \right] d\tau = \int_{\bar{t}}^{t} [\mathcal{L}(q/q'/\tau)] d\tau \,,$$

where t and $\bar{t}$ are two time-instants, $q' \equiv dq/d\tau$, $\mathcal{H}$ and $\mathcal{L}$ the Hamiltonian and the Lagrangian functions, and the integral has to be taken along the actual trajectory of the dynamical system. The right-hand side of the above equation does indeed satisfy the Hamilton–Jacobi equation and also the additional equation[32]

$$\frac{\partial S}{\partial \bar{t}} - \mathcal{H} = 0 \,.$$

Remark 9 *For conservative systems, S depends actually only on the difference $t - \bar{t}$, so that*

$$\frac{\partial S}{\partial \bar{t}} - \mathcal{H} = 0 \Leftrightarrow \frac{\partial S}{\partial t} + \mathcal{H} = 0 \,.$$

¶Paul Adrien Maurice Dirac was born in Bristol in 1902, and died in 1984. After his degree, obtained at Bristol University in 1921, he moved to Cambridge University. In this university, he was Lucasian professor, a chair already covered by Newton, from the year 1932. Dirac has been one of the most important physicist of our age and can be considered the father of modern physics. We just need to mention the *Dirac equation* predicting the existence of the positron and more generally of antiparticles, the *Fermi-Dirac statistics* and the *constraints method*, which is an essential tool for the Hamiltonian formulation of Einstein's equation, considered then as a step towards a quantum theory of gravity. The constraints method has been also a fundamental step for the quantization of gauge theories. His books are now considered as classical works. Together with Schrödinger, Dirac was appointed to the Nobel Prize in 1933.

4.5 The Hamilton–Jacobi Equation for the Kepler Potential

In terms of spherical-polar coordinates (r, ϑ, φ), the Cartesian coordinates (x, y, z) of a point are expressed as follows:

$$\begin{cases} x = r \sin\vartheta \cos\varphi \,, \\ y = r \sin\vartheta \sin\varphi \,, \\ z = r \cos\theta \,. \end{cases}$$

The line length $ds = (dx^2 + dy^2 + dz^2)^{\frac{1}{2}}$, representing the infinitesimal distance between two points of coordinates (x, y, z) and $(x+dx, y+dy, z+dz)$, is given by

$$ds^2 = dr^2 + r^2 d\vartheta^2 + r^2 \sin^2\vartheta d\varphi^2 \,.$$

The kinetic energy $\mathcal{T}$ of a massive particle will be written as

$$\begin{aligned} \mathcal{T} &= \frac{1}{2} m v^2 = \frac{1}{2m} \left(\frac{ds}{dt} \right)^2 \\ &= \frac{1}{2} m (\dot{r}^2 + r^2 \dot{\vartheta}^2 + r^2 \sin^2\vartheta \dot{\varphi}^2) \,, \end{aligned}$$

so that the Lagrangian of a particle in the potential $\mathcal{U}(\vec{r})$ can be written as follows:

$$\mathcal{L} = \frac{1}{2} m (\dot{r}^2 + r^2 \dot{\vartheta}^2 + r^2 \sin^2\vartheta \dot{\varphi}^2) - \mathcal{U}(\vec{r}) \,.$$

By introducing the conjugate momenta $(p_r, p_\vartheta, p_\varphi)$ of (r, ϑ, φ),

$$p_r = m\dot{r} \,, \quad p_\vartheta = mr\dot{\vartheta} \,, \quad p_\varphi = mr^2 \sin^2\vartheta \dot{\varphi} \,,$$

the corresponding Hamiltonian will be given by

$$\mathcal{H} = \frac{1}{2m} \left(p_r^2 + \frac{p_\vartheta^2}{r^2} + \frac{p_\varphi^2}{r^2 \sin^2\vartheta} \right) + \mathcal{U}(\vec{r}) \,.$$

Since the Hamiltonian does not depend explicitly on the time, the Hamilton–Jacobi equation

$$\frac{\partial V}{\partial t} + \frac{1}{2m} \left(\left(\frac{\partial V}{\partial r} \right)^2 + \frac{1}{r^2} \left(\frac{\partial V}{\partial \vartheta} \right)^2 + \frac{1}{r^2 \sin^2\vartheta} \left(\frac{\partial V}{\partial \varphi} \right)^2 \right) + \mathcal{U}(\vec{r}) = 0$$

can be reduced, with $V = W - Et$, to the form

$$\frac{1}{2m}\left(\left(\frac{\partial W}{\partial r}\right)^2 + \frac{1}{r^2}\left(\frac{\partial W}{\partial \vartheta}\right)^2 + \frac{1}{r^2 \sin^2\vartheta}\left(\frac{\partial W}{\partial \varphi}\right)^2\right) + \mathcal{U}(\vec{r}) = E\,.$$

For a central potential $\mathcal{U}(r)$, it is possible to find a complete integral of previous equation by using the *method of separation of variables*, which consists of searching for a solution $W(r, \vartheta, \varphi)$ of the form

$$W(r, \vartheta, \varphi) = W_r(r) + W_\vartheta(\vartheta) + W_\varphi(\varphi)\,;$$

that is, for a solution that is the sum of three different functions W_r, W_ϑ and W_φ, each one depends only on one of the variables r, ϑ and φ.

In this way, the Hamilton–Jacobi equation for W becomes

$$\frac{1}{2m}\left(\left(\frac{dW_r}{dr}\right)^2 + \frac{1}{r^2}\left(\frac{dW_\vartheta}{d\vartheta}\right)^2 + \frac{1}{r^2 \sin^2\vartheta}\left(\frac{dW_\varphi}{d\varphi}\right)^2\right) + \mathcal{U}(r) = E\,,$$

and with a simple manipulation, it can be written in the form

$$\left(\frac{dW_\varphi}{\partial \varphi}\right)^2 = r^2 \sin^2\vartheta \left(2m[E - \mathcal{U}(r)] - \left(\frac{dW_r}{\partial r}\right)^2 - \frac{1}{r^2}\left(\frac{dW_\vartheta}{\partial \vartheta}\right)^2\right).$$

The left-hand side of the above equation depends only on φ, while the right-hand side depends only on r and ϑ. Since the variables r, ϑ and φ are independent, each side must be equal to a constant, namely π_φ^2.

Thus, we obtain

$$\begin{cases} \dfrac{dW_\varphi}{d\varphi} = \pi_\varphi\,, \\ \left(\dfrac{dW_\vartheta}{d\vartheta}\right)^2 + \dfrac{\pi_\varphi^2}{\sin^2\vartheta} = r^2\left(2m[E - \mathcal{U}(r)] - \left(\dfrac{dW_r}{dr}\right)^2\right). \end{cases}$$

Once again we observe that the left-hand side depends only on ϑ, while the right-hand side depends only on r. Since r and ϑ are independent variables, both sides must be equal to a constant, namely π_ϑ^2.

Remark 10 *The constant π_φ has a clear physical meaning: it simply corresponds to the component p_φ of the angular momentum along the z axis. Thus, it expresses the uniformity of the spanning, by the projected vector radius $\vec{R} = \vec{r} - z\vec{k}$, of the areas in the (x, y) plane.*

The constant π_ϑ corresponds to the modulus of the angular momentum

$$|\vec{L}|^2 = |m\vec{r} \wedge \vec{v}|^2 = p_\vartheta^2 + \frac{p_\varphi^2}{\sin^2 \vartheta},$$

so that, if α denotes the angle between the orbit plane and the (x, y) plane, we have

$$p_\varphi = |\vec{L}| \cos\alpha,$$

and also

$$\pi_\varphi = \pi_\vartheta \cos\alpha. \tag{4.25}$$

Therefore, the Hamilton–Jacobi equation for W is, for this solution, equivalent to

$$\begin{cases} \dfrac{dW_\varphi}{d\varphi} = \pi_\varphi, \\ \left(\dfrac{dW_\vartheta}{d\vartheta}\right)^2 + \dfrac{\pi_\varphi^2}{\sin^2\vartheta} = \pi_\vartheta^2, \\ \left(\dfrac{dW_r}{dr}\right)^2 = 2m[E - \mathcal{U}(r)] - \dfrac{\pi_\vartheta^2}{r^2}. \end{cases}$$

In the case of the Kepler dynamics we choose $\mathcal{U}(r) = -k/r$, so that the above equations can be written as follows:

$$\begin{cases} \dfrac{dW_\varphi}{d\varphi} = \pi_\varphi, \\ \left(\dfrac{dW_\vartheta}{d\vartheta}\right)^2 = \pi_\vartheta^2 - \dfrac{\pi_\varphi^2}{\sin^2\vartheta}, \\ \left(\dfrac{dW_r}{dr}\right)^2 = 2m\left[E + \dfrac{k}{r}\right] - \dfrac{\pi_\vartheta^2}{r^2}. \end{cases}$$

Thus, we have

$$V = -Et + \pi_\varphi\varphi + \int d\vartheta \sqrt{\pi_\vartheta^2 - \frac{\pi_\varphi^2}{\sin^2\vartheta}} + \int dr \sqrt{2mE + \frac{2mk}{r} - \frac{\pi_\vartheta^2}{r^2}}.$$

Problems

1. Find a complete integral of the Hamilton–Jacobi equation for the harmonic oscillator with *3* degrees of freedom,

$$\mathcal{H} = \frac{1}{2m}\sum_{i=1}^{3}(p_i^2 + m^2\omega_i^2 q_i^2),$$

by separating the variables in spherical-polar coordinates.

2. Find a complete integral of the Hamilton–Jacobi equation for the Kepler dynamics,

$$\mathcal{H} = \frac{p^2}{2m} - \frac{k}{r}$$

by using parabolic coordinates (ξ, η, φ), defined by

$$\begin{cases} x = \sqrt{\xi\eta}\cos\varphi\,, \\ y = \sqrt{\xi\eta}\sin\varphi\,, \\ z = \dfrac{1}{2}(\xi - \eta)\,, \end{cases}$$

with $\xi \in [0, +\infty[$, $\eta \in [0, +\infty[$. The surfaces, defined by $\xi = constant$, $\eta = constant$, define two families of revolution paraboloids having the z axis as a symmetry axis.

3. Find a complete integral of the Hamilton–Jacobi equation for the Hamiltonian with *3* degrees of freedom,

$$\mathcal{H} = \frac{p^2}{2m} + \frac{e_1}{r_1} + \frac{e_1}{r_1}\,,$$

describing the dynamics of a particle in the field (Newtonian or Coulombian‖) generated by two particles, located at distance a at

‖Charles Augustin Coulomb, was born in Angoulême in 1736, and died in Paris in 1806. He worked as an engineer and was a member of the *Institute de France*. During the last years of his life he has been general overseer of Paris University. His contributions to friction laws and to electromagnetism can be considered of basic importance.

positions $\vec{r}_1, \vec{r}_2$. Use elliptic coordinates (ξ, η, φ), defined by

$$\begin{cases} x = \sigma\sqrt{(\xi^2-1)(\eta^2-1)}\cos\varphi\,, \\ y = \sigma\sqrt{(\xi^2-1)(\eta^2-1)}\sin\varphi\,, \\ z = \sigma\xi\eta\,, \end{cases}$$

where σ is an arbitrary parameter and $\xi \in [1, +\infty[$, $\eta \in [-1, 1[$. (Hint: choose $\sigma = a$.)

4.6 The Liouville Theorem on the Complete Integrability

4.6.1 *Reduction*

The knowledge of a first integral f for a given dynamical system, described by the equations

$$\dot{x}^i = X^i(x/t)\,, \quad \forall i \in \{1, 2, \ldots, m\}\,,$$

simplifies the integration problem, since the relation

$$f(x(t)/t) = \text{constant} \equiv f_0$$

must be satisfied by any solution $x(t)$ of the equations of motion; of course, for a suitable choice of the constant, which depends on the initial conditions. All $m-1$ hypersurfaces obtained by varying the constant f_0 will *foliate* the whole space, and each trajectory will belong to one and only one of them. The foliation is called *regular*, if the hypersurfaces have the same dimension; in this way each hypersurface is called a *leaf.*

Furthermore, if an additional functionally independent first integral g is known, a given trajectory also belongs to the hypersurface

$$g(x(t)/t) = g_0\,,$$

defined by g. Each trajectory thus lies on the generically $m-2$ dimensional intersection of leaves of the two foliations. It follows that the knowledge of $m-1$, functionally independent first integrals, defining regular foliations, completely solves the integration problem, since the 1-dimensional intersections of leaves just correspond to curves representing trajectories.

Remark 11 *The previous picture is just the description of a virtual case and is given as a motivation for introducing the Lie theorem below. It almost* *never realizes, even for a simple system, as the one of a harmonic oscillator with 2 degrees of freedom, described by the Hamiltonian*

$$\mathcal{H} = \frac{1}{2}\sum_{i=1}^{2}(p_i^2 + \omega_i^2 q_i^2),$$

with ω_1/ω_2, an irrational number. A fine global coordinate analysis can be found in the Wintner book,[58] *in which the notion of isolating integral (also less correctly called uniform) is introduced. An integral $f(x)$ is said to be isolating if it "can enable one to make predictions concerning the possible future (or past) positions of the points $x = x(t)$ of the solution path which goes at $t = 0$ through x_0 (the case $x(t) = x_0$ of an equilibrium solution being not excluded)."*

Thus, it is naturally expected that, for a canonical system, the knowledge of $2n - 1$, functionally independent first integrals, defining regular foliations, completely solves the integration problem.

Actually, for a canonical system with n degrees of freedom (and with a $2n$-dimensional phase space Φ), it turns out that the integration problem is completely solved knowing only n functionally independent first integrals f_i, which are in *involution*; that is, such that

$$\{f_i, f_j\} = 0, \quad \forall i, j \in \{1, 2, \ldots, n\}.$$

More precisely, the following remarkable theorem was proven by Sophus Lie.[134,25]

Theorem 15 (Lie) *If for a canonical system of rank $2n$, m functionally independent and involutive first integrals are known, which can be solved with respect to m of the p's, the integration problem reduces to the integration of a new canonical system of rank $n - m$.*

In other words, while for a generic first order differential system the knowledge of m first integrals reduces the rank by m units, for a Hamiltonian system the rank is reduced by $2m$ units.

It is interesting, for concrete applications, the case in which $m = n$; that is, the case in which the number of such first integrals is just equal to the number n of degrees of freedom.

4.6.2 *The Liouville theorem*

Theorem 16 (Liouville) *If for a Hamiltonian system, n functionally independent and involutive first integrals are known, which can be solved with respect to the p's, the integration problem reduces to pure quadratures; that is, the equations of motion can be solved simply by evaluating integrals.*

The proof will be carried out by means of the Hamilton–Jacobi integration method. According to this method, in order to have the general integral of a canonical system

$$\begin{cases} \dot{p}_h = -\dfrac{\partial \mathcal{H}}{\partial q_h}, \\ \dot{q}_h = \dfrac{\partial \mathcal{H}}{\partial p_h}, \end{cases} \quad \forall h \in \{1, \ldots, n\},$$

it is sufficient to find a complete integral V of the partial differential equation

$$\frac{\partial V}{\partial t} + \mathcal{H}\left(\frac{\partial V}{\partial q} \Big/ q/t\right) = 0. \tag{4.26}$$

Then the Liouville statement will be proven if we show that the knowledge of n first integrals $f_r(p/q/t)$, with $r \in \{1, \ldots, n\}$ of the canonical system, satisfying the following properties:

(i) are functionally independent; i.e. $\sum_{h=1}^n \lambda_h df_h = 0 \Rightarrow \lambda_h = 0$,
(ii) are in involution; i.e. $\{f_r, f_s\} = 0$, $r, s \in \{1, \ldots, n\}$,
(iii) and define an algebraic system $f_r(p/q/t) = \pi_h$ solvable with respect to the n variables p_r; i.e.

$$\frac{\partial(f_1, \ldots, f_n)}{\partial(p_1, \ldots, p_n)} \neq 0,$$

allows us to determine, by pure quadratures, a complete integral of Eq. (4.26).

Let us first consider the case in which the Hamiltonian $\mathcal{H}$ and the functions f_r do not explicitly depend on the time t.

The equations

$$f_r(p/q) = \pi_r \quad r \in \{1, \ldots, n\}, \tag{4.27}$$

for fixed π's, define a submanifold $M_{f(\pi)}$, known as *the level manifold*, of the phase space Φ. By using hypothesis (iii), Eq. (4.27) can be solved with respect to p's in the form

$$p_\alpha = \varphi_\alpha(q/\pi), \quad \alpha \in \{1, \ldots, n\}.$$

Of course, on the level manifold M_π the differences $p_\alpha - \varphi_\alpha(q/\pi)$, and consequently, the Poisson brackets $\{p_\alpha - \varphi_\alpha(q/\pi), p_\beta - \varphi_\beta(q/\pi)\}$ vanish identically.

Furthermore, at the end of the section, we will show that the involutivity of the f's on the whole phase space Φ implies the vanishing of the Poisson brackets $\{p_\alpha - \varphi_\alpha(q/\pi), p_\beta - \varphi_\beta(q/\pi)\}$ on the whole phase space Φ, namely

$$\{f_r, f_s\} = 0\,, \quad \forall r,s \in \{1,\ldots,n\}$$
$$\Rightarrow \{p_\alpha - \varphi_\alpha, p_\beta - \varphi_\beta\} = 0\,, \quad \forall \alpha,\beta \in \{1,\ldots,n\}\,.$$

On the other hand,

$$\{p_\alpha - \varphi_\alpha, p_\beta - \varphi_\beta\} = \frac{\partial \varphi_\beta}{\partial q_\alpha} - \frac{\partial \varphi_\alpha}{\partial q_\beta}\,,$$

so that

$$\{f_r, f_s\} = 0\,, \quad \forall r,s \in \{1,\ldots,n\} \Rightarrow \frac{\partial \varphi_\beta}{\partial q_\alpha} = \frac{\partial \varphi_\alpha}{\partial q_\beta}\,, \quad \forall \alpha,\beta \in \{1,\ldots,n\}\,,$$

which implies that the differential form

$$\sum_{h=1}^{n} \varphi_h(q/\pi)dq_h\,,$$

is closed, or which is the same, locally exact. In other words, in any simply connected part of M_π, a function $W(q/\pi)$ exists, such that

$$\varphi_h(q/\pi/t) = \frac{\partial W}{\partial q_h}\,. \tag{4.28}$$

In this way, it has been shown that the involutivity of the f's implies that the solutions of the Eq. (4.27), with respect to the p's, can be locally expressed as follows:

$$p_h = \frac{\partial W}{\partial q_h}(q/\pi)\,.$$

Let us now define n new coordinates, namely $\chi_1, \chi_2, \ldots, \chi_n$, by

$$\chi_h = \frac{\partial W}{\partial \pi_h}(q/\pi/t)\,.$$

Then, we easily check that the relations

$$\begin{cases} p_h = \dfrac{\partial W}{\partial q_h}(q/\pi)\,, \\ \chi_h = \dfrac{\partial W}{\partial \pi_h}(q/\pi)\,, \end{cases}$$

implicitly define an invertible transformation between the variables (p/q) and (π/χ), which turns out to be canonical for the results obtained in subsection (3.1.2). The invertibility of the transformation easily follows by observing that

$$\frac{\partial^2 W}{\partial q_h \partial \pi_k} = \frac{\partial}{\partial \pi_k}\frac{\partial W}{\partial q_h} = \frac{\partial \varphi_h}{\partial \pi_k}\,,$$

that is, by noting that $\partial^2 W/\partial q_h \partial \pi_k$ is just the generic element of the Jacobian matrix of the φ's with respect to the π's, so that the Jacobian determinant $\det(\partial^2 W/\partial q_h \partial \pi_k)$ is the inverse of the Jacobian $\det(\partial f_i/\partial p_j)$, which is supposed (hypothesis (iii)) to be different from zero.

Therefore, we come to the new canonical system

$$\begin{cases} \dfrac{d}{dt}\pi_h = -\dfrac{\partial \mathcal{K}}{\partial \chi_h}\,, \\ \dfrac{d}{dt}\chi_h = \dfrac{\partial \mathcal{K}}{\partial \pi_h}\,, \end{cases} \tag{4.29}$$

with a new characteristic function $\mathcal{K}$ given by

$$\mathcal{K}(\pi/\chi) = \mathcal{H}^*\,,$$

where the $*$ indicates, as usual, that the transformation has been performed.

The canonical Eq. (4.29) show that $\mathcal{K}$ really does not depend on the χ's, since $\dot{\pi}_h = 0$. As a consequence, the derivatives $\partial\mathcal{K}/\partial\pi_h$ will not depend on time t, so that the system (4.29) is trivially integrated in the following form:

$$\begin{cases} \pi_h = \text{constant}\,, \\ \chi_h = \nu_h t + \delta_h\,, \end{cases}$$

where $\nu_h \equiv \partial\mathcal{K}/\partial\pi_h$, and the δ's are arbitrary constants.

The general case can be treated by introducing two additional auxiliary parameters p_0, q_0, and the *double bracket*

$$\begin{aligned}\{\{u,v\}\} &= \sum_{h=0}^{n}\left(\frac{\partial u}{\partial q_h}\frac{\partial v}{\partial p_h} - \frac{\partial u}{\partial p_h}\frac{\partial v}{\partial q_h}\right)\\ &= \frac{\partial u}{\partial q_0}\frac{\partial v}{\partial p_0} - \frac{\partial u}{\partial p_0}\frac{\partial v}{\partial q_0} + \{u,v\}.\end{aligned}$$

Of course, if u and v do not depend on p_0,

$$\{\{u,v\}\} = \{u,v\},$$

and then

$$\{\{f_r, f_s\}\} = \{f_r, f_s\} = 0.$$

Furthermore, by choosing $q_0 = t$, we have

$$\begin{aligned}\{\{p_0 + H, f_r\}\} &= 1 \cdot \frac{\partial f_r}{\partial t} - 0 + \{p_0 + H, f_r\}\\ &= \frac{\partial f_r}{\partial t} + \{H, f_r\}.\end{aligned}$$

In terms of the double bracket, Liouville's hypotheses on the knowledge of n involutive first integrals $f_r(p/q/t)$, $r \in \{1,\ldots,n\}$ can be expressed as follows:

$$\{\{f_\mu, f_\nu\}\} = 0, \quad \mu,\nu \in \{0,1,\ldots,n\},$$

with $f_0 = p_0 + \mathcal{H}$.

On the other hand, from the same hypotheses it follows that the algebraic system

$$\begin{cases} f_r(p/q/t) = \pi_r, \\ p_0 + \mathcal{H} = 0, \end{cases}$$

can be solved in the form

$$p_\alpha - \varphi_\alpha(q/\pi/t) = 0, \quad \alpha \in \{0,1,\ldots,n\}, \tag{4.30}$$

with

$$\varphi_0(q/\pi/t) = -\mathcal{H}(\varphi/q/t).$$

As before, it turns out that the vanishing of $\{\{f_\mu, f_v\}\}$ for all $\mu, v \in \{0, 1, \ldots, n\}$ implies that

$$\{\{p_\alpha - \varphi_\alpha, p_\beta - \varphi_\beta\}\} = 0\,, \quad \forall\, \alpha, \beta \in \{0, \ldots, n\}\,,$$

or, more explicitly,

$$\{\{p_\alpha - \varphi_\alpha, p_\beta - \varphi_\beta\}\} = \frac{\partial \varphi_\beta}{\partial q_\alpha} - \frac{\partial \varphi_\alpha}{\partial q_\beta}\,.$$

Since

$$\{\{f_\mu, f_v\}\} = 0\,, \quad \forall\, \mu, v \in \{0, 1, \ldots, n\}$$
$$\Rightarrow \frac{\partial \varphi_\beta}{\partial q_\alpha} = \frac{\partial \varphi_\alpha}{\partial q_\beta}\,, \quad \forall\, \alpha, \beta \in \{0, 1, \ldots, n\}\,,$$

the differential form

$$\sum_{h=0}^{n} \varphi_\alpha(q/\pi/t) dq_\alpha$$

is closed or, which is the same, locally exact. In other words, in a sufficiently small region a function $V(q/\pi/t)$ exists such that

$$\varphi_\alpha(q/\pi/t) = \frac{\partial V}{\partial q_\alpha}\,, \quad \forall\, \alpha \in \{0, 1, \ldots, n\}\,.$$

It has thus been shown that the involutivity of the f's implies that the solutions of the equations $f_r(p/q/t) = \pi_r$, with respect to the p's, can be locally expressed as follows:

$$p_r = \frac{\partial V}{\partial q_r}(q/\pi/t)\,, \quad r \in \{1, \ldots, n\}\,,$$

where V is such that

$$\frac{\partial V}{\partial t} = \varphi_0(q/\pi/t) = -\mathcal{H}(\varphi/q/t) = -\mathcal{H}\left(\frac{\partial V}{\partial q}\Big/ q/t\right)\,.$$

This is equivalent to the fact that V is a complete integral of the Hamilton–Jacobi equation

$$\frac{\partial V}{\partial t} + \mathcal{H}\left(\frac{\partial V}{\partial q}\Big/ q/t\right) = 0\,.$$

Therefore, by defining n new coordinates, namely $\chi_1, \chi_2, \ldots, \chi_n$, by

$$\chi_h = \frac{\partial V}{\partial \pi_h}(q/\pi/t),$$

the relations

$$\begin{cases} p_h = \dfrac{\partial V}{\partial q_h}(q/\pi/t), \\ \chi_h = \dfrac{\partial V}{\partial \pi_h}(q/\pi/t), \end{cases}$$

implicitly define an invertible transformation between the variables (p/q) and (π/χ). For the results at the subsection (3.1.2), this transformation is canonical and leads to the, trivially integrable, new canonical system

$$\begin{cases} \dot{\pi}_h = 0, \\ \dot{\chi}_h = 0. \end{cases}$$

Lemma 17 (Involutive relations) *Given n functions g_r on the phase space Φ such that the algebraic system of equations*

$$g_r(p/q) = 0, \quad r \in \{1, \ldots, n\}, \tag{4.31}$$

can be solved with respect to the p's in the form

$$p_i = \varphi_i(q), \quad i \in \{1, \ldots, n\}, \tag{4.32}$$

then on the whole phase space Φ, we have

$$\{g_r, g_s\} = 0, \quad \forall r, s \in \{1, \ldots, n\}$$
$$\Rightarrow \{p_\alpha - \varphi_\alpha, p_\beta - \varphi_\beta\} = 0, \quad \forall \alpha, \beta \in \{1, \ldots, n\}.$$

Proof. *Equation (4.32), when replaced in Eq. (4.31), give identically*

$$\tilde{g}_r(q) \equiv g_r(\varphi(q)/q) = 0, \quad r \in \{1, \ldots, n\},$$

and then on Φ,

$$\frac{d}{dq_h} g_r(\varphi(q)/q) = 0.$$

The last relation explicitly gives

$$0 = \frac{\partial}{\partial q_h}\tilde{g}_s = \left|\frac{\partial g_s}{\partial q_h} + \sum_{i=1}^{n} \frac{\partial g_s}{\partial p_i}\frac{\partial \varphi_i}{\partial q_h}\right|_{p=\varphi(q)}, \quad r, h \in \{1, \ldots, n\}.$$

Therefore, the partial derivatives $\partial g_s/\partial q_h$ can be expressed in the form

$$\frac{\partial g_s}{\partial q_h} = \sum_{i=1}^{n} \frac{\partial g_s}{\partial p_i}\frac{\partial (p_i - \varphi_i)}{\partial q_h}.$$

On the other hand, since $\partial(p_j - \varphi_j)/\partial p_h = \delta_{jh}$, we have

$$\frac{\partial g_r}{\partial p_h} = \sum_{j=1}^{n} \frac{\partial g_r}{\partial p_j}\frac{\partial (p_j - \varphi_j)}{\partial p_h},$$

so that

$$\frac{\partial g_r}{\partial p_h}\frac{\partial g_s}{\partial q_h} = \sum_{i,j=1}^{n} \frac{\partial g_r}{\partial p_i}\frac{\partial g_s}{\partial p_j}\frac{\partial (p_i - \varphi_i)}{\partial p_h}\frac{\partial (p_j - \varphi_j)}{\partial q_h},$$

$$\frac{\partial g_r}{\partial p_h}\frac{\partial g_s}{\partial q_h} = \sum_{i,j=1}^{n} \frac{\partial g_r}{\partial p_i}\frac{\partial g_s}{\partial p_j}\frac{\partial (p_i - \varphi_i)}{\partial q_h}\frac{\partial (p_j - \varphi_j)}{\partial p_h}.$$

It thus follows that

$$\begin{aligned} \{g_r, g_s\} &= \sum_{h=1}^{n} \left(\frac{\partial g_r}{\partial p_h}\frac{\partial g_s}{\partial q_h} - \frac{\partial g_r}{\partial p_h}\frac{\partial g_s}{\partial q_h}\right) \\ &= \sum_{i,j=1}^{n} \frac{\partial g_r}{\partial p_i}\frac{\partial g_s}{\partial p_j}\{p_i - \varphi_i, p_j - \varphi_j\}. \end{aligned}$$

The above relation can be written in the form

$$\{g_r, g_s\} = \sum_{i=1}^{n} \frac{\partial g_r}{\partial p_i} X_i^{(s)}, \tag{4.33}$$

with

$$X_i^{(s)} = \sum_{j=1}^{n} \frac{\partial g_s}{\partial p_j}\{p_i - \varphi_i, p_j - \varphi_j\}. \tag{4.34}$$

Since the Jacobian determinant $\partial(g_1,\dots,g_n)/\partial(p_1,\dots,p_n)$ is nonvanishing, from Eq. (4.33) it follows that

$$\{g_r, g_s\} = 0 \Rightarrow X_i^{(s)} = 0\,.$$

Similarly, by using for Eq. (4.34) the same argument, we finally have

$$X_i^{(s)} = 0 \Rightarrow \{p_i - \varphi_i, p_j - \varphi_j\} = 0\,.$$

4.6.3 *Remarks on the Liouville theorem*

Let us consider a canonical system with a Hamiltonian function $\mathcal{H}$, which does not explicitly depend on the time. It has been shown that the knowledge of n first integrals $f_r(p/q)$, $r \in \{1,\dots,n\}$, which are functionally independent; i.e. $\sum_{h=1}^{n} \lambda_h df_h = 0 \Rightarrow \lambda_h = 0$, in involution; i.e. $\{f_r, f_s\} = 0$, $r, s \in \{1,\dots,n\}$, and such that the algebraic system defined by $f_r(p/q/t) = \pi_h$ can be solved with respect to the n variables p_r, allows us to trivially integrate the equations of the motion.

Arnold has given a global formulation of the theorem by requiring that the level manifold M_π be compact and connected. This will be treated in details in Part III. Here we shall limit ourselves to the following considerations.

Let us first observe that W, in Eq. (4.28), is defined only locally. As a consequence, the coordinates χ are not uniquely defined on $M_{f(\pi)}$. They will be continuous multivalued functions of the point $p \in M_{f(\pi)}$:

$$\chi : p \in M_{f(\pi)} \to p' = \chi(p) \in \Re^n\,.$$

Therefore, to each point $p \in M_{f(\pi)}$ we can associate a point $p' \in \Re^n$ whose coordinates are just given by $\chi^1, \chi^2, \dots, \chi^n$. Really, as the χ's are not uniquely defined, we can associate to p infinitely many points, one for any chosen determination of the χ's:

$$p \in M_{f(\pi)} \to p'_1, p'_2, p'_3, \dots\,.$$

It is clear that, as the χ's change continuously, all points p', associated with all the points p of $M_{f(\pi)}$, will cover the whole space $\Re^n$.

Let us investigate more closely the multivalued structure of the χ's.

A vector $\vec{a}$ will be called a *period* if $\forall\, \vec{\chi}$, $\vec{\chi}$ and, $\vec{\chi} + \vec{a}$ represent the same point in $M_{f(\pi)}$. Of course, it will be independent on $\vec{\chi}$, as $\vec{\chi}$ and $\vec{\chi}+\vec{a}$ are both solutions of Eq. (4.29). Moreover, the modulus $|\vec{a}|$ of $\vec{a}$ cannot be arbitrarily small since, in a sufficiently small region, $\vec{\chi}$ is single-valued.

If $\vec{a}_1$ is a period with a minimal modulus, then $m_1\vec{a}_1$, with $m_1 \in N$, is still a period. Furthermore, any period which is parallel to $\vec{a}_1$ must be an integer multiple of $\vec{a}_1$. In fact, if $\vec{a}' = \lambda\vec{a}_1$, with $\lambda \in (\Re - N)$, then by denoting with $[\lambda]$ the maximal integer lower than λ, $\vec{a}' - [\lambda]\vec{a}_1 = (\lambda - [\lambda])\vec{a}_1$ would be a period with modulus lower than $|\vec{a}_1|$. As a consequence, any new period $\vec{a}$ will have a component which is orthogonal to $\vec{a}_1$. By choosing among them, the period $\vec{a}_2$ whose component orthogonal to $\vec{a}_1$ has the lowest modulus, it turns out that the vectors $m_1\vec{a}_1 + m_2\vec{a}_2$ are periods. Moreover, in the plane spanned by $\vec{a}_1$, $\vec{a}_2$ there are no periods of different form. It follows, inductively, that all periods are of the form

$$m_1\vec{a}_1 + m_2\vec{a}_2 + \cdots + m_r\vec{a}_r\,, \quad r \in \{0, \ldots, n\}, m_i \in N\,,$$

where $r = 0$ corresponds to the absence of periods and $r \leq n$, since all vectors $\vec{a}_i$ are, by construction, linearly independent. Thus, each point $p \in M_{f(\pi)}$ will have just one image in each parallelepiped with sides $\vec{a}_i$ (if $r < i \leq n$, then $\vec{a}_i = \infty$).

The motion region is bounded if $r = n$ and unbounded if $r \neq n$.

If $n = 2$, three cases can occur:

- $r = 0 \Rightarrow M_{f(\pi)}$ is topologically equivalent to a plane;
- $r = 1 \Rightarrow M_{f(\pi)}$ is topologically equivalent to a cylinder;
- $r = 2 \Rightarrow M_{f(\pi)}$ is topologically equivalent to a torus.

Only in the last case the motion region is compact.

More generally, if $\dim M_{f(\pi)} = n$, the compact hypersurface corresponding to $r = n$ is called an n-torus T^n. In any case, the motion develops on $M_{f(\pi)} \subset \Phi$, which is *invariant.*

4.6.4 *Action-angle coordinates*

Let us consider more closely the case of the torus. By denoting with γ_l the closed curves, which on $M_{f(\pi)}$ are images of segments $\lambda\vec{a}_l$ with $0 \leq \lambda \leq 1$, let us define

$$J_l = \frac{1}{2\pi}\oint_{\gamma_l}\sum_{h=1}^{n}\frac{\partial W}{\partial q_h}dq_h\,, \quad l \in \{1, \ldots, n\}\,.$$

The J's will be first integrals, as they are functions of the f's and

$$J_l = \frac{1}{2\pi}\Delta_l W\,,$$

where $\Delta_l W$ represents the variation of W along the curve γ_l. Moreover, they will be independent and involutive since

$$\{J_h, J_k\} = \sum_{r,s=1}^{n} \frac{\partial J_h}{\partial f_r}\frac{\partial J_h}{\partial f_s}\{f_r, f_s\} = 0\,.$$

Therefore, starting from the very beginning with the J's instead of the f's, we can introduce their conjugate variables φ_h in the same way as the χ's were introduced as conjugate variables to the f's. Along a cycle γ_h, we will have

$$\Delta_k \varphi_h = \Delta_k \frac{\partial W}{\partial J_h} = \frac{\partial}{\partial J_h}\Delta_k W = 2\pi\delta_{hk}\,. \tag{4.35}$$

According to the above equation, the φ's are *angle* variables, since their variation is 2π along any closed walk, turning the torus just one time. Their conjugate momenta J give, apart from a constant factor the variation of the action W along a cycle in which all the φ's, but one, are constant. For this reason they are called *action variables.* The Hamiltonian function $\mathcal{K} = \mathcal{H}^*$ will be function of the action variables J alone, and the angle variables satisfy the equations

$$\dot{\varphi}_h = \frac{\partial \mathcal{K}}{\partial J_h} = \nu_h(J)\,,$$

whose integration give

$$\varphi_h = \nu_h(J)t + \delta_h\,.$$

The motion described by them is called *a multiperiodic motion* with *frequencies* ν_h.

Let us finally observe that the action-angle variables are not uniquely defined, since any linear transformation of the φ's, with integer coefficients and determinant of the associated matrix equal to 1, will again give angle variables, whose conjugate variables will still be action variables.

4.6.5 *The action-angle coordinates for the harmonic oscillator*

The Hamiltonian of the harmonic oscillator, with $n = 1$ degree of freedom, is given by

$$H = \frac{1}{2m}(p^2 + m^2\omega^2 q^2)\,.$$

The system has just one first integral which, of course, is in involution with itself. Thus the system is completely integrable à la Liouville. The level manifold

$$M_E \equiv \left\{ (p/q) \in \Phi : \frac{1}{2m}(p^2 + m^2\omega^2 q^2) = E \right\}$$

is, in the phase space Φ, an ellipse having $a = \sqrt{2mE}$ and $b = (1/\omega)\sqrt{2E/m}$ as semi-axis. The action variable

$$J = \frac{1}{2\pi}\oint_\gamma p dq$$

must be evaluated along the curve γ determined by the values of q_* at turning points; i.e. by the values

$$q_* = \pm(1/\omega)\sqrt{\frac{2E}{m}},$$

whose corresponding momenta vanish.

The corresponding integral can be easily performed. It can also be evaluated more simply by observing that, apart from the factor $1/2\pi$, the action variable is the area πab of the mentioned ellipse. Thus, we obtain

$$J = \frac{E}{\omega}.$$

The Hamiltonian of the harmonic oscillator in terms of action variables is then given by

$$\mathcal{K} = \omega J.$$

The angle coordinate can be evaluated by using Eq. (4.24).

The same procedure, applied to the harmonic oscillator with n degrees of freedom,

$$H = \frac{1}{2m}\sum_{h=1}^{n}(p_h^2 + m^2\omega_h^2 q_h^2),$$

leads to

$$K = \sum_{h=1}^{n} \omega_h J_h.$$

4.6.6 The Kepler dynamics in action-angle variables

In the previous section, it was shown that the Hamiltonian function for the Kepler dynamics, in spherical-polar coordinates, reads

$$\mathcal{H} = \frac{1}{2m}\left(p_r^2 + \frac{p_\vartheta^2}{r^2} + \frac{p_\varphi^2}{r^2 \sin^2\vartheta}\right) - \frac{k}{r},$$

and that the corresponding reduced Hamilton–Jacobi equation has the form

$$\frac{1}{2m}\left(\left(\frac{\partial W}{\partial r}\right)^2 + \frac{1}{r^2}\left(\frac{\partial W}{\partial \vartheta}\right)^2 + \frac{1}{r^2 \sin^2\vartheta}\left(\frac{\partial W}{\partial \varphi}\right)^2\right) - \frac{k}{r} = E.$$

It was also shown that, by using the method of separation of variables, a complete integral of the equation takes the following form:

$$W(r, \vartheta, \varphi) = W_r(r) + W_\vartheta(\vartheta) + W_\varphi(\varphi),$$

where the functions $W_r(r)$, $W_\vartheta(\vartheta)$, $W_\varphi(\varphi)$ satisfy the equations

$$\begin{cases} \dfrac{dW_\varphi}{d\varphi} = \pi_\varphi, \\ \left(\dfrac{dW_\vartheta}{d\vartheta}\right)^2 + \dfrac{\pi_\varphi^2}{\sin^2\vartheta} = \pi_\vartheta^2, \\ \left(\dfrac{dW_r}{dr}\right)^2 = 2m\left[E + \dfrac{k}{r}\right] - \dfrac{\pi_\vartheta^2}{r^2}. \end{cases}$$

In the compact case, characterized by $E < 0$, we can introduce action variables J_r, J_ϑ, and J_φ by writing

$$\begin{cases} J_\varphi = \dfrac{1}{2\pi}\displaystyle\oint \pi_\varphi d\varphi, \\ J_\vartheta = \dfrac{1}{2\pi}\displaystyle\oint d\vartheta \sqrt{\pi_\vartheta^2 - \dfrac{\pi_\varphi^2}{\sin^2\vartheta}}, \\ J_r = \dfrac{1}{2\pi}\displaystyle\oint dr \sqrt{2mE + \dfrac{2mk}{r} - \dfrac{\pi_\vartheta^2}{r^2}}. \end{cases}$$

Since π_φ is constant and $0 \le \varphi \le 2\pi$, we have

$$J_\varphi = \pi_\varphi. \tag{4.36}$$

The remaining two closed curves of integration are fixed by requiring the vanishing of the corresponding velocities, or better, of the corresponding momenta p_ϑ and p_r expressed, of course, in terms of variables π_ϑ and π_φ. In this way, the integration limits are fixed by

$$\begin{cases} p_\vartheta^2 \equiv \pi_\vartheta^2 - \dfrac{\pi_\varphi^2}{\sin^2\vartheta} = 0\,, \\ p_r^2 \equiv 2mE + \dfrac{2mk}{r} - \dfrac{\pi_\vartheta^2}{r^2} = 0\,. \end{cases}$$

Therefore, the "ϑ" integration must be performed between the limits ϑ_1 and ϑ_2 given by the solutions of

$$\sin^2\vartheta = \frac{\pi_\varphi^2}{\pi_\vartheta^2} = \cos^2\alpha\,,$$

where Eq. (4.25) has been used. Since ϑ itself always lies between 0 and π, where $\sin\vartheta > 0$, we have $\sin\vartheta_1 = \sin\vartheta_2 = \cos\alpha$. Thus, the integration goes from $\vartheta_1 = \pi/2 - \alpha$ to $\pi/2$ to $\vartheta_2 = \pi/2 + \alpha$ and again back to ϑ_1, so that the $\sin\vartheta$ goes from $\cos\alpha$ to 1, then to $\cos\alpha$. In this way, we obtain (see Appendix C)

$$J_\vartheta = \pi_\vartheta - \pi_\varphi\,. \tag{4.37}$$

The "r" integration, which is also performed in Appendix C by using the *method of residues*, gives

$$J_r = -\pi_\vartheta + \frac{mk}{\sqrt{-2mE}}\,. \tag{4.38}$$

From Eqs. (4.36), (4.37) and (4.38), we have

$$J_r + J_\vartheta + J_\varphi = \frac{mk}{\sqrt{-2mE}}\,.$$

The above equation allows us to write directly the new Hamiltonian function $\mathcal{K} = \mathcal{H}^*$ as follows:

$$\mathcal{K}(J) = -\frac{mk^2}{2(J_r + J_\vartheta + J_\varphi)^2}\,.$$

4.6.7 *The perturbations of integrable systems and the KAM theorem*

There exist very few dynamical systems which satisfy Liouville's theorem hypotheses. Generally, for Hamiltonian function not depending explicitly on the time t, there exists just the first integral given by the Hamiltonian. Some classical integrable systems are given by *systems with a central symmetry, a particle in a Newtonian gravitational field generated by two fixed points, the spherical top in a Newtonian gravitational field.* There exist some more dynamical systems with finitely many degrees of freedom, recently found in connection with integrability problem in field theory, where much more examples occur.[161,44]

For the applications, it is interesting to elaborate methods which will allow us to study a given Hamiltonian dynamics by separating its Hamiltonian function $\mathcal{H}$ as the sum of two parts. The first part, namely $\mathcal{H}_0$, is required to be a completely integrable one, so that

$$\mathcal{H} = \mathcal{H}_0 + \lambda \mathcal{H}_1 \,,$$

where λ is a "small" parameter and $\mathcal{H}_1$ is an analytic function of $2n$ variables. By using action-angle variables (J, φ), the above equation can be written as follows:

$$\mathcal{H}(J, \varphi) = \mathcal{H}_0(J) + \lambda \mathcal{H}_1(J, \varphi) \,.$$

For $\lambda = 0$, the phase space is foliated, according to Liouville's theorem, in n-dimensional invariant tori $M_{J(\pi)}$, defined by $J_h = \pi_h$, on which the curves,

$$\varphi_h = \nu_h(J)t + \varphi_h(0) \,,$$

completely wound.

It was common opinion, before 1954, that $\lambda \neq 0$ completely destroys the foliation in invariant tori and the beautiful geometrical structure underlying integrable systems, giving rise to the ergodic behavior; that is, to orbits densely filling the submanifold $\mathcal{H} =$ constant. Therefore, the question is to know what remains of this geometrical structure when $\lambda \neq 0$. This opinion was supported by the fact that in the perturbative series there appear denominator terms like $\vec{\nu} \cdot \vec{k}$, where $\vec{k} \equiv (k_1, k_2, \ldots, k_n)$ are integer numbers. Therefore, when the ratios ν_i/ν_j are rational numbers, the series diverges. To say that the ratios ν_i/ν_j are rational numbers is equivalent to say that there exists a period T, which is a multiple of all period $\tau_j = 1/\nu_j$, so that the orbit on the torus T^n is closed. In such cases the torus is said to be *resonant*. Beyond this case, close to the resonance there will appear, however, terms too large (little divisors), since the rational numbers Q are dense in $\Re$. This happens, for instance, in the case of Jupiter and Saturnus, which move along their orbits each day by 299″1′ and 120″5′ degrees, respectively, so that $2\nu_1 - 5\nu_2 \simeq 0$. The existence of a strong perturbation, with a large period, of the motion of planets, connected with the little denominator $2\nu_1 - 5\nu_2$, was already known to Laplace.

The presence of $\vec{\nu} \cdot \vec{k}$ in the denominators can be easily understood, by considering that the terms in the Fourier's expansion of $\mathcal{H}_1$:

$$\mathcal{H}_1(J, \varphi) = \sum_{\vec{k}} \mathcal{H}_1^k(J) \exp i\vec{k} \cdot \vec{\varphi},$$

in a perturbative scheme, will be derived or integrated in t.

Finally, in 1954, a positive answer about the applicability of perturbative methods and the role of the parameter λ in the convergence of corresponding series was offered by Kolmogorov. His theorem, extended and formalized by Arnold (1963) and Moser (1967), is today known as KAM theorem.

The theorem[2] proves that, for *small* values of $|\lambda|$ and nonvanishing Hessian of the Hamiltonian, only few invariant tori are destroyed. A large number of them are only deformed by the perturbation. On such deformed tori the orbits are still dense and almost periodic with n frequencies everywhere. Such invariant deformed tori correspond to unperturbed initial conditions for which

$$|\vec{\nu} \cdot \vec{k}| \geq a|\vec{k}|^{-b},$$

with a and b positive constants. It is shown that, for sufficiently large b, the constant a is of order $\bigcirc(\lambda)$ and gives the measure of the lost tori.

We will not go into more details and refer the interested reader to the literature (besides Arnold's book,[2] see for instance Refs. 45 and 3).

4.6.8 *The Poincaré representation*

It is possible to concretely *see* what happens by using the so called *Poincaré*** *map*.

Let us consider a Hamiltonian system with $n = 2$ degrees of freedom whose Hamiltonian function $\mathcal{H}$ does not depend explicitly on the time t. Let us also

**Henry Poincaré was born in Nancy in 1854, and died in Paris in 1912. He was a professor at Paris University and École Polytechnique. An analysis, by Hadamard, Langevin, Boutroux, and Volterra, of his basic contribution to mathematics and theoretical physics can be found in La Nouvelle Collection Scientifique (Paris: Alcan, 1914).

suppose that the 3-dimensional manifold $M_E = \{p \in \Phi : \mathcal{H}(p/q) = E\}$, defined by the first integral of the energy, is compact. We know that the existence of a second first integral, namely f, will ensure the complete integrability, and that the manifold M_E will be foliated in 2-dimensional invariant tori T^2. For a given initial condition, the motion will be represented by a helix belonging just to one torus and densely winding on it, never returning, provided the torus is not resonant, exactly at the same point.

Let us now consider the 2-dimensional manifold Σ defined by the equation

$$\mathcal{H}(p_1, p_2, q_1, 0) = E\,,$$

representing the intersection of M_E with the hyperplane defined by $q_2 = 0$, and a point $\sigma_0 \in \Sigma$, which can be fixed by giving a point in the plane $S \equiv (p_1, q_1)$.

By considering σ_0 as the initial condition at the time t_0 of the flow of $\mathcal{H}$, there will exist a time instant, namely t_1, in which the trajectory will again meet Σ in a point $\sigma_1 \in \Sigma$. Thus, recursively, there will be a sequence of time instants t_k in which the trajectory will meet Σ at points $\sigma_k \in \Sigma$. Since the sequence of points $\sigma_k \in \Sigma$ is the image of a sequence $s_k = (p_1^{(k)}, q_1^{(k)})$ of points in the plane S, it is possible to describe the evolution by giving the sequence $s_k \in S$. The map

$$\Sigma \to S\,,$$

which describes the evolution, is called the *Poincaré map.*

Furthermore, as it winds on the torus T^2, the trajectory meets Σ on the 1-dimensional submanifold determined by the ulterior equation $f(p_1, p_2, q_1, q_2) = \pi$; that is, on a smooth closed curve, which is also the image of a similar curve of S. Therefore, for a not resonant torus, the points s_k will dispose along a regular curve, while for a resonant torus the sequence will stop; that is, there will be an integer number r such that $s_{k+r} = s_k$, and so on.

The above description can be also applied to *noncompletely integrable* dynamics. In this cases the trajectory will not meet Σ along a regular curve but in points σ_k covering a 2-dimensional region, which is an image of a 2-dimensional region of S (chaotic behavior).

The KAM theorem predicts that, by increasing the value of the parameter λ, it is possible to observe, in S, a transition from a picture composed by regular curves to a picture composed by a large part of previous curves together with extra points, and then to a picture composed by few curves and too many isolated points covering the whole interested region.

Computer analysis gave, of course, exactly what was expected and was allowed to discover new remarkable completely integrable systems.

Further Readings

- G. dell' Antonio, *Elementi di Meccanica* (Liguori, Naples, 1996).
- A. Lichnerowicz, "Les variétés de Poisson et leurs algèbres de Lie associées," *J. Diff. Geom.* **12**, 253 (1977).
- A. Romano, *Lezioni di Meccanica Razionale* (Liguori, Naples, 1990).
- E. C. G. Sudarshan and N. Mukunda, *Classical Dynamic: A Modern Perspective* (John Wiley, 1974).
- W. Yourgrau and S. Mandelstam, *Variational Principles in Dynamics and Quantum theory* (Dover, 1968).

Part II

Basic Ideas of Differential Geometry

Differential geometry is the differential calculus that does not depend on the coordinate system and so, it is the best language for a science (Physics) deputed to describe phenomena, which do not depend on the observer.

Part II is devoted to a summary survey of useful geometric concepts, as *differential manifold*, *tangent space*, *fiber bundles*, *Lie derivative*, *differential forms*, and *exterior derivative.*

Differential forms describe all the relevant physical entities as *the work*, *the heat*, *the internal energy*, *the electromagnetic field*, and so on.

Fiber bundles are used, in elementary particle physics, to describe particles with internal structures (isotopic spin, etc.) and to construct *instantonic solutions* in field theory.

Henceforth, the sum over repeated, upper and lower indices is understood according to the Einstein convention: $\sum_i a_i b^i \equiv a_i b^i$.

Chapter 5

Manifolds and Tangent Spaces

5.1 Differential Manifolds

A manifold $\mathcal{M}$ is a separable topological space such that every point of it is representable at least on one chart; a chart $(\mathcal{U}, \varphi)$ of a manifold $\mathcal{M}$ is an open set $\mathcal{U} \subseteq \mathcal{M}$, called domain of the chart, with a homeomorphism

$$\varphi : \mathcal{U} \subset \mathcal{M} \rightarrow A \subset \Re^n$$

from $\mathcal{U}$ to an open set A in $\Re^n$.

A map $f : U \rightarrow V$ between two topological spaces U e V is called a *homeomorphism* if it is one-to-one and both f and f^{-1} are continuous maps.

Let p be an element in $\mathcal{M}$ representable on the chart $(\mathcal{U}, \varphi)$. The coordinates $\varphi^i(p)$ of the image of the point p are called coordinates of p in the chart $(\mathcal{U}, \varphi)$, or local coordinates of p. Thus, a chart is essentially a local coordinates system.

A C^k atlas on a manifold $\mathcal{M}$ is a collection $\{(\mathcal{U}_\alpha, \varphi_\alpha)\}$ of charts such as the domains $\mathcal{U}_\alpha$ form a covering of $\mathcal{M}$, and the homeomorphisms

$$\varphi_\beta \circ \varphi_\alpha^{-1} : \varphi_\alpha(\mathcal{U}_\alpha \cap \mathcal{U}_\beta) \rightarrow \varphi_\beta(\mathcal{U}_\alpha \cap \mathcal{U}_\beta)$$

are C^k maps between open sets of $\Re^n$.

Two C^k atlases are equivalent if their union is a C^k atlas.

A manifold $\mathcal{M}$ with a C^k atlas equivalence class is called a C^k *differential manifold*; its dimension is n.

A manifold is said to be *orientable* if an atlas can be chosen in such a way that for all α, β the Jacobian determinants $\det(\partial\varphi^i_\alpha/\partial\varphi^j_\beta)$ have the same sign. An intrinsic definition will be given in Sec. 2.5.

The sphere as a differential manifold

The n-dimensional sphere S^n is defined by

$$S^n = \left\{(x_1, \ldots, x_{n+1}) \in \Re^{n+1} \;:\; \sum_{i=1}^{n+1} x_i^2 = 1\right\}.$$

It is a separable topological space, and associated with $\mathcal{U} \subset S^n$ and $\mathcal{V} \subset S^n$, two charts $(\mathcal{U}, \varphi)$ and $(\mathcal{V}, \psi)$ can be introduced as follows:

$$\mathcal{U} \equiv S^n - N\,, \quad \mathcal{V} \equiv S^n - S\,,$$

where N and S are the *north* and *south poles* defined by

$$N \equiv (0, \ldots, 0, 1)\,, \quad S \equiv (0, \ldots, 0, -1)\,.$$

By using the notation $\vec{x} \equiv (x_1, \ldots, x_n)$, a point $p \in \mathcal{U}$ will have coordinates $(\vec{x}, x_{n+1})$.

The map

$$\varphi : p \in \mathcal{U} \to \varphi(p) = \vec{t} = \frac{\vec{x}}{1 - x_{n+1}} \in \Re^n$$

is called the *stereographic projection* of S^n *with respect to the north pole* N. It is a one-to-one map that is continuous together with its inverse, which is given by

$$\varphi^{-1} \;:\; \vec{t} \in \Re^n \to \varphi^{-1}(\vec{t}) = \left(\frac{2\vec{t}}{|t|^2 + 1}, \frac{|t|^2 - 1}{|t|^2 + 1}\right) \in \mathcal{U}\,.$$

Similarly, the *stereographic projection* of S^n *with respect to the south pole* S is defined as

$$\psi : p \in \mathcal{V} \to \psi(p) = \vec{u} = \frac{\vec{x}}{1 + x_{n+1}} \in \Re^n\,,$$

and it is a one-to-one map continuous together with its inverse given by

$$\psi^{-1} \ : \ \vec{u} \in \Re^n \to \psi^{-1}(\vec{u}) = \left(\frac{2\vec{u}}{1+|u|^2}, \frac{1-|u|^2}{1+|u|^2} \right) \in \mathcal{V}\,.$$

Since the images of the intersection

$$\mathcal{U} \cap \mathcal{V} = S^n - \{N, S\}\,,$$

for both φ and ψ, is $(\Re^n - \{0\})$; i.e.

$$\varphi(\mathcal{U} \cap \mathcal{V}) = \psi(\mathcal{U} \cap \mathcal{V}) = \Re^n - \{0\}\,,$$

the map

$$\varphi \circ \psi^{-1} \ : \ \vec{u} \in (\Re^n - \{0\}) \to \vec{t} = (\varphi \circ \psi^{-1})(\vec{u}) = \frac{\vec{u}}{|u|^2} \in (\Re^n - \{0\})$$

is a C^∞ map.

Thus, the two compatible charts $(\mathcal{U}, \varphi)$ and $(\mathcal{V}, \psi)$ are a C^∞ atlas for S^n, and $(t_1, \ldots, t_n)$, $(u_1, \ldots, u_n)$ represent the local coordinates of a point $p \in S^n$ with respect to them.

Exercise 5.1.1. *Show that the sphere S^2 is orientable.*

5.2 Curves on a Differential Manifold

Differentiable functions on a manifold

If

$$f : \mathcal{M} \to \Re$$

is a function defined on a differential manifold $\mathcal{M}$ and p is an element in $\mathcal{M}$ representable on the chart $(\mathcal{U}, \varphi)$, the map

$$f \circ \varphi^{-1} : \varphi(\mathcal{U}) \subset \Re^n \to \Re$$

transforms open set of $\Re^n$ to open set of $\Re$. As the coordinates $\varphi^i(p)$ of $\varphi(p)$ represent the point p in the local chart $(\mathcal{U}, \varphi)$, so the map $\tilde{f} = f \circ \varphi^{-1}$ represents the function f in the local chart.

The function f is *said to be differentiable at* $p \in \mathcal{M}$, if in a given chart $(\mathcal{U}, \varphi)$, $\tilde{f} = f \circ \varphi^{-1}$ is differentiable at $\varphi(p)$, or equivalently, if expressed in local coordinates, gives rise to a differentiable function.

The given definition is independent from the choice of the chart. Indeed, if the point p is representable on two charts $(\mathcal{U}, \varphi)$ and $(\mathcal{V}, \psi)$, we have

$$f \circ \psi^{-1} = (f \circ \varphi^{-1}) \circ (\varphi \circ \psi^{-1}).$$

If $(f \circ \varphi^{-1})$ is differentiable at $\varphi(p)$, then $f \circ \psi^{-1}$ is differentiable at $\psi(p)$, since $\varphi \circ \psi^{-1}$ is differentiable at $\varphi(p)$ and $(\varphi \circ \psi^{-1})(\psi(p)) = \varphi(p)$.

Let $\mathcal{M}$ and $\mathcal{N}$ be two differential manifolds, with m and n dimensions, respectively, and let f be a map

$$f : p \in \mathcal{M} \to f(p) \in \mathcal{N} \tag{5.1}$$

from $\mathcal{M}$ to $\mathcal{N}$.

If p and $f(p)$ are representable on the charts $(\mathcal{U}, \varphi)$ and $(\mathcal{V}, \psi)$, respectively, the function

$$\psi \circ f \circ \varphi^{-1} : \varphi(\mathcal{U}) \subset \Re^n \to \psi(f(\mathcal{U})) \subset \Re^n$$

represents the map f in the local charts $(\mathcal{U}, \varphi)$ and $(\mathcal{V}, \psi)$, and we will say that *f is differentiable at the point $p \in \mathcal{M}$*, if $\tilde{f} = \psi \circ f \circ \varphi^{-1}$ is differentiable at $\varphi(p)$.

The map (5.1) is said to be *differentiable in $\mathcal{M}$* if it is differentiable at every point $p \in \mathcal{M}$; if f is one-to-one and f and f^{-1} are differentiable, then f is said to be a *diffeomorphism.*

Curves on a manifold

A curve γ on a differential manifold $\mathcal{M}$ is a homeomorphism

$$\gamma : \tau \in I \subset \Re \to \gamma(\tau) \in \mathcal{M}$$

from an open set $I \subseteq \Re$ (open interval) to an open set in $\mathcal{M}$.

The curve γ is said to be *differentiable at* $\tau = 0$ if the map

$$\varphi \circ \gamma : \tau \in I \subset \Re \to \varphi(\gamma(\tau)) \in \Re^n$$

is differentiable at $\tau = 0$, where φ is the homeomorphism of the chart on which $p \equiv \gamma(0)$ is representable.

Two curves, γ and γ' on $\mathcal{M}$ are called *equivalent* if

$$\gamma(0) = \gamma'(0) = p\,, \quad \lim_{\tau \to 0} \frac{\varphi(\gamma(\tau)) - \varphi(\gamma'(\tau))}{\tau} = 0 \tag{5.2}$$

in a chart $(\mathcal{U}, \varphi)$. Of course, the relation (5.2) is true in any other chart.[1,2,7]

5.3 Tangent Space at a Point

5.3.1 *Tangent vectors to a curve on a manifold*

Definition 18 It is called tangent vector X_p to a differentiable curve $\gamma = \gamma(\tau)$ on a manifold $\mathcal{M}$, at the point $p = \gamma(\tau_0)$, the directional derivative operator along the curve $\gamma = \gamma(\tau)$, at the point p:

$$X_p = \left.\frac{d}{d\tau}\right|_{\gamma(\tau_0)} .$$

If $p \in \mathcal{U}_\alpha \subset \mathcal{M}$, $\tau \in I \in \Re$ and f is a differentiable function on $\mathcal{M}$, we have

$$(X_p f)(p) = \left.\frac{df(\gamma(\tau))}{d\tau}\right|_{\gamma(\tau_0)} = \left.\frac{d(f \circ \gamma)(\tau)}{d\tau}\right|_{\gamma(\tau_0)} = \left.\frac{d(f \circ \varphi_\alpha^{-1} \circ \varphi_\alpha \circ \gamma)(\tau)}{d\tau}\right|_{\gamma(\tau_0)} ,$$

where φ_α is the homeomorphism of the chart $(\mathcal{U}_\alpha, \varphi_\alpha)$ on which p is represented. Since

$$\tilde{f} \equiv f \circ \varphi_\alpha^{-1} : (x^1, \ldots, x^n) \in A \subseteq \Re^n \to \tilde{f}(x^1, \ldots, x^n) \in \Re ,$$
$$\tilde{\gamma} \equiv \varphi_\alpha \circ \gamma : \tau \in \Re \to \tilde{\gamma}(\tau) = (x^1(\tau), \ldots, x^n(\tau)) \in A \subseteq \Re^n ,$$

we have

$$(X_p f)(p) = \left.\frac{d(\tilde{f} \circ \tilde{\gamma})(\tau)}{d\tau}\right|_{\gamma(\tau_0)} = \left.\frac{\partial \tilde{f}}{\partial x^i}\right|_p \left.\frac{d\tilde{\gamma}^i}{d\tau}\right|_{\tau_0} = \left(X_p^i \frac{\partial}{\partial x_p^i}\right) \tilde{f} ,$$

where $X_p^i = d\tilde{\gamma}^i/d\tau|_{\tau_0}$.

From the above formula it follows that every tangent vector to a given curve, at a given point, is a linear combination of the n partial derivatives $\{\partial/\partial x_p^i\}$, which are linearly independent.

Of course, every tangent vector is tangent to an infinite number of different curves through p for two different reasons. The first is that there are many curves which are tangent to one another and have the same tangent vector at p, and the second is that the same path may be reparametrized in such a way as to give the same tangent at p.

So it becomes natural to give the alternative following definition.

5.3.2 *Tangent vectors to a manifold*

The equivalence class of the curves $\gamma(\tau)$ on $\mathcal{M}$ through p is called *tangent vector to the manifold $\mathcal{M}$ at the point p*. The set of all the tangent vectors to the manifold $\mathcal{M}$ at the point p, with the sum and the product by scalars defined by

$$(aX_p + bY_p)(f) = aX_p(f) + bY_p(f)\,,$$

is a vector space called *tangent space to $\mathcal{M}$ at p* and it will be denoted with $\mathcal{T}_p\mathcal{M}$. This space has the same dimension of the manifold, and the components of every tangent vector $X_p \in \mathcal{T}_p\mathcal{M}$, in the local chart $(\mathcal{U}, \varphi)$ on which p is represented, are given by the relation

$$X_p^i = \frac{d}{d\tau}\varphi^i(\gamma(\tau))\bigg|_{\tau=0}\,.$$

Remark 12 *A tangent vector to a manifold $\mathcal{M}$ at a point p, could equivalently, be defined as a linear function from the space $\mathcal{F}(\mathcal{U})$ of differentiable functions, defined on a neighborhood $\mathcal{U}$ of p, to $\Re$:*

$$\mathcal{F}(\mathcal{U}) \to \Re\,,$$

satisfying the Leibnitz rule; i.e.

$$X_p(af + bg) = aX_p(f) + bX_p(g)\,,$$
$$X_p(fg) = f(p)X_p(g) + g(p)X_p(f)\,,$$

with $a, b \in \Re$, and f and g differentiable functions defined on a neighborhood $\mathcal{U}$ of p.

The value of X_p in f is

$$X_p(f) = \left(\frac{\partial \tilde{f}}{\partial x^i}\right)_p X_p^i\,, \tag{5.3}$$

where $x^i = \varphi^i(p)$ and X_p^i are the components of X_p in the given chart $(\mathcal{U}, \varphi)$. The previous formula is often written in the form

$$X_p = \sum_{i=1}^n X_p^i \left(\frac{\partial}{\partial x^i}\right)_p\,. \tag{5.4}$$

The tangent vectors

$$\left(\frac{\partial}{\partial x^1}\right)_p, \ldots, \left(\frac{\partial}{\partial x^n}\right)_p \tag{5.5}$$

constitute a basis for the tangent space $\mathcal{T}_p\mathcal{M}$, called *the natural basis.*

Since a chart $(\mathcal{U}, \varphi)$ at p induces an isomorphism; that is, an invertible linear map between the spaces $\mathcal{T}_p\mathcal{M}$ and $\Re^n$, the dimension of $\mathcal{T}_p\mathcal{M}$ coincides with the dimension of the manifold $\mathcal{M}$.

Transformation laws

If p belongs to the intersection $\mathcal{U}_i \cap \mathcal{U}_j$, then two sets of local coordinates can be introduced, namely $(x^1, x^2, \ldots, x^n)$ and $(x'^1, x'^2, \ldots, x'^n)$, and a vector V_p can be locally written in two different forms according to the chosen coordinate basis

$$\left(\frac{\partial}{\partial x^1}\right)_p, \ldots, \left(\frac{\partial}{\partial x^n}\right)_p, \quad \text{or} \quad \left(\frac{\partial}{\partial x'^1}\right)_p, \ldots, \left(\frac{\partial}{\partial x'^n}\right)_p.$$

Thus, we have

$$V_p = V_p^{\prime j}\left(\frac{\partial}{\partial x'^j}\right)_p = V_p^i\left(\frac{\partial}{\partial x^i}\right)_p.$$

The above relation, once applied to the functions $x'^k : \Re^n \to \Re$, gives the familiar transformation law for the components of a vector

$$V_p^{\prime j} = V_p^i\left(\frac{\partial x'^k}{\partial x^i}\right)_p.$$

5.4 A Digression on Vectors and Covectors

5.4.1 *Vector space*

Let us recall that a set E, whose elements we are denoting with capital Latin letters $X, Y, Z, \ldots$, is called a *vector space* (over the real numbers $\Re$) if

- an internal composition law

  $$+ : (X, Y) \in E \times E \mapsto E$$

  can be defined, with respect to which E is an Abelian group;

- a multiplication by real numbers $c \in \Re$ is defined to satisfy the following properties:

$$c \cdot (X+Y) = c \cdot X + c \cdot Y\,,$$
$$(c_1 c_2) \cdot X = c_1 \cdot (c_2 \cdot X)\,,$$
$$1 \cdot X = X\,.$$

  It is common use to drop the multiplication dot and the parentheses used in the previous properties.

The elements of E are called *vectors*, and the vector corresponding to the identity element in E is denoted with 0.

A set of vectors $X_1, X_2, \ldots, X_k$, in E is said to be *linearly independent* if

$$\sum_{i=1}^{k} c_i X_i = 0 \Rightarrow c_i = 0\,, \quad \forall i \in \{1, 2, \ldots, k\}\,,$$

and *linearly dependent*, otherwise. The set is said to be *maximal linearly independent*, if the set, obtained by adding to it any other vector of E, is linearly dependent. This means that any other vector in E can be expressed as a linear combination of a maximal set, which for this reason, is called a *basis* for E. The number of vectors of a basis is called *the dimension of* E. If $\{e_i\}$ denotes a basis for the vector space E, then any vector V can be written as

$$V = V^i e_i\,,$$

and the coefficients $V^i \in \Re$ are called *the components of* V in the given basis $\{e_i\}V^i \in \Re$.

5.4.2 *Dual vector space*

The space of all linear maps from E to $\Re$ is denoted with

$$E^* = \mathrm{Lin}(E, \Re)\,,$$

and is called the *dual space of* E. The elements of E^* will be denoted with small Greek letters $\alpha, \beta, \gamma, \ldots$.

For an arbitrary element $\alpha \in E^*$, we have

$$\alpha : X \in E \mapsto \alpha(X) \in \Re\,,$$

$$\alpha(X+Y) = \alpha(X) + \alpha(Y), \quad \forall X, Y \in \Re,$$
$$\alpha(cX) = c\alpha(X), \quad \forall c \in \Re.$$

It is easy to see that

- E^* is also a vector space, the internal composition law and the multiplication by real numbers being trivially defined as

$$(\alpha+\beta)(X) = \alpha(X) + \beta(X),$$
$$(c\alpha)(X) = c\alpha(X).$$

The vectors of E^* are called *covectors* in order to distinguish them from the vectors of E;

- E^* has the same dimension of E.

If $e_1, e_2, \ldots, e_n$ is a basis of E, a given vector $X \in E$ can be expressed as

$$X = X^i e_i, \quad i \in \{1, 2, \ldots, n\}.$$

Associated with the given basis $\{e_i\}$ of E, let us introduce the elements of E^*, namely $\vartheta^1, \vartheta^2, \ldots, \vartheta^n$, defined by

$$\vartheta^i(X) = X^i, \quad i \in \{1, 2, \ldots, n\},$$

or equivalently

$$\vartheta^i(e_j) = \delta^i_j, \quad i, j \in \{1, 2, \ldots, n\}.$$

Since for any elements $\alpha \in E^*$ and $X \in E$, we have

$$\alpha(X) = \alpha(X^i e_i) = X^i \alpha(e_i) = \alpha_i \vartheta^i(X),$$

with $\alpha_i = \alpha(e_i) \in \Re$, the element α can be expressed as

$$\alpha = \alpha_i \vartheta^i.$$

The set $\vartheta^1, \vartheta^2, \ldots, \vartheta^n$ is then a basis for E^*.

It is also possible to consider the dual $E^{**} = (E^*)^*$ of the dual of a given vector space E. The reader can easily prove that if the dimension of E is finite, then E^{**} is isomorphic to E and they can be identified.

Thus, E and E^* are duals of each other; in order to underline the reciprocal duality, the value that a covector α takes on a vector X is also denoted by the bracket $\langle\cdot,\cdot\rangle$, so that

$$\alpha(X) \equiv \langle\alpha, X\rangle\,.$$

If $\{e'_i\}$ and $\{e_i\}$ are two different bases and $\{\vartheta'^i\}$ and $\{\vartheta^i\}$ are their dual bases, respectively, a given vector V can be represented in two different forms. Then, we have

$$V = V'^i e'_i = V^i e_i\,.$$

Thus, the value of a generic element ϑ'^k of $\{\vartheta'^i\}$ on V will be

$$\vartheta'^k(V) = V'^i \vartheta'^k(e'_i) = V^i \vartheta'^k(e_i)\,,$$

and then

$$V'^k = M^k{}_i V^i\,,$$

M denoting the matrix whose elements are $\vartheta'^k(e_i)$.

5.5 Cotangent Space at a Point

As for any vector space E, we can introduce the dual $\mathcal{T}^*_p\mathcal{M}$ of the vector space $\mathcal{T}_p\mathcal{M}$. The vector space $\mathcal{T}^*_p\mathcal{M}$ is called the *cotangent space* to the manifold $\mathcal{M}$ at the point p. Special covectors of $\mathcal{T}^*_p\mathcal{M}$ can be associated with any differentiable function f defined on $\mathcal{M}$, as follows:

If $f:\mathcal{M}\to\Re$ is a differentiable function at the point $p\in\mathcal{M}$, its *differential* (or gradient, or exterior derivative) df_p *at the point* p is the linear map

$$df_p : \mathcal{T}_p\mathcal{M} \to \Re$$

of the tangent space $\mathcal{T}_p\mathcal{M}$ in $\Re$ defined by

$$\langle df_p, X_p\rangle = X_p f\,,\quad \forall\, X_p \in \mathcal{T}_p\mathcal{M}\,.$$

More explicitly, let X_p be a tangent vector to $\mathcal{M}$ in p, and $\gamma(\tau)$ a curve, belonging to the equivalence class of curves through p, ($\gamma(0) = p$), which represents X_p, then

$$\langle df_p, X_p\rangle = \frac{d}{d\tau}(f\circ\gamma)(\tau)\bigg|_{\tau=0}\,. \tag{5.6}$$

If p is representable on the chart $(\mathcal{U}, \varphi)$, we have

$$\begin{aligned}\langle df_p, X_p\rangle &= \frac{d}{d\tau}(f\circ\gamma)(\tau)\bigg|_{\tau=0} = \frac{d}{d\tau}(\tilde{f}\circ(\varphi\circ\gamma))(\tau)\bigg|_{\tau=0} \\ &= \left(\frac{\partial\tilde{f}}{\partial x^i}\right)_p \left(\frac{d\gamma^i(\tau)}{d\tau}\right)_{\tau=0} = \left(\frac{\partial\tilde{f}}{\partial x^i}\right)_p X_p^i\,. \qquad (5.7)\end{aligned}$$

By choosing for f the coordinate functions $x_p^j : p \in \mathcal{U} \to \varphi^j(p)$, we obtain

$$\langle dx_p^j, X_p\rangle = X_p^j\,.$$

Then, Eq. (5.7) can also be written in the form

$$\langle df_p, X_p\rangle = \left(\frac{\partial\tilde{f}}{\partial x^i}\right)_p \langle dx_p^i, X_p\rangle\,,$$

or equivalently, in the form

$$df_p = \left(\frac{\partial\tilde{f}}{\partial x^i}\right)_p dx_p^i\,.$$

It follows that the set of covectors $\{dx_p^j\}$ is the dual basis of the natural basis $\{\partial/\partial x_p^i\}$ of $\mathcal{T}_p\mathcal{M}$.

5.6 Maps Between Manifolds

If $\phi : \mathcal{M} \to \mathcal{N}$ is a map from a manifold $\mathcal{M}$ to another manifold $\mathcal{N}$, the image of a curve γ on $\mathcal{M}$ will be a curve $\gamma' = \phi\circ\gamma$ on $\mathcal{N}$, and a tangent vector X_p to γ at the point p, will have, as an image, a tangent vector to γ' at the transformed point $\phi(p)$. Thus, ϕ induces the linear map

$$\phi_{*p} : \mathcal{T}_p\mathcal{M} \to \mathcal{T}_{\phi(p)}\mathcal{N}$$

between the tangent spaces $\mathcal{T}_p\mathcal{M}$ and $\mathcal{T}_{\phi(p)}\mathcal{N}$ defined, formally, as in Eq. (5.6),

$$\phi_{*p}X_p = \frac{d}{d\tau}(\phi\circ\gamma)(\tau)\bigg|_{\tau=0} \qquad \forall\, X_p \in \mathcal{T}_p\mathcal{M}\,.$$

However, here $\phi_{*p}X_p$ is a vector belonging to $\mathcal{T}_{\phi(p)}\mathcal{N}$. Indeed, if p is representable in the chart $(\mathcal{U}, \varphi)$, and $q = \phi(p)$ in the chart $(\mathcal{V}, \psi)$, then we have

$$(\phi_{*p}X_p)^i = \left(\frac{\partial \tilde{\phi}^i}{\partial x^j}\right)_p \left(\frac{d\gamma^j(\tau)}{d\tau}\right)_{\tau=0} = \left(\frac{\partial \tilde{\phi}^i}{\partial x^j}\right)_p X_p^j,$$

where $\tilde{\phi} = \psi \circ \phi \circ \varphi^{-1}$. Vice versa we can associate with a given covector $\alpha_q \in \mathcal{T}_q^*\mathcal{N}$, the covector $\beta_p \in \mathcal{T}_p^*\mathcal{M}$ defined by

$$\langle \beta_p, X_p \rangle \equiv \langle \alpha_q, \phi_{*p}X_p \rangle ,$$

with $q = \phi(p)$.

Thus, $\phi_{*p}X_p$ is called the *push-forward* of X_p, while $\beta_p \equiv \phi_p^*\alpha_{\phi(p)}$ is called *the pull-back* of $\alpha_{\phi(p)}$.

5.7 Vector Fields

A *vector field* X on a manifold $\mathcal{M}$ is a rule which to every point $p \in \mathcal{M}$ associates a tangent vector $X_p \in \mathcal{T}_p\mathcal{M}$; i.e. a map

$$X : p \in \mathcal{M} \to X(p) = X_p \in \mathcal{T}_p\mathcal{M}$$

is defined on $\mathcal{M}$. In a local coordinate system, a vector field X can be written in the form

$$X(p) = X^i(p) \left(\frac{\partial}{\partial x^i}\right)_p .$$

The vector field $X(p)$ is said to be C^h *differentiable* on a C^k manifold $\mathcal{M}$, with $h \leq k-1$, if the functions $X^i(p)$ are C^h differentiable on the manifold $\mathcal{M}$.

5.7.1 *Holonomic and anholonomic basis of vector fields*

The natural basis $\{\partial/\partial x^i\}$ is not, of course, the sole possible basis for vector fields. By taking n, point-wise linearly independent, combinations e_i of its elements; i.e.

$$e_i = a_i^j(x)\frac{\partial}{\partial x^j}, \quad i = 1, \ldots, n, \tag{5.8}$$

with $\det(a_i^j(x)) \neq 0$ everywhere, we obtain a new basis $\{e_i(x)\}$. In fact, provi ous relations can be solved with respect to $\partial/\partial x^j$ in the form

$$\frac{\partial}{\partial x^i} = b_i^j(x) e_j \,, \quad i = 1, \ldots, n \,,$$

where $(b_i^j(x))$ is the inverse matrix of $(a_i^j(x))$.

Thus, an arbitrary vector field X, that in the original basis was written as $X = X^i(\partial/\partial x^i)$, takes the form

$$X = X'^i e_i \,,$$

where $X'^j = b_i^j(x) X^i$ are the new components.

In other words, a basis is given by n arbitrary point-wise linearly independent vector fields.

If the matrix $(a_i^j(x))$ is a Jacobian matrix, coordinates y^i exist such that

$$e_i = \frac{\partial}{\partial y^i} \,, \quad i = 1, \ldots, n \,.$$

Of course, such coordinates are the solutions of the differential system

$$\frac{\partial y^j}{\partial x^i} = b_i^j(x) \,, \tag{5.9}$$

which defines, then, a coordinate transformation between the x's and the y's.

The natural basis, as well the ones related to it by a coordinate transformation, has the following obvious property:

$$\left[\frac{\partial}{\partial x^i}, \frac{\partial}{\partial x^j}\right] = 0 \,, \quad i, j = 1, \ldots, n \,.$$

For an arbitrary linear combination, as the one in Eq. (5.8), it will happen, generally, that

$$[e_i, e_j] \neq 0 \,,$$

at least for some values of the indices i, j.

If this linear combination is made with a Jacobian matrix, the previous property is again satisfied, since

$$[e_i, e_j] = \left[\frac{\partial}{\partial y^i}, \frac{\partial}{\partial y^j}\right] = 0 \,, \quad i, j = 1, \ldots, n \,.$$

It is true also the converse, namely: *If the elements of a basis $\{e_i\}$ of vector fields fulfill the property*

$$[e_i, e_j] = 0\,, \quad i,j = 1,\ldots,n\,,$$

then coordinates y^i exist such that

$$e_i = \frac{\partial}{\partial y^i}\,, \quad i = 1,\ldots,n\,.$$

This can be understood by observing that the elements e_i of a basis, however, this one is chosen, are linear combination (with functions as coefficients) of the partial derivatives $\partial/\partial x^i$:

$$e_i = a_i^j(x)\frac{\partial}{\partial x^j}\,, \quad i = 1,\ldots,n\,.$$

Thus,

$$[e_i, e_j] = \left[a_i^h(x)\frac{\partial}{\partial x^h}, a_j^k(x)\frac{\partial}{\partial x^k}\right] = \left(a_i^h(x)\frac{\partial a_j^k(x)}{\partial x^h} - a_j^h(x)\frac{\partial a_i^k(x)}{\partial x^h}\right)\frac{\partial}{\partial x^k}\,.$$

Then,

$$[e_i, e_j] = 0 \Leftrightarrow a_i^h(x)\frac{\partial a_j^k(x)}{\partial x^h} = a_j^h(x)\frac{\partial a_i^k(x)}{\partial x^h}\,.$$

If $(b_i^j(x))$ is the inverse matrix of $(a_i^j(x))$, we have

$$\begin{aligned}
b_l^i a_i^k &= \delta_l^k \Rightarrow (\partial_h b_l^i)a_i^k = -b_l^i(\partial_h a_i^k) \Rightarrow (\partial_h a_j^k) = -a_j^l(\partial_h b_l^i)a_i^k \\
&\Rightarrow a_i^h(\partial_h a_j^k) - a_j^h(\partial_h a_i^k) = a_j^h a_i^l(\partial_h b_l^p)a_p^k - a_i^h a_j^l(\partial_h b_l^p)a_p^k \\
&= a_j^l a_i^h(\partial_l b_h^p)a_p^k - a_i^h a_j^l(\partial_h b_l^p)a_p^k = a_j^l a_i^h a_p^k((\partial_l b_h^p) - (\partial_h b_l^p))\,,
\end{aligned}$$

where for convenience, the shorthand notation $\partial_h \equiv \partial/\partial x^h$ has been introduced.

Thus,

$$[e_i, e_j] = a_j^l a_i^h\left(\frac{\partial b_h^p}{\partial x^l} - \frac{\partial b_l^p}{\partial x^h}\right)e_p\,, \tag{5.10}$$

and, since $\det(a_j^l) \neq 0$,

$$[e_i, e_j] = 0 \Leftrightarrow \frac{\partial b_h^p}{\partial x^l} = \frac{\partial b_l^p}{\partial x^h}\,,$$

which is just the compatibility conditions to be satisfied, in order a differential system, like Eq. (5.9), admit solutions. In other words, the conditions

$$\frac{\partial b_h^p}{\partial x^l} = \frac{\partial b_l^p}{\partial x^h}$$

say that the differential forms

$$\vartheta^p = b_h^p dx^k , \quad p = 1, \ldots, n$$

are closed; that is, functions y^p, at least locally, exist such that $\vartheta^p = dy^p$.

Equation (5.10) shows that the commutation relations between elements of an arbitrary basis are of the form

$$[e_i, e_j] = c_{ij}^p e_p \, .$$

A basis for which $c_{ij}^p = 0$, $\forall\, i,\, j,\, p$, will be called *holonomic* (integrable), or *anholonomic* (nonintegrable), otherwise.

5.8 The Tangent Bundle

The union $\cup_{p \in \mathcal{M}} \mathcal{T}_p\mathcal{M}$ of all tangent spaces $\mathcal{T}_p\mathcal{M}$ to a manifold $\mathcal{M}$ can be endowed with a structure of a differential manifold in a very natural way. An element of it is a pair (p, X_p) with $p \in \mathcal{M}$ and $X_p \in \mathcal{T}_p\mathcal{M}$. Thus, a system of local coordinates $(x^1, x^2, \ldots, x^n, X^1, X^2, \ldots, X^n)$ for $\cup_{p \in \mathcal{M}} \mathcal{T}_p\mathcal{M}$ can be introduced by using the local coordinates $(x^1, x^2, \ldots, x^n)$ of p and the ones $(X^1, X^2, \ldots, X^n)$ of X_p. Regarded as a differential manifold, the set $\cup_{p \in \mathcal{M}} \mathcal{T}_p\mathcal{M}$ is denoted with $\mathcal{TM}$ and it is called the *tangent bundle of* $\mathcal{M}$, while a single tangent space $\mathcal{T}_p\mathcal{M}$ is called a *fiber* of $\mathcal{TM}$. Of course, the dimension of $\mathcal{TM}$ is twice that of $\mathcal{M}$.

A curve on $\mathcal{TM}$ identifies a vector at each point of $\mathcal{M}$, and so it defines a vector field on $\mathcal{M}$. Such a curve; that is, a curve *transversal* to the fibers, is called a *cross-section of* $\mathcal{TM}$.

The names *bundle, fiber, and cross-section* refer, however, to general structures that our "special" differential manifold $\mathcal{TM}$ shares with a general class of differential manifolds called *fiber bundles*. A general fiber bundle consists of a *base manifold*, which in our case is $\mathcal{M}$, and of one *fiber* attached to each point of the base space. If the dimension of the base space is n and the one of each fiber is m, the bundle has $m + n$ dimensions. The points of each fiber are related to one another, while points on different fibers are not.

The fibers need not be related to the differential structure of the base manifold $\mathcal{M}$. In *elementary particle physics*, bundles are considered, whose fibers are isospin spaces attached to points in the space-time, which is the base manifold. Such a bundle will describe, besides the coordinates (t, x, y, z), also the isospin of an elementary particle.

5.9 General Definition of Fiber Bundle

Let E, $\mathcal{M}$, F be differential manifolds and π a differentiable map

$$\pi : E \to \mathcal{M}$$

from E to $\mathcal{M}$.

Let $\{\mathcal{U}_j\}_{j=1,\dots,n}$ be a covering of $\mathcal{M}$, made of open subsets of $\mathcal{M}$, which are compatible charts' domains

$$\mathcal{M} = \bigcup_{j=1}^{n} \mathcal{U}_j \, .$$

Let us suppose that, for every open set $\mathcal{U}_j$, there is a homeomorphism φ_j from $\pi^{-1}(\mathcal{U}_j)$ to the Cartesian product $\mathcal{U}_j \times F$ of the form

$$\varphi_j : y \in \pi^{-1}(\mathcal{U}_j) \to \varphi_j(y) = \left(\pi(y), \overset{\triangle}{\varphi}_j\,(y)\right) \in \mathcal{U}_j \times F \, ,$$

where $\overset{\triangle}{\varphi}_j \colon \pi^{-1}(\mathcal{U}_j) \to F$ and the restriction $\overset{\triangle}{\varphi}_j\,|_{\pi^{-1}(p)}$ of $\overset{\triangle}{\varphi}_j$ to $\pi^{-1}(p)$,

$$\overset{\triangle}{\varphi}_j \Bigg|_{\pi^{-1}(p)} \equiv \overset{\triangle}{\varphi}_{j,p} \colon \pi^{-1}(p) \to F \, ,$$

is a homeomorphism of $\pi^{-1}(p)$ onto F, such that the diagram

$$\begin{array}{ccc} & \pi^{-1}(\mathcal{U}_j) & \\ \pi \swarrow & & \searrow \varphi_j \\ \mathcal{U}_j & \overset{\rho}{\longleftarrow} & \mathcal{U}_j \times F \end{array}$$

commutes; i.e.

$$\pi = \rho \circ \varphi_j \, ,$$

ρ denoting the canonical projection from $\mathcal{U}_j \times F$ onto $\mathcal{U}_j$.

Of course, the set of maps $\overset{\wedge}{\varphi}_{k,p} \circ \overset{\triangle}{\varphi}{}^{-1}_{j,p}: F \to F$ for all $p \in \mathcal{U}_j \cap \mathcal{U}_k$ and for all $j, k \in \{1, \ldots, n\}$ is a group G in a natural way. If this group is a Lie group and the maps

$$g_{jk} : p \in \mathcal{U}_j \cap \mathcal{U}_k \to g_{jk}(p) = \overset{\triangle}{\varphi}_{k,p} \circ \overset{\triangle}{\varphi}{}^{-1}_{j,p} \in G$$

are differentiable, we will say that $(E, \mathcal{M}, \pi, F, G)$ is a *differentiable fiber bundle*; $\mathcal{M}$ is called the *base* of the *bundle*, F is called *the typical fiber*, and $\pi^{-1}(p)$ is called *the fiber* in p. The group is called *the structure group.* Usually the bundle $(E, \mathcal{M}, \pi, F, G)$ is simply indicated by E.

Locally, a fiber bundle is always a Cartesian product; i.e., $\pi^{-1}(\mathcal{U}_j) \simeq \mathcal{U}_j \times F$. A fiber bundle E, which is *globally* a Cartesian product; that is, if $E = \mathcal{M} \times F$, is called *trivial.* For more details see Ref. 19.

5.9.1 *More on the tangent bundle*

Let us consider again an n-dimensional differential manifold $\mathcal{M}$ and the set of the pairs (p, X_p) where $p \in \mathcal{M}$ and $X_p \in \mathcal{T}_p\mathcal{M}$. Such a set, denoted by the symbol $\mathcal{TM}$, can be provided of a differential bundle structure in the following way: as a base we take the differential manifold $\mathcal{M}$ and the map π is the projection

$$\pi : (p, X_p) \in \mathcal{TM} \to \pi(p, X_p) = p \in \mathcal{M}\,.$$

The typical fiber F is the Euclidean space $\Re^n$, and for every $p \in \mathcal{M}$, the fiber $\pi^{-1}(p)$ is $\mathcal{T}_p\mathcal{M}$, the tangent space to $\mathcal{M}$ at p. The covering of $\mathcal{M}$ is made by the domains $\mathcal{U}_j$ of compatible charts $(\mathcal{U}_j, \psi_j)$ such that

$$\mathcal{M} = \bigcup_{j=1}^{n} \mathcal{U}_j\,.$$

The homeomorphisms φ_j are defined as follows:

$$\varphi_j : (p, X_p) \in \pi^{-1}(\mathcal{U}_j) \to \varphi_j(p, X_p) = (\pi(p, X_p), (\psi_j)_{*p}(\pi_2(p, X_p))) \in \mathcal{U}_j \times \Re^n\,,$$

where $\pi_2(p, X_p) = X_p$.

The coordinates of the point (p, X_p) in $\mathcal{TM}$ are then

$$(x^1, \ldots, x^n, X^1_p, \ldots, X^n_p)\,,$$

where x^i are the coordinates of p and

$$X_p^i = [(\psi_j)_{*p}(X_p)]^i$$

are the coordinates of X_p in the chart $(\mathcal{U}_j, \psi_j)$.

The linear map

$$(\psi_j)_{*p} : \mathcal{T}_p\mathcal{M} \to \Re^n$$

is an isomorphism between the spaces $\mathcal{T}_p\mathcal{M}$ and $\Re^n$.

If p is representable on two charts $(\mathcal{U}_j, \psi_j)$ and $(\mathcal{U}_k, \psi_k)$, the map

$$(\psi_k)_{*p} \circ ((\psi_j)_{*p})^{-1} : \Re^n \to \Re^n$$

is an isomorphism of $\Re^n$ onto itself. Thus, the structure group G is $GL(n, \Re)$.

As we have already noted, the space $\mathcal{TM}$ is usually called the tangent bundle.

The cotangent bundle

The cotangent bundle is built exactly as the tangent bundle $\mathcal{TM}$, by simply replacing the tangent spaces $\mathcal{T}_p\mathcal{M}$ with the cotangent spaces $\mathcal{T}_p^*\mathcal{M}$.

5.9.2 *Analysis of two bundles with S^1 as base manifold*

The tangent bundle $\mathcal{T}S^1$ of the circle S^1 can be visualized as a cylinder, which *globally* is the Cartesian product $S^1 \times \Re$, and so it is a trivial bundle.

A cylinder, once cut along a directrix, "becomes" an infinite rectangle belonging to $\Re^2$; a part of it is a finite rectangle.

Let a and a' be the upper corners of the rectangle and b and b' the down ones. Before the cut, $a - a'$ was identified with the same point of the cylinder and $a' - b'$ with a different point along the same directrix.

Thus, by gluing a with a' and b with b', we obtain an upper- and down-bounded part of the cylinder, while by gluing a with b' and b with a', we obtain the so-called *Möbius band.*

Then, with the same base space S^1 and the same fibers $\Re$, we can build two globally different bundles. The first, $\mathcal{T}S^1$, is a trivial bundle, the second, a nontrivial one; both show the same local properties. Their difference, which is of global type, is described by the *structure group.*

The bundle $\mathcal{T}S^1$

Let $\{\mathcal{U}_j\}_{j=1,\ldots,n}$ be an open covering of, $S^1 = \bigcup_{j=1}^{n} \mathcal{U}_j$. Every $\mathcal{U}_j$ has, as coordinate system, a parameter τ_j along S^1, and for $p \in \mathcal{U}_j$, a basis of $\mathcal{T}_p S^1 \simeq F = \Re$ will be given by the vector $d/d\tau_j$. Thus, a given vector $V \in \mathcal{T}_p S^1$ will be represented by $v^j d/d\tau_j$, where v^j is a real number, and since j is fixed, there is no sum over j. If p belongs to the intersection of two neighborhoods $\mathcal{U}_j$ and $\mathcal{U}_i$, the vector V will have two representations, $v^j d/d\tau_j$ and $v^i d/d\tau_i$, where since τ_j and τ_j are unrelated, v^j and v^i are two nonzero real numbers.

Since the typical fiber is $\Re$, the homeomorphism $\overset{\triangle}{\varphi}_{i,p} \circ \overset{\triangle}{\varphi}{}^{-1}_{j,p}: \Re \to \Re$ maps v^j in v^i, so that it reduces simply to the multiplication by the real number $r_{ij} = v^i/v^j$.

Thus, the structure group of $\mathcal{T}S^1$, since r_{ij} are nonzero arbitrary real numbers, is $GL(1, \Re) \equiv (\Re - \{0\}, \times)$; that is, $\Re - \{0\}$ with the composition law given by the multiplication.

We observe now that, for any j, the parameters τ_j can be chosen to be *concordant*; that is, in such a way that any two of them, namely τ_j and τ_j, increase in the same direction of S^1 $(d\tau_i/d\tau_j > 0)$, in the intersection $\mathcal{U}_j \cap \mathcal{U}_i$. With this choice, $r_{ij} > 0$ and the structure group reduces to $\Re^+$, the positive real numbers with the composition law given by the multiplication.

Moreover, the Jacobians $d\tau_i/d\tau_j$ could be chosen in such a way that, in the intersection $\mathcal{U}_j \cap \mathcal{U}_i$, we have $d\tau_i/d\tau_j = 1$. Thus, the structure group reduces, finally, to $(1, \times)$, a trivial group as trivial as the bundle $\mathcal{T}S^1$.

The Möbius fiber bundle

It is easy to see that the Möbius strip is not an orientable manifold, so that at least one of the real numbers $r_{ij} = v^i/v^j$ will be -1. In this way, the structure group reduces to $(\{1, -1\}, \times)$.

The bundle of frames and the principal fiber bundle

The *frame bundle* $Fr\mathcal{M}$ of an n-dimensional differential manifold $\mathcal{M}$ is a bundle having $\mathcal{M}$ as base space, $GL(n, \Re)$ as structure group (the same of the tangent bundle $\mathcal{TM}$), and the set of all bases of $\Re^n$ as fiber. Since the set of all bases of $\Re^n$ is homeomorphic to $GL(n, \Re)$, the typical fiber F is just $GL(n, \Re)$.

Then $\mathcal{TM}$ has $G = GL(n, \Re)$ and $F = \Re^n$, while $Fr\mathcal{M}$ has $G = GL(n, \Re)$ and $F = GL(n, \Re)$.

The frame bundle $Fr\mathcal{M}$ is just an example of a bundle in which the structure group (not necessarily $GL(n, \Re)$) is homeomorphic to the fiber. Such a bundle, that is, a bundle in which the structure group is homeomorphic to the fiber, is called a *principal fiber bundle.*

5.10 Integral Curves of a Vector Field

As it has already been said, given a vector $X_p \in \mathcal{T}_p\mathcal{M}$, there exist infinitely many differentiable curves on $\mathcal{M}$ which in p admit X_p as tangent vector. Given two vectors, X_p and X_q, it is also easy to find curves on $\mathcal{M}$ admitting X_p and X_q as tangent vectors at p and q, respectively. It is also clear that the search for such curves becomes more and more difficult as the number of the given vectors increases. It is in a sense surprising that, given a vector field X on $\mathcal{M}$, and then the assignment of infinitely many tangent vectors (one in each tangent space $\mathcal{T}_p\mathcal{M}$, $\forall p \in \mathcal{M}$), there exists always a curve $p = p(\tau), \tau \in]a, b[\subseteq \Re$ on $\mathcal{M}$, whose tangent vector in a point $p_0 = p(\tau_0)$ coincides with the value X_{p_0} of the field at the point, $\forall \tau_0 \in]a, b[$.

As a matter of the fact, if $x^i = x^i(\tau)$ is the local parametric representation of the unknown curve, the derivatives $dx^i/d\tau|_{\tau=\tau_0}$ will represent the components of the tangent vector at τ_0, while, if $X = X^i(\partial/\partial x^i)$ is the local representation of the vector field, $X^i(p_0)$ will be the components of the vector corresponding to the value of the vector field X at p_0. Thus, the unknown curve will be the solution of the system of differential equations

$$\frac{dx^i}{d\tau} = X^i(x), \quad \forall i \in (1, 2, \ldots, n).$$

We know that, for *smooth* X^i, the above system always admits, locally, a unique solution $x^i = x^i(\tau)$, assuming at $\tau = \tau_0$ a prefixed value x_0^i. Such a solution is called an *integral curve of the vector field* X.

Such curves are well known in physics as *force lines*, a name given by M. Faraday who introduced them for the electric and magnetic vector fields E and H.

Example 19 *Let*

$$X = x\frac{\partial}{\partial y} - y\frac{\partial}{\partial x}$$

be the harmonic oscillator vector field. Its integral curves will be given by solutions of the differential system

$$\begin{cases} \dfrac{dx}{d\tau} = -y\,, \\ \dfrac{dy}{d\tau} = \quad x\,, \end{cases}$$

which are circles

$$x^2 + y^2 = r^2$$

of arbitrary radius r.

Example 20 *The integral curves of the vector field*

$$X = \left(x + \frac{y}{r}\right)\frac{\partial}{\partial y} - \left(y + \frac{x}{r}\right)\frac{\partial}{\partial x}\,,$$

where $r = (x^2 + y^2)^{1/2}$, will be given by the solutions of the differential system

$$\begin{cases} \dfrac{dx}{d\tau} = -\left(y + \frac{x}{r}\right)\,, \\ \dfrac{dy}{d\tau} = \quad \left(x + \frac{y}{r}\right)\,. \end{cases}$$

By multiplying the first equation by x and the second by y, we obtain

$$\begin{cases} x\dfrac{dx}{d\tau} = -xy + \frac{x^2}{r}\,, \\ y\dfrac{dy}{d\tau} = \quad yx + \frac{y^2}{r}\,. \end{cases}$$

Thus,

$$\frac{1}{2}\frac{d}{d\tau}(x^2 + y^2) = r\,,$$

or

$$\frac{dr}{d\tau} = 1\,,$$

which gives

$$r = \tau + c\,,$$

represents infinitely many spirals, one for each value of the constant c.

We can summarize saying that, for every point p in $\mathcal{M}$ there exists an integral curve of X; i.e. a curve on $\mathcal{M}$

$$\gamma : \tau \in I \subset \Re \rightarrow \gamma(\tau) \in \mathcal{M}\,,$$

which satisfies the differential equations

$$\frac{d\gamma(\tau)}{d\tau} = X(\gamma(\tau))\,, \quad \forall \tau \in I\,.$$

More precisely, for every $p \in \mathcal{M}$, there exists an interval $I_p \subset \Re$ and an integral curve $\tau \in I_p \rightarrow \gamma(\tau, p)$ of class C^{h+1} in I_p, such that $\gamma(0,p) = p$. Moreover, this integral curve is uniquely determined.

It follows that, if σ and τ are elements of I_p such that $\sigma + \tau \in I_p$, we have

$$\gamma(\tau, \gamma(\sigma, p)) = \gamma(\tau + \sigma, p)\,, \quad \forall p \in \mathcal{M}\,. \tag{5.11}$$

If $p_0 \in \mathcal{M}$, there exist an open set $\mathcal{U}(p_0) \subset \mathcal{M}$ containing p_0 and an interval $I_{p_0} \subset \Re$ such that γ is defined on $I_{p_0} \times \mathcal{U}(p_0)$

$$\gamma : (\tau, p) \in I_{p_0} \times \mathcal{U}(p_0) \rightarrow \gamma(\tau, p) \in \mathcal{M}\,.$$

For every $\tau \in I_{p_0}$, the map

$$\gamma^\tau : p \in \mathcal{U}(p_0) \rightarrow \gamma^\tau(p) = \gamma(\tau, p) \in \mathcal{M} \tag{5.12}$$

is a diffeomorphism between open subsets of $\mathcal{M}$; a point p in $\mathcal{U}(p_0)$ goes to a point $\gamma^\tau(p) \in \mathcal{M}$ along the integral curve of X at p. The position of $\gamma^\tau(p)$ is determined by τ.

Let $\{\mathcal{U}(p_0)\}$ be a covering of $\mathcal{M}$. The intersection I of the intervals I_{p_0} corresponding to the open sets $\mathcal{U}(p_0)$ can be empty, but if the manifold $\mathcal{M}$ is compact, the covering $\{\mathcal{U}(p_0)\}$ contains a finite subcovering and the intersection I of the corresponding intervals is certainly not empty. In such a case the diffeomorphisms (5.12) can be extended to the entire manifold $\mathcal{M}$, with $\tau \in I$ and the vector field X said to be *complete.*[29] From Eq. (5.11), we have

$$\gamma(\tau + \sigma, p) = \gamma^{\tau+\sigma}(p)\,, \quad \gamma(\tau, \gamma(\sigma, p)) = \gamma^\tau(\gamma(\sigma, p)) = \gamma^\tau \circ \gamma^\sigma(p)\,,$$

and then

$$\gamma^{\tau+\sigma} = \gamma^{\tau} \circ \gamma^{\sigma} .$$

Moreover, every γ^τ is supplied by an inverse; that is by the diffeomorphism $\gamma^{-\tau}$, so that we can define γ^τ for every $\tau \in \Re$.

What has been said above can be summarized by the following theorem.[29]

Theorem 21 *With every C^h differentiable vector field $X(p)$, on a C^k ($h \leq k-1$) compact, differential manifold $\mathcal{M}$, a one parameter group is associated*

$$\gamma^\tau : \mathcal{M} \to \mathcal{M} . \tag{5.13}$$

This group of diffeomorphisms of $\mathcal{M}$ in itself is such that

$$\left.\frac{d}{d\tau}\gamma^\tau(p)\right|_{\tau=0} = X(p) , \quad \forall p \in \mathcal{M} .$$

The group γ^τ is also called the *flow of the vector field* $X(p)$, and it is also denoted by γ^τ_X.

The group (5.13) is then well defined if the manifold $\mathcal{M}$ is compact. In the general case, the γ^τ are defined just like in Eq. (5.12) only in neighborhoods of a point $p_0 \in \mathcal{M}$ and for small τ.

5.11 The Lie Derivative

Let $X(p)$ and $Y(p)$ be two vector fields on an n-dimensional differential manifold $\mathcal{M}$, and ϕ^τ_X be the flow of the vector field X.

The Lie derivative $L_X Y$ of the vector field Y is the vector field defined by the relation

$$(L_X Y)(p) = \lim_{\tau \to 0} \frac{(\phi^{-\tau})_{*\phi^\tau(p)}(Y(\phi^\tau(p))) - Y(p)}{\tau} , \tag{5.14}$$

where

$$(\phi^{-\tau})_{*\phi^\tau(p)} : \mathcal{T}_{\phi^\tau(p)}\mathcal{M} \to \mathcal{T}_p\mathcal{M} .$$

Let us calculate the Lie derivative of the basis vectors $\{\partial/\partial x^i\}$.

In order to simplify the notation, let us indicate with $x \in \Re^n$ the coordinates of the points of $\mathcal{M}$ and let us set $\phi^\tau(x) = y$. If $\phi^i(\tau, x)$ are the coordinates of $\phi^\tau(x)$, and $\tilde{\phi}^i(\tau, x)$ those of $\phi^{-\tau}(y)$, then $\phi^i(0, p) = x^i$ and $\tilde{\phi}^i(0, y) = y^i$.

From Eq. (5.14), we have

$$\left(L_X\frac{\partial}{\partial x^i}\right)(p)=\lim_{\tau\to 0}\frac{1}{\tau}\left[(\phi^{-\tau})_{*y}\left(\frac{\partial}{\partial x^i}\right)_y-\left(\frac{\partial}{\partial x^i}\right)_x\right].$$

The components of $\partial/\partial x^i$, in the natural coordinates system (5.5), are δ_i^j, since

$$\left(\frac{\partial}{\partial x^i}\right)_p=\delta_i^j\left(\frac{\partial}{\partial x^j}\right)_p,$$

while the components of $(\phi^{-\tau})_{*y}(\frac{\partial}{\partial x^i})_y$ are given by

$$\frac{\partial\tilde{\phi}^j(\tau,y)}{\partial y^h}\delta_i^h=\frac{\partial\tilde{\phi}^j(\tau,y)}{\partial y^i}.$$

Then, we can write

$$\left(L_X\frac{\partial}{\partial x^i}\right)(p)=\lim_{\tau\to 0}\frac{1}{\tau}\left[\frac{\partial\tilde{\phi}^j(\tau,y)}{\partial y^i}-\delta_i^j\right]\left(\frac{\partial}{\partial x^i}\right)_p, \tag{5.15}$$

with

$$\tilde{\phi}^j(0,y)=y^j\,,\quad \frac{\partial\phi^j(0,y)}{\partial y^i}=\delta_i^j\,. \tag{5.16}$$

By using Eq. (5.16), Eq. (5.15) can be rewritten in the following form:

$$\left(L_X\frac{\partial}{\partial x^i}\right)=\frac{d}{d\tau}\left(\frac{\partial\tilde{\phi}^j}{\partial y^i}\right)_{\tau=0}\left(\frac{\partial}{\partial x^j}\right)_p.$$

Since

$$\frac{d}{d\tau}\left(\frac{\partial\tilde{\phi}^j}{\partial y^h}\frac{\partial\phi^h}{\partial x^i}\right)_{\tau=0}=\frac{d}{d\tau}\delta_i^j=0\,,$$

we have

$$\left(\frac{d}{d\tau}\frac{\partial\tilde{\phi}^j(\tau,y)}{\partial y^h}\right)_{\tau=0}\frac{\partial\phi^h(0,x)}{\partial x^i}+\frac{\partial\tilde{\phi}^j(0,y)}{\partial y^h}\left(\frac{d}{d\tau}\frac{\partial\phi^h(\tau,x)}{\partial x^i}\right)_{\tau=0}=0\,,$$

and

$$\left(\frac{d}{d\tau}\frac{\partial\tilde{\phi}^j}{\partial y^h}\right)_{\tau=0}\delta_i^h+\delta_h^j\left(\frac{d}{d\tau}\frac{\partial\phi^h}{\partial x^i}\right)_{\tau=0}=0\,,$$

so that

$$\left(\frac{d}{d\tau}\frac{\partial\tilde{\phi}^j}{\partial y^i}\right)_{\tau=0} = -\left(\frac{d}{d\tau}\frac{\partial\phi^j}{\partial x^i}\right)_{\tau=0},$$

and

$$\left(L_X\frac{\partial}{\partial x^i}\right)(p) = -\left(\frac{d}{d\tau}\frac{\partial\phi^j}{\partial x^i}\right)_{\tau=0}\frac{\partial}{\partial x^j}\bigg|_p = -\frac{\partial X^j}{\partial x^i}\left(\frac{\partial}{\partial x^j}\right)_p. \qquad (5.17)$$

The Lie derivative is an additive operator; i.e.

$$L_X(U+V) = L_XU + L_XV\,,$$

where X, U, and V are vector fields on the manifold $\mathcal{M}$. Moreover, it satisfies the Leibnitz rule

$$L_X(U\otimes V) = (L_XU)\otimes V + U\otimes L_XV\,,$$

where the symbol $\otimes$ denotes the *tensor product* defined in the next chapter.

By using the Eq. (5.17) and the relations $\langle dx^i, \partial/\partial x^j\rangle = \delta^i_j$, we can calculate the Lie derivative $L_X dx^i$ of the basis differential 1-forms $\{dx^i\}$. Indeed

$$L_X\left\langle dx^i \frac{\partial}{\partial x^j}\right\rangle = L_X\delta^i_j = 0\,,$$

so that

$$\begin{aligned}
\left\langle L_Xdx^i, \frac{\partial}{\partial x^j}\right\rangle &= -\left\langle dx^i, L_X\frac{\partial}{\partial x^j}\right\rangle = \left\langle dx^i, \frac{\partial X^k}{\partial x^j}\frac{\partial}{\partial x^k}\right\rangle \\
&= \frac{\partial X^k}{\partial x^j}\delta^i_k = \frac{\partial X^i}{\partial x^j} = \frac{\partial X^i}{\partial x^k}\delta^k_j \\
&= \frac{\partial X^i}{\partial x^k}\left\langle dx^k, \frac{\partial}{\partial x^j}\right\rangle = \left\langle \frac{\partial X^i}{\partial x^k}dx^k, \frac{\partial}{\partial x^j}\right\rangle \\
&= \left\langle \frac{\partial X^i}{\partial x^k}dx^k, \frac{\partial}{\partial x^j}\right\rangle = \left\langle dX^i, \frac{\partial}{\partial x^j}\right\rangle.
\end{aligned}$$

Therefore, we obtain

$$L_Xdx^i = dX^i\,.$$

The Lie derivative of a differentiable function f on the manifold $\mathcal{M}$ has the following expression:

$$(L_X f)(x) = \frac{\partial \tilde{f}}{\partial x^j} X^j ,$$

where $\tilde{f} = f \circ \psi^{-1}$ is the function representing f in the chart $(\mathcal{U}, \psi)$ in which p is represented.

5.12 Submanifolds

Examples of submanifolds are given by a sphere S^2 or a curve γ in the space $\Re^3$. In some neighborhood $\mathcal{U} \subseteq \Re^3$ of any point $p \in S^2$, a coordinate system (x, y, z) can be introduced for $\Re^3$, such that the points of $S^2 \cap \mathcal{U}$ are characterized by $z = 0$.

Similarly, in some neighborhood $\mathcal{U} \subseteq \Re^3$, of any point $p \in \gamma$, a coordinate system (x, y, z) can be introduced for $\Re^3$, such that the points of $\gamma \cap \mathcal{U}$ are characterized by $y = z = 0$.

A sphere S^2 (or a curve γ) is said to be 2-dimensional (1-dimensional) submanifold S of the manifold $\mathcal{M} = \Re^3$.

Thus, it is natural to say that an m-dimensional *submanifold* S, of an n-dimensional manifold $\mathcal{M}$, is a set of points of $\mathcal{M}$ such that, in some neighborhood $\mathcal{U} \subseteq \mathcal{M}$ of any point $p \in S$, a coordinate system $(x^1, \ldots, x^n)$ can be introduced for $\mathcal{M}$ in which the points of $S \cap \mathcal{U}$ are characterized by $x^{m+1} = x^{m+2} = \cdots = x^n = 0$.

More formally, a one-to-one map $f : Q \to \mathcal{M}$ is said to be an *embedding* of the m-dimensional manifold Q in an n-dimensional manifold $\mathcal{M}$, $(m \leq n)$, if at every $q \in Q$ there is a neighborhood $\mathcal{V} \subseteq Q$ of q and a chart $(\mathcal{U}, \varphi)$ of $\mathcal{M}$ at $p = f(q)$, such that $(\mathcal{V}, \varphi \circ f|_{\mathcal{V}})$ is a chart of Q; that is, $\varphi \circ f|_{\mathcal{V}} : Q \to \Re^m$ are coordinates on $\mathcal{V}$ for Q. The manifold Q is said to be *embedded* in the manifold $\mathcal{M}$. The image $S = f(Q)$ is called a *submanifold* of the manifold $\mathcal{M}$, provided with the manifold structure for which $f : Q \to S \subseteq \mathcal{M}$ is a diffeomorphism.

If f is not one-to-one, we shall speak of *immersion*. In other words, a map $f : Q \to \mathcal{M}$ is said to be an *immersion* of the manifold Q in a manifold $\mathcal{M}$, if at every $p \in Q$, there is a neighborhood $\mathcal{V} \subseteq Q$ of p and a chart $(\mathcal{U}, \varphi)$ of $\mathcal{M}$ at $f(p)$, such that $(\mathcal{V}, \varphi \circ f)$ is a chart of Q; that is, $\varphi \circ f : Q \to \Re^n$ are coordinates on $\mathcal{V}$ for Q. The manifold Q is said to be *immersed* in the manifold $\mathcal{M}$.

By recalling what has been said in Sec. 5.6, concerning maps between manifolds, a vector field defined on a submanifold S is also a vector field on $\mathcal{M}$, and a *covector field* on $\mathcal{M}$ is also a *covector field* on S. A suggested reading on the subject and its applications is given by the Marmo, Saletan, Simoni, Vitale book.[41]

5.12.1 *The Frobenius theorem*

It has been shown that, given a smooth vector field X on an n-dimensional manifold $\mathcal{M}$, one can find a curve (*integral curve*) that, at every point $p \in \mathcal{M}$, the value X_p of the vector field X coincides with the tangent vector to the curve at the same point.

In other words, since a vector field X is an assignment at every point $p \in \mathcal{M}$ of a vector X_p in the tangent space $\mathcal{T}_p\mathcal{M}$, we can paraphrase the previous statement saying:

Given, at every point $p \in \mathcal{M}$, *a* 1*-dimensional subspace* D_p *of the tangent space* $\mathcal{T}_p\mathcal{M}$, *one can find a* 1*-dimensional submanifold* $\mathcal{N}$ *such that* $D_p = \mathcal{T}_p\mathcal{N}, \forall\, p \in \mathcal{M}$.

It is interesting to have an answer to the analogous problem:

Given, at every point $p \in \mathcal{M}$, *a* 2*-dimensional subspace* D_p *of the tangent space* $\mathcal{T}_p\mathcal{M}$ (i.e. *a plane*), *does a* 2*-dimensional submanifold* $\mathcal{N}$, *such that* $D_p = \mathcal{T}_p\mathcal{N}$, *exist* $\forall\, p \in \mathcal{M}$?)

The answer is generally: No.

In order to discuss the general case, it is advisable to introduce the following useful definitions:

- An assignment D at every point $p \in \mathcal{M}$, of a h-dimensional subspace D_p of the tangent space $\mathcal{T}_p\mathcal{M}$, that is, a hyperplane, is called a *h-dimensional distribution on* $\mathcal{M}$, or also, *a differential systems of h-planes on* $\mathcal{M}$.
- A h-dimensional distribution D is said to be C^∞ if, at every point $p \in \mathcal{M}$, there exists a neighborhood $\mathcal{U}$ of p and h C^∞-vector fields, namely $X_1, \ldots, X_h$, defined in $\mathcal{U}$ and defining, at every point $q \in \mathcal{U}$, a basis $X_1(q), \ldots, X_h(q)$ for D_q. The vector fields $X_1, \ldots, X_h$ are then called a *local basis* for D.
- A vector field X is said *to belong* to D if $X_p \in D_p$ at every point $p \in \mathcal{M}$.

- A $\mathcal{C}^\infty$ distribution D is called *involutive* if

$$X \in D, Y \in D \Rightarrow [X, Y] \in D.$$

The above relation is equivalent to say that a local basis $\{X_1, \ldots, X_h\}$ of a involutive distribution has the following property:

$$[X_i, X_j] = c_{ij}^k X_k,$$

since the Lie bracket of any two vector fields X and Y, which are their f-linear (i.e. the coefficients are functions) combinations

$$X = f^i(p) X_i, \quad Y = g^i(p) X_i,$$

will be linear combinations of X_i:

$$\begin{aligned} [X, Y] &= [f^i X_i, g^j X_j] \\ &= f^i g^j [X_i, X_j] + f^i Y_i(g^k) X_k - g^i Y_i(f^k) X_k \\ &= (f^i g^j c_{ij}^k + f^i X_i(g^k) - g^i X_i(f^k)) X_k = d_{ij}^k X_k. \end{aligned}$$

- A connected submanifold $\mathcal{N}$ of $\mathcal{M}$ is called an *integral manifold of the distribution* D if $f_*(\mathcal{T}_p\mathcal{N}) = D_q$ for all $p \in \mathcal{N}$, where f is the embedding of $\mathcal{N}$ into $\mathcal{M}$. The submanifold $\mathcal{N}$ is called a *maximal integral manifold of* D, when no other integral manifold of D, containing $\mathcal{N}$, exists.

It can be proven (see for instance Refs. 11, 29 and 50) that

Theorem 22 (Frobenius) *If D is an involutive distribution on a differential manifold $\mathcal{M}$, through every point $p \in \mathcal{M}$, there passes a unique maximal integral manifold $\mathcal{N}(p)$ of D. Any integral manifold through p is an open submanifold of $\mathcal{N}(p)$.*

In other words, if $X_1, \ldots, X_h$ are $h(< n)$ vector fields defined on a region $\mathcal{U}$ of an n-dimensional manifold $\mathcal{M}$ such that

$$[X_i, X_j] = c_{ij}^k X_k,$$

the integral curves of vector fields mesh to form a family of submanifolds.

Each submanifold has dimension equal to the dimension of the vector space these fields define at any point, which is at most h. Each point of $\mathcal{U}$ belongs to one and only one submanifold, provided that the dimension of the vector space defined by the fields is the same everywhere in $\mathcal{U}$. This family of submanifolds is called a *foliation* of $\mathcal{U}$, and each submanifold a *leaf* of $\mathcal{U}$.

The central idea underlying the Frobenius theorem is that, if the integral curves, of the vector fields $X_1, \ldots, X_h$ defining a distribution, are to define a submanifold to which the vector fields must be tangent, they have to mesh one another as cotton threads in a web. In other words the flows $\varphi^t_{X_i}$ of the vector field X_i have to transform an integral curve of a vector field X_j, in the integral curve (of the image) of a vector field constructed as linear combination (with functions) of $X_1, \ldots, X_h$. This will be guaranteed if all their Lie brackets $[X_i, X_j]$ are themselves tangent; that is, belong to the distribution $[X_i, X_j] = c^k_{ij} X_k$. This just means that the distribution has to be involutive.

Chapter 6

Differential Forms

6.1 The Tensors

In previous sections it has been shown how to construct the dual space E^* of a given vector space E. The elements of such spaces constitute the simplest examples of *tensors*.

A more interesting example is given by the *area* $A(U, V)$ of a given parallelogram constructed by two vectors U, V. Its most important property is expressed as follows:

$$A(X + Y, Z) = A(X, Z) + A(Y, Z) .$$

Thus, the area of a parallelogram is a rule

$$A : (U, V) \in E \times E \longrightarrow A(U, V) \in \Re ,$$

which associates a real number with two vectors *linearly* in the entries U, V.

Any *bilinear map* T, from the Cartesian product $E \times E$ to $\Re$, is called a *tensor of* $(0, 2)$*-type*.

The space of all such tensors is denoted with

$$\mathcal{T}_2^0(E) \equiv \mathrm{Lin}(E \times E, \Re) ,$$

and can be endowed naturally with a vector space structure, defined by

$$(T_1 + T_2)(X,Y) = T_1(X,Y) + T_2(X,Y),$$

$$(kT)(X,Y) = k(T(X,Y)), \quad \forall k \in \Re.$$

A basis of such vector space can be easily constructed by using a basis $\{e_i\}$ of E and its dual basis $\{\vartheta^i\}$.

In the given basis $\{e_i\}$, the vectors X and Y can be written as

$$X = X^i e_i, \quad Y = Y^j e_j,$$

and we have

$$T(X,Y) = T(X^i e_i, Y^j e_j) = X^i Y^j T(e_i, e_j) = T_{ij} X^i Y^j,$$

with $T_{ij} \equiv T(e_i, e_j) \in \Re$.

Since, by definition, for all $X \in E$, $\vartheta^i(X) = X^i$, the previous relation can also be written in the form

$$T(X,Y) = T_{ij}\vartheta^i(X)\vartheta^j(Y). \tag{6.1}$$

Thus, by introducing the *tensor product* $\otimes$ of two covectors, $\alpha \in E^*$, $\beta \in E^*$, by

$$(\alpha \otimes \beta)(X,Y) := \alpha(X)\beta(Y),$$

the relation (6.1) becomes

$$T(X,Y) = (T_{ij}\vartheta^i \otimes \vartheta^j)(X,Y),$$

or for the arbitrariness of X, Y,

$$T = T_{ij}\vartheta^i \otimes \vartheta^j.$$

Since the tensor T is an arbitrary element of the vector space $\mathcal{T}_2^0(E)$, the last relation shows that a basis for this space is given by the n^2 elements $\{\vartheta^i \otimes \vartheta^j\}$. Thus, a basis in E will fix a basis in its dual space, E^*, and also a basis in the vector space of $(0,2)$ tensors. For this reason the n^2 elements of $\Re$ are called the *components of the tensor T in the given basis.*

Similarly, any *bilinear map* R, from the Cartesian product $E^* \times E^*$ to $\Re$,

$$R : (\alpha, \beta) \in E^* \times E^* \longrightarrow R(\alpha,\beta) \in \Re,$$

is called a *tensor of* $(2, 0)$*-type*. The space of all such tensors is denoted with

$$\mathcal{T}_0^2(E) \equiv \text{Lin}(E^* \times E^*, \Re),$$

and can be endowed naturally with a vector space structure defined by

$$(R_1 + R_2)(X, Y) = R_1(X, Y) + R_2(X, Y),$$
$$(kR)(X, Y) = k(R(X, Y)), \quad \forall k \in \Re.$$

By defining the tensor product $X \otimes Y$, of two vectors X, Y of E, to be the $(2, 0)$ tensor given by

$$(X \otimes Y)(\alpha, \beta) = \alpha(X)\beta(Y), \quad \forall \alpha \in E^*, \ \beta \in E^*,$$

a basis $\{e_i \otimes e_j\}$ of $\mathcal{T}_0^2(E)$ is fixed in terms of a chosen basis $\{e_i\}$ of E.

Once more, any *bilinear map* S, from the Cartesian product $E^* \times E$ to $\Re$,

$$S : (\alpha, X) \in E^* \times E \longrightarrow S(\alpha, X) \in \Re,$$

is called a *tensor of* $(1, 1)$*-type*. The space of all such tensors is denoted with

$$\mathcal{T}_1^1(E) \equiv \text{Lin}(E^* \times E, \Re),$$

and can be endowed naturally with a vector space structure defined by

$$(S_1 + S_2)(\alpha, X) = S_1(\alpha, X) + S_2(\alpha, X),$$
$$(kS)(\alpha, X) = k(S(\alpha, X)), \quad \forall k \in \Re.$$

Exercise 6.1.1 *Show that a basis of* $\mathcal{T}_1^1(E)$ *can be given by* $\{\vartheta^i \otimes e_j\}$, *with an obvious definition for this tensor product.*

Previous examples exhaust the concepts of *tensor of rank* 2.

More generally, any *multilinear map*

$$T : \underbrace{E \times E \times \cdots \times E}_{q \text{ times}} \times \underbrace{E^* \times E^* \times \cdots \times E^*}_{p \text{ times}} \longrightarrow \Re,$$

is called a *tensor of* (p, q)*-type*. The tensor T is also said to be *of rank* $p + q$.

The space of all such tensors is denoted by

$$\mathcal{T}_q^p(E) \equiv \text{Lin}(\underbrace{E \times E \times \cdots \times E}_{q \text{ times}} \times \underbrace{E^* \times E^* \times \cdots \times E^*}_{p \text{ times}}, \Re),$$

and can be endowed naturally with a vector space structure defined by

$$(T_1 + T_2)(X, Y, \ldots, Z, \alpha, \beta, \ldots, \gamma) = T_1(X, Y, \ldots, Z, \alpha, \beta, \ldots, \gamma) + T_2(X, Y, \ldots, Z, \alpha, \beta, \ldots, \gamma),$$

$$(kT)(X, Y, \ldots, Z, \alpha, \beta, \ldots, \gamma) = k(T(X, Y, \ldots, Z, \alpha, \beta, \ldots, \gamma)), \quad \forall k \in \Re.$$

A basis of $\mathcal{T}_q^p(E)$ is easily found by using the same procedure used for $\mathcal{T}_2^0(E)$, $\mathcal{T}_0^2(E)$, or $\mathcal{T}_1^1(E)$. Indeed, let T be a (p, q) tensor, $\{e_i\}$ be a basis of E and $\{\vartheta^i\}$ be its dual basis. We have

$$\begin{aligned} T(X, Y, \ldots, Z, \alpha, \beta, \ldots, \gamma) &= X^i Y^j \cdots Z^k \alpha_p \beta_q \cdots \gamma_r T \\ &\quad \times (e_i, e_j, \ldots, e_k, \alpha^p, \beta^q, \ldots, \gamma^r) \\ &= T_{ij\cdots k}^{pq\cdots r} X^i Y^j \cdots Z^k \alpha_p \beta_q \cdots \gamma_r \\ &= T_{ij\cdots k}^{pq\cdots r} \vartheta^i(X) \vartheta^j(Y) \cdots \vartheta^k(Z) e_p(\alpha) e_q(\beta) \cdots e_r(\gamma) \\ &= \left(T_{ij\cdots k}^{pq\cdots r} \vartheta^i \otimes \vartheta^j \otimes \cdots \otimes \vartheta^k \otimes e_p \otimes e_q \otimes \cdots \otimes e_r \right) \\ &\quad \times (X, Y, \ldots, Z, \alpha, \beta, \ldots, \gamma). \end{aligned}$$

Since $X, Y, \ldots, Z, \alpha, \beta, \ldots, \gamma$ are arbitrary vectors and covectors, we can write

$$T = T_{ij\cdots k}^{pq\cdots r} \vartheta^i \otimes \vartheta^j \otimes \cdots \otimes \vartheta^k \otimes e_p \otimes e_q \otimes \cdots \otimes e_r, \tag{6.2}$$

which shows that a basis for the vector space $\mathcal{T}_q^p(E)$ is given by

$$\underbrace{\vartheta^i \otimes \vartheta^j \otimes \cdots \otimes \vartheta^k}_{q \text{ times}} \otimes \underbrace{e_p \otimes e_q \otimes \cdots \otimes e_r}_{p \text{ times}}.$$

Remark 13 *According to the previous definition we can say that*

- *A covector is a tensor of* $(0, 1)$*-type. The corresponding vector space* E^* *is also denoted, besides* $\mathcal{T}_1^0(E)$, *with* $\Lambda(E)$, *or simply* Λ. *So* $\Lambda(E) = \mathcal{T}_1^0(E) = E^*$.
- *A vector is a tensor of* $(1, 0)$*-type. The corresponding vector space* E *could be also denoted with* $\mathcal{T}_0^1(E)$.
- *The elements of* $\Re$ *are called tensors of* $(0, 0)$*-type.*

A tensor T of $(0,2)$-type is said to be

- *symmetric* if $T(X,Y) = T(Y,X)$
- *antisymmetric* if $T(X,Y) = -T(Y,X)$

The same definition can be given for tensors of $(2,0)$-type and, more generally, for tensors of $(0,p)$ or $(q,0)$-type. As for a tensor of $(1,1)$-type, no meaning can be given to the interchange of a vector with a covector.

The set of all antisymmetric tensor of $(0,2)$ type is, of course, a vector subspace $\Lambda^2(E)$ of the vector space $\mathcal{T}_2^0(E)$. A basis can be easily found by considering a generic element A of $\Lambda^2(E)$. In a given basis $\{e_i\}$ of E, the antisymmetric $(0,2)$ tensor A can be written as

$$A = A_{ij}\vartheta^i \otimes \vartheta^j ,$$

where the $\frac{1}{2}n(n-1)$ distinct numbers $A_{ij} = A(e_i, e_j)$ are antisymmetric for the interchange $i \leftrightarrow j$. Thus

$$A = \frac{1}{2}A_{ij}\vartheta^i \otimes \vartheta^j - \frac{1}{2}A_{ji}\vartheta^i \otimes \vartheta^j = \frac{1}{2}A_{ij}\left(\vartheta^i \otimes \vartheta^j - \vartheta^j \otimes \vartheta^i\right) .$$

Then, by introducing the *exterior* (*or wedge*) *product* $\vartheta^i \wedge \vartheta^j$, of the basis elements ϑ^i and ϑ^j by

$$\vartheta^i \wedge \vartheta^j := \vartheta^i \otimes \vartheta^j - \vartheta^j \otimes \vartheta^i ,$$

the antisymmetric $(0,2)$ tensor A can also be written in the form

$$A = \frac{1}{2}A_{ij}\vartheta^i \wedge \vartheta^j .$$

Thus, a basis for $\Lambda^2(E)$ is given by the $\frac{1}{2}n(n-1)$ elements $\{\vartheta^i \wedge \vartheta^j\}$.

More generally, a tensor T of $(0,q)$-type is said to be

- *symmetric* if $T(X_a, X_b, \ldots X_c) = T(X_1, X_2, \ldots X_q)$ for all permutations $(a, b, \ldots, c)$ of $(1, 2, \ldots, q)$
- *antisymmetric* if $T(X_a, X_b, \ldots X_c) = -T(X_1, X_2, \ldots X_q)$ for all odd permutations $(a, b, \ldots, c)$ of $(1, 2, \ldots, q)$

A similar definition can be given for tensors of $(q,0)$-type.

6.1.1 *The p-covectors*

Antisymmetric $(0,p)$ tensors are called *p-covectors* or *p-forms* and usually denoted with small Greek letters. So a *p-form* ω is a function

$$\omega : (X_1, \ldots, X_p) \in E \times \cdots \times E \to \omega(X_1, \ldots, X_p) \in \Re\,,$$

which is

- p-times linear; that is, for all $i = 1, 2, \ldots, p$

$$\begin{aligned} &\omega(X_1, \ldots, X_{i-1}, aY_i + bZ_i, X_{i+1}, \ldots, X_p) \\ &\quad = a\omega(X_1, \ldots, X_{i-1}, Y_i, X_{i+1}, \ldots, X_p) \\ &\qquad + b\omega^k(X_1, \ldots, X_{i-1}, Z_i, X_{i+1}, \ldots, X_p)\,; \end{aligned} \tag{6.3}$$

- completely antisymmetric

$$\omega(X_1, \ldots, X_k) = (-1)^{|\sigma|}\omega(X_{i_1}, \ldots, X_{i_p})\,, \tag{6.4}$$

  where $|\sigma| = 0$ or 1 according to the parity (even or odd, respectively) of the permutation $\sigma = (i_1, \ldots, i_p)$ of $(1, 2, \ldots, p)$.

The set of p-covectors is a subspace $\Lambda^p(E)$ of the vector space $\mathcal{T}_1^0(E)$. A basis of $\Lambda^p(E)$ can be found by applying the usual procedures which require, however, the notion of *wedge* or *exterior product of p* covectors.

6.1.2 *The exterior product*

Let $\alpha_1, \alpha_2, \ldots, \alpha_p$ be p-covectors on a vector space E. Their exterior product $\alpha_1 \wedge \alpha_2 \wedge \cdots \wedge \alpha_p$ is the p-form on E defined by

$$(\alpha_1 \wedge \alpha_2 \wedge \cdots \wedge \alpha_p)(X_1, \ldots, X_p) = \det \begin{pmatrix} \alpha_1(X_1)\,, & \cdots & \alpha_1(X_p) \\ : & : & : \\ \alpha_p(X_1) & \cdots & \alpha_1(X_p) \end{pmatrix}, \tag{6.5}$$

that is, by the determinant of the matrix $(\alpha_i(X_j))$.

The properties of the determinant show that the exterior product defined by (6.5) is a p-form.

By using the procedure already used for 2-forms, it is easy to check that any p-form ω can be written, in a given basis $\{e_i\}$ of E, as follows

$$\omega = \frac{1}{p!} \sum \omega_{i_1 \dots i_p} \vartheta^{i_1} \wedge \dots \wedge \vartheta^{i_p} ,$$

with $\omega_{i_1 \dots i_p} = \omega(e_{i_1}, \dots, e_{i_p})$ and $\{\vartheta^i\}$ the dual basis of $\{e_i\}$.

Thus, the $\binom{n}{p} = n!/p!(n-p)!$ distinct elements

$$\vartheta^{i_1} \wedge \dots \wedge \vartheta^{i_p} , \quad p \leq n , \tag{6.6}$$

make *a basis* in the vector space, $\Lambda^p(E)$, of the p-forms on E; that is, any p-form ω can be expressed in terms of them, and then

$$\dim \Lambda^p(E) = \binom{n}{p} .$$

The *exterior product* between a p-form $\alpha \in \Lambda^p(E)$ and a q-form $\beta \in \Lambda^q(E)$ is the $(p+q)$-form $\alpha \wedge \beta \in \Lambda^{p+q}(E)$ defined as

$$(\alpha \wedge \in)(X_1, \dots, X_{k+l}) = \sum_\sigma (-1)^{|\sigma|} \alpha(X_{i_1}, \dots, X_{i_k}) \beta(X_{j_1}, \dots, X_{jl}) ,$$

where the sum is over all permutation $\sigma = (i_1, \dots, i_k, j_1, \dots, j_l)$ of $(1, \dots, k+l)$ and $|\sigma| = 0$ or 1, according to the parity (even or odd, respectively) of the permutation.

It is easy to verify that

- $\alpha \wedge \beta$ is truly a $(p+q)$-form,
- that the product is
  - ◇ anticommutative: $\alpha \wedge \beta = (-1)^{pq} \beta \wedge \alpha$,
  - ◇ distributive with respect to the sum: $(a\alpha + b\alpha) \wedge \beta = a\alpha \wedge \beta + b\alpha \wedge \beta$,
  - ◇ associative: $(\alpha \wedge \beta) \wedge \gamma = \alpha \wedge (\beta \wedge \gamma)$,
  - ◇ coincides with the product defined by Eq. (6.5) if α and β are monomials; that is, if α is the exterior product of p covectors $\alpha_1, \dots, \alpha_p$ and β is the exterior product of q covectors $\beta_1, \dots, \beta_q$, respectively:

$$\alpha = \alpha_1 \wedge \dots \wedge \alpha_p , \quad \beta = \beta_1 \wedge \dots \wedge \beta_q .$$

The pair

$$\left(\Lambda = \bigcup_{p=1}^{\infty} \Lambda^p(E), \wedge\right)$$

of all p-covectors with arbitrary p, endowed with the exterior product $\wedge$, is called a *Grassmann algebra.*

6.1.3 *The metric tensor on a vector space*

A metric tensor on an n-dimensional vector space E is a $(0,2)$ tensor g satisfying the following requirement:

- *symmetry*: $g(X,Y) = g(Y,X)$, $X, Y \in E$
- *not degenerate*: $g(X,Y) = 0$, $\forall X \in E \Longleftrightarrow Y = 0$.

Exercise 6.1.2 *Show that previous requirements are equivalent, in a given basis* $\{e_i\}$, *to*

- *symmetry*: $g_{ij} = g_{ji}$
- *not degenerate*: $\det(g_{ij}) \neq 0$,

where $g_{ij} = g(e_i, e_j)$ *are the components of* g *in the basis* $\{e_i\}$.

Since the matrix $\tilde{g} \equiv (g_{ij})$ is symmetric, there exists a basis $\{\varepsilon_i = U_i^j e_j\}$, with $U = (U_i^j)$ an orthogonal matrix, such that

$$g(\varepsilon_i, \varepsilon_j) = \lambda_i \delta_{ij} \,,$$

where the eigenvalues λ_i, $i = 1, \ldots, n$ are not vanishing by the hypothesis that g is not degenerate.

Thus, if g is a metric tensor, a basis

$$\left\{ e_i' = \varepsilon_i / \sqrt{\lambda_i} \right\}$$

exists such that

$$g_{ij}' \equiv g\left(e_i', e_j'\right) = \pm \delta_{ij} \,.$$

A metric tensor provides an isomorphism between vectors and covectors. In fact, with any vector $X \in E$, we can associate the covector $\chi = i_X g$ defined by

$$\chi = (i_X g)(Y) = g(X,Y) \,,$$

whose components in a given basis are

$$\chi_j \equiv (i_X g)_j = X^i g_{ij} \,.$$

Since $\det(g_{ij}) \neq 0$, the previous map is invertible, so that

$$X^i = g^{ij} \chi_j \,,$$

where (g^{ij}) is the inverse of (g_{ij}) defined by

$$g^{ih} g_{hj} = \delta_{ij} \,.$$

6.2 The Tensor Fields

A (m,n)*-type tensor field* S on a manifold $\mathcal{M}$ is a rule that associates, with every point $p \in \mathcal{M}$, a (m,n)-type tensor $S_p \in \mathcal{T}^m_n(\mathcal{T}_p\mathcal{M})$; i.e. a map

$$S : p \in \mathcal{U} \subseteq \mathcal{M} \to S(p) = S_p \in \mathcal{T}^m_n(\mathcal{T}_p\mathcal{M}) \,.$$

By applying the same algebraic procedure used to obtain the Eq. (6.2), it is easy to see that, in a local coordinate system, a tensor field S can be written in the form

$$S(p) = S^{ab\cdots c}_{ef\cdots g}(p) dx^e \otimes dx^f \otimes \cdots \otimes dx^g \otimes \frac{\partial}{\partial x^a} \otimes \frac{\partial}{\partial x^b} \otimes \cdots \otimes \frac{\partial}{\partial x^c} \,, \tag{6.7}$$

where $x^i = \varphi^i(p)$ are the coordinate of p in the chart $(\mathcal{U}, \varphi)$. The tensor field $S(p)$ is said to be C^h *differentiable* on a C^k manifold $\mathcal{M}$, with $h \leq k-1$, if the functions $S^{ab\cdots c}_{ef\cdots g}(p)$ are C^h differentiable on the manifold $\mathcal{M}$.

6.2.1 *The Lie derivative of a tensor field*

Since the Lie derivative has been defined on functions, differential 1-forms and vector field, it is also defined, by the Leibnitz rule, on a general tensor field as the one given in Eq. (6.7).

One of the most important use of the Lie derivative in physics is to check if a tensor field is invariant under some transformation. If the transformation is generated by some vector field Δ, then the invariance of the tensor field S is expressed by

$$L_\Delta S = 0 \,.$$

The invariance condition preserves its elegance, also locally.

For instance, let the *mixed* tensor field

$$S : (X, \alpha) \to S(X, \alpha)$$

be locally represented by

$$S = S_i^j dx^i \otimes \frac{\partial}{\partial x^j}\,.$$

Its Lie derivative, with respect to the vector field Δ, is given by

$$\begin{aligned} L_\Delta S &= L_\Delta \left(S_i^j dx^i \otimes \frac{\partial}{\partial x^j} \right) \\ &= (L_\Delta S_i^j) dx^i \otimes \frac{\partial}{\partial x^j} + S_i^j (L_\Delta dx^i) \otimes \frac{\partial}{\partial x^j} + S_i^j dx^i \otimes \left(L_\Delta \frac{\partial}{\partial x^j} \right) \\ &= \frac{\partial S_i^j}{\partial x^k} \Delta^k dx^i \otimes \frac{\partial}{\partial x^j} + S_i^j \frac{\partial \Delta^i}{\partial x^k} dx^k \otimes \frac{\partial}{\partial x^j} - S_i^j \frac{\partial \Delta^k}{\partial x^j} dx^i \otimes \frac{\partial}{\partial x^k} \\ &= \left(\frac{\partial S_i^j}{\partial x^k} \Delta^k + S_k^j \frac{\partial \Delta^k}{\partial x^i} - S_i^k \frac{\partial \Delta^j}{\partial x^k} \right) dx^i \otimes \frac{\partial}{\partial x^j}\,. \end{aligned} \tag{6.8}$$

Thus, the invariance of S is expressed by

$$\frac{\partial S_i^j}{\partial x^k} \Delta^k = S_i^k \frac{\partial \Delta^j}{\partial x^k} - \frac{\partial \Delta^k}{\partial x^i} S_k^j\,.$$

In terms of the matrices

$$\underset{\sim}{S} = (S_i^j) \quad \Delta' = \left(\Delta'^j_k = \frac{\partial \Delta^j}{\partial x^k} \right),$$

it can be written as follows:

$$\frac{d}{d\tau} \underset{\sim}{S} = \left[\underset{\sim}{S}, \Delta' \right].$$

For the interested reader, an intrinsic definition can be given as in the case of a vector field.

The Lie derivative, with respect to X of a tensor field S, is defined by

$$(L_X S)(p) = \lim_{\tau \to 0} \frac{(\tilde{\phi}^{-\tau})_{*\phi^\tau(p)} (S(\tilde{\phi}^\tau(p))) - S(p)}{\tau}\,, \tag{6.9}$$

where

- ϕ^τ is the flow of the vector field X

$$\phi^\tau : p \in \mathcal{M} \to \phi^\tau(p) \in \mathcal{M}\,,$$

- $(\phi^\tau)_*$ its derivative

$$(\phi^\tau)_{*p} : \mathcal{T}_p\mathcal{M} \mapsto \mathcal{T}_{\phi^\tau(p)}\mathcal{M}\,,$$

- $(\tilde{\phi}^\tau)_*$ the *extension*[29] of $(\phi^\tau)_*$

$$(\tilde{\phi}^\tau)_{*p} : \mathcal{T}(p) \to \mathcal{T}(\phi^\tau(p))\,,$$

  to the whole tensor algebra

$$\mathcal{T}(p) \equiv \sum_{m,n=0}^{\infty} \mathcal{T}_n^m(\mathcal{T}_p\mathcal{M})\,.$$

Thus, in our case

$$(\tilde{\phi}^{-\tau})_{*\phi^\tau(p)} : \mathcal{T}(\phi^\tau(p)) \to \mathcal{T}(p)\,.$$

This definition can appear formally complicated. In reality it is very simple from a geometrical point of view. As a matter of fact, a map ϕ between two manifolds $\mathcal{M}$ and $\mathcal{N}$, transforms a curve trough $p \in \mathcal{M}$ to a curve trough the point $\phi(p) \in \mathcal{N}$; then, it also transforms a tangent vector to $\mathcal{M}$ at p in a tangent vector to $\mathcal{N}$ at $\phi(p)$. So ϕ induces a map, ϕ_*, between the corresponding tangent spaces $\mathcal{T}_p\mathcal{M}$ and $\mathcal{T}_{\phi(p)}\mathcal{N}$ at corresponding points. Of course, it also induces maps between the tensor spaces $\mathcal{T}_n^m(\mathcal{T}_p\mathcal{M})$ and $\mathcal{T}_n^m(\mathcal{T}_{\phi^\tau(p)}\mathcal{M})$ at corresponding points and, finally, between $\mathcal{T}(p) = \sum_{m,n=0}^{\infty} \mathcal{T}_n^m(\mathcal{T}_p\mathcal{M})$ and $\mathcal{T}(\phi^\tau(p)) = \sum_{m,n=0}^{\infty} \mathcal{T}_n^m(\mathcal{T}_{\phi^\tau(p)}\mathcal{M})$.

If S is a $(1,r)$-tensor field, the relation

$$\langle \alpha, \tilde{S}(Y^1, Y^2, \ldots, Y^r)\rangle \equiv S(\alpha, Y^1, Y^2, \ldots, Y^r)$$

defines a vector field $\tilde{S}(Y^1, Y^2, \ldots, Y^r)$.

It can be easily proven that, for any vector field X, we have

$$(L_X\tilde{S})(Y^1, Y^2, \ldots, Y^r) = [X, \tilde{S}(Y^1, Y^2, \ldots, Y^r)] - \sum_{i=1}^{r} \tilde{S}(Y^1, \ldots, [X, Y^i], \ldots, Y^r)\,. \quad (6.10)$$

The Leibnitz rule gives the following general properties of the Lie derivative:

$$L_X(R\otimes S) = (L_X R)\otimes S + R\otimes (L_X S)\,, \tag{6.11}$$

$$\begin{aligned} L_X(T(\alpha_1,\ldots,\alpha_p,X_1,\ldots,X_q)) &= (L_X T)(\alpha_1,\ldots,\alpha_p,X_1,\ldots,X_q) \\ &\quad + \sum_{i=1}^{p} T(\alpha_1,\ldots,L_X\alpha_i,\ldots,\alpha_p,X_1,\ldots,X_q) \\ &\quad + \sum_{i=1}^{p} T(\alpha_1,\ldots,\alpha_p,X_1,\ldots,L_X X_i,\ldots,X_q)\,. \end{aligned} \tag{6.12}$$

Equation (6.10) is just a particular case of Eq. (6.12).

Exercise 6.2.1 *Show that for any vector fields X and Y :*

$$L_{[X,Y]} = [L_X, L_Y]\,.$$

6.2.2 *The differential p-forms*

A *differential* 1-*form* α on the manifold $\mathcal{M}$ is a regular map

$$\alpha : \mathcal{TM} \to \Re \tag{6.13}$$

of the tangent bundle of the manifold $\mathcal{M}$ in $\Re$, linear in every tangent space $\mathcal{T}_p\mathcal{M}$:

$$\alpha_p(aX+bY) = a\alpha_p(X) + b\alpha_p(Y)\,, \quad \forall\, a,b\in\Re\,, \quad \forall\, X,Y\in\mathcal{T}_p\mathcal{M}\,.$$

In this way, a differential 1-form on $\mathcal{M}$ is a covector on $\mathcal{T}_p\mathcal{M}$ differentiable in p. Let us suppose that the functions $x^1,\ldots,x^n$ are a system of local coordinates in a given domain $\mathcal{U}$ of the manifold $\mathcal{M}$,

$$x^i : p_0\in\mathcal{U} \to x^i(p_0) = x_0^i \in \Re \quad \forall\, i=1,\ldots,n\,.$$

These functions are differentiable, and their differentials $dx^i_{p_0}$ at the point p_0,

$$dx^i_{p_0} : X\in\mathcal{T}_{p_0}\mathcal{M} \to dx^i_{p_0}(X)\in\Re \quad \forall\, i=1,\ldots,n\,,$$

are covectors on $\mathcal{T}_{p_0}\mathcal{M}$.

The values of the differentials $dx^1_{p_0}, \ldots, dx^n_{p_0}$ on the vector X are the components $X^1, \ldots, X^n$ of the vector.

If α_{p_0} is any covector on $\mathcal{T}_{p_0}\mathcal{M}$, because of the linearity of α_{p_0}, we have

$$\begin{aligned}
\langle \alpha_{p_0}, X \rangle &= \left\langle \alpha_{p_0}, X^i \left(\frac{\partial}{\partial x^i}\right)_{p_0} \right\rangle = X^i \left\langle \alpha_{p_0}, \left(\frac{\partial}{\partial x^i}\right) \right\rangle_{p_0} \\
&= \left\langle \alpha_{p_0}, \left(\frac{\partial}{\partial x^i}\right)_{p_0} \right\rangle dx^i_{p_0}(X) = \alpha_i\,(p_0)\,dx^i_{p_0}(X) \\
&= \langle \alpha_i(p_0) dx^i_{p_0}, X \rangle \,,
\end{aligned}$$

with

$$\alpha_i(p_0) = \left\langle \alpha_{p_0}, \left(\frac{\partial}{\partial x^i}\right)_{p_0} \right\rangle .$$

The covector α_{p_0} can thus be expressed locally in the form

$$\alpha_{p_0} = \alpha_1(p_0) dx^1_{p_0} + \cdots + \alpha_n(p_0) dx^n_{p_0} \,.$$

Therefore, every differential 1-form α (Eq. (6.13)) in the domain $\mathcal{U}$ can be locally expressed as

$$\alpha = \alpha_1(p) dx^1 + \cdots + \alpha_n(p) dx^n \,.$$

A k-covector ω_p at the point $p \in \mathcal{M}$ is a k-times linear (Eq. (6.3)) and antisymmetric (Eq. (6.4)) function,

$$\omega_p : (X_1, \ldots, X_k) \in \mathcal{T}_p\mathcal{M} \times \cdots \times \mathcal{T}_p\mathcal{M} \to \omega_p(X_1, \ldots, X_k) \in \Re \,. \tag{6.14}$$

A differential k-form ω is defined on the manifold $\mathcal{M}$ if the form (Eq. (6.14)) is given at every point p in $\mathcal{M}$ and, moreover, if it is differentiable.

Every differential k-form ω can be uniquely expressed in a domain with local coordinates $x^1, \ldots, x^n$ as

$$\omega = \frac{1}{k!} \omega_{i_1 \cdots i_k}(x^1, \ldots, x^n) dx^{i_1} \wedge \cdots \wedge dx^{i_k} \,, \tag{6.15}$$

where $dx^{i_1} \wedge \cdots \wedge dx^{i_k}$ are the exterior products of the basis 1-forms $dx^1, \ldots, dx^n$.

Operations such as the sum of k-forms, the product with real numbers, the exterior product between forms are always point-wise possible; that is, at every point $p \in \mathcal{M}$ the corresponding exterior forms on the tangent spaces $\mathcal{T}_p\mathcal{M}$ can be summed and multiplied with numbers or exteriorly.

Lie derivative of a differential k-form

From Eq. (6.9), defining the Lie derivative of a tensor field, we obtain, for a differential k-form ω, the useful formula

$$(L_X\omega)(Y^1, Y^2, \ldots, Y^k) = [X, \omega(Y^1, Y^2, \ldots, Y^k)] - \sum_{i=1}^{r} \omega(Y^1, \ldots, [X, Y^i], \ldots, Y^k), \quad (6.16)$$

which is similar to the one given, for a $(1, k)$-tensor field, by the Eq. (6.10).

6.2.3 *The exterior derivative*

On the space of differential k-forms we can define an operator d, called *exterior derivative*, having the following properties:

If $\alpha \in \Lambda^k(\mathcal{M})$, $\beta \in \Lambda^k(\mathcal{M})$, $\gamma \in \Lambda^h(\mathcal{M})$,

(1) $d(\alpha + \beta) = d\alpha + d\beta$;
(2) $d(\alpha \wedge \gamma) = d\alpha \wedge \gamma + (-1)^k \alpha \wedge d\gamma$;
(3) $d^2\alpha = 0$;
(4) On the differential 0-forms; that is, on functions, the operator d coincides with the differential defined in Sec. 5.5.

The operator d, as it easily follows from its properties, transforms differential k-forms in differential $(k+1)$-forms.

By using the properties (1), (2), (3) and (4), we can easily calculate the exterior derivative of a k-form in a coordinate basis. For ω given by Eq. (6.15), we obtain

$$d\omega = \frac{1}{(k+1)!} d\omega_{i_1 \cdots i_k} \wedge dx^{i_1} \wedge \cdots \wedge dx^{i_k},$$

because $ddx^i = 0$.

6.2.4 *Closed and exact differential forms*

A differential k-form is said to be *closed* if

$$d\omega = 0\,,$$

and to be *exact* if there exists $\alpha \in \Lambda^{k-1}(\mathcal{M})$, such that

$$\omega = d\alpha\,.$$

Since $d^2 = 0$, an exact p-form is also closed. The converse is not true and the following is a classical example in $\Re^2$.

Example 23 *Consider the differential* 1*-form*

$$\omega = \frac{xdy - ydx}{x^2 + y^2}\,,$$

which it is easy to see to be closed,

$$d\omega = 0\,.$$

In polar coordinates

$$\begin{cases} x = r\cos\vartheta\,, \\ y = r\sin\vartheta\,, \end{cases}$$

it becomes

$$\omega = d\vartheta\,.$$

Thus, one is tempted to say that ω is an exact differential form also. But the angles do not exist really!

The misunderstanding is solved by observing that ω is not defined at the point $(0,0)$, *as well as the transformation from Cartesian to polar coordinates.*

6.2.5 *The contraction operator i_X*

Let E be an n-dimensional vector space and $\Lambda^r(E)$ be the vector space of r-covectors defined on it.

If $\omega \in \Lambda^r(E)$ is an antisymmetric multilinear map from $\underbrace{E \times E \times \cdots \times E}$ to $\Re$, and $X_1, X_2, \ldots X_r$ are vectors of E, then

$$\omega(X_1, X_2 \ldots X_r) \tag{6.17}$$

is a real number antisymmetric under the interchange of any two vectors. Therefore, an r-covector is defined once the number (Eq. (6.17)) is given $\forall X_i$, $i = 1, \ldots, r$.

It is natural, starting from any r-covector $\omega \in \Lambda^r(E)$ and a vector $X \in E$, to *define* a $(r-1)$-covector, namely $i_X\omega \in \Lambda^{r-1}(E)$ by the following equality:

$$(i_X\omega)(X_1, X_2, \ldots X_{r-1}) := \omega(X, X_1, X_2, \ldots X_{r-1}) .$$

In this way, $(i_X\omega)$ is the $(r-1)$-covector, built from $\omega \in \Lambda^r(E)$ and $X \in E$, which evaluated on $(r-1)$ vectors $X_1, X_2, \ldots X_{r-1}$, gives the same real number given by ω on the r vectors $X, X_1, X_2, \ldots X_{r-1}$.

The operator i_X is called the *contraction operator with respect to* X. We already met it in the case in which ω is a simple covector. In fact, by denoting with α an element of $\Lambda^1(E) \equiv \Lambda(E) \equiv E^*$, the previous definition simply reduces to

$$i_X\alpha = \alpha(X) \equiv \langle \alpha, X \rangle .$$

In order to illustrate the given definition, let us represent the r-covector ω in a basis $\{\vartheta^i\}$, as follows:

$$\omega = \frac{1}{r!}\omega_{ij\cdots k}\vartheta^i \wedge \vartheta^j \wedge \cdots \wedge \vartheta^k . \tag{6.18}$$

Thus,

$$(i_{X_1}\omega)(X_2, \ldots, X_r) = \frac{1}{r!}\omega_{i_1 i_2 \cdots i_r} \det \begin{pmatrix} \vartheta^{i_1}(X_1), \ldots, \vartheta^{i_1}(X_r) \\ \vartheta^{i_2}(X_1), \ldots, \vartheta^{i_2}(X_r) \\ \vdots \qquad\quad \vdots \\ \vartheta^{i_r}(X_1), \ldots, \vartheta^{i_r}(X_r) \end{pmatrix} , \tag{6.19}$$

and then, by using the Laplace expansion (first column) of the determinant, $i_X\omega$ is represented by

$$\begin{aligned}(i_{X_1}\omega) = &\frac{1}{r!}\omega_{i_1 i_2 \cdots i_r}\vartheta^{i_1}(X_1)\vartheta^{i_2} \wedge \vartheta^{i_3} \wedge \cdots \wedge \vartheta^{i_r} \\ &- \frac{1}{r!}\omega_{i_1 i_2 \cdots i_r}\vartheta^{i_2}(X_1)\vartheta^{i_1} \wedge \vartheta^{i_3} \wedge \cdots \wedge \vartheta^{i_r} \\ &+ \frac{1}{r!}\omega_{i_1 i_2 \cdots i_r}\vartheta^{i_3}(X_1)\vartheta^{i_1} \wedge \vartheta^{i_2} \wedge \vartheta^{i_4} \cdots \wedge \vartheta^{i_r} + \cdots\end{aligned}$$

$$+\frac{1}{r!}(-1)^{r-1}\omega_{i_1 i_2 \cdots i_r}\vartheta^{i_r}(X_1)\vartheta^{i_1}\wedge\vartheta^{i_2}\wedge\cdots\wedge\vartheta^{i_{r-1}} \tag{6.20}$$

$$=\frac{1}{(r-1)!}\omega_{i_1 i_2 \cdots i_r}\vartheta^{i_1}(X_1)\vartheta^{i_2}\wedge\vartheta^{i_3}\wedge\cdots\wedge\vartheta^{i_r}\,. \tag{6.21}$$

What are the properties of the operator i_X?

- It is easy to see that

$$i_X i_Y = -i_Y i_X\,. \tag{6.22}$$

This easily follows by observing that, $\forall\,\omega\in\Lambda^r(E)$,

$$\begin{aligned}\omega(X,Y,X_1,X_2,\dots X_{r-2}) &= (i_X\omega)(Y,X_1,X_2,\dots X_{r-2})\\ &= (i_Y i_X\omega)(X_1,X_2,\dots X_{r-2})\\ \omega(Y,X,X_1,X_2,\dots X_{r-2}) &= (i_Y\omega)(X,X_1,X_2,\dots X_{r-2})\\ &= (i_X i_Y\omega)(X_1,X_2,\dots X_{r-2})\,.\end{aligned}$$

As a particular case, it follows that

$$i_X^2 = 0\,. \tag{6.23}$$

- If $\alpha\in\Lambda^r(E)$, and $\beta\in\Lambda^s(E)$, with $r+s\leq n$, then

$$i_X(\alpha\wedge\beta) = (i_X\alpha)\wedge\beta + (-1)^r\alpha\wedge i_X\beta\,. \tag{6.24}$$

This easily follows from the following formula:

$$\begin{aligned}(\alpha\wedge\beta)(X_1,\dots,X_{r+s}) &= \frac{1}{(r+s)!}\sum(-1)^\sigma\alpha(X_{j_1},\dots,X_{j_r})\\ &\quad\times\beta(X_{j_{r+1}},\dots,X_{j_{r+s}})\,,\end{aligned}$$

where the sum is over all permutation $(j_1,j_2,\dots,j_{r+s})$ of $(1,2,\dots,r+s)$ and $\sigma=0$ or 1, according to its parity (even or odd, respectively).

Properties of Eqs. (6.23) and (6.24) extend in a natural way, with the only additional requirement that on 0-forms $f\in\mathcal{F}(M)$, $i_X f=0$, to r-forms on a differential manifold $\mathcal{M}$.

Let X be a vector field on $\mathcal{M}$ and $\alpha\in\Lambda^k(\mathcal{M})$.

The operator i_X which, acting on the differential k-form α, transforms it in a differential $(k-1)$-form $i_X\alpha$ (also called *interior product* between X and α), is defined point-wise as

$$(i_X\alpha)_p(X_1,\ldots X_{k-1}) = \alpha_p(X(p), X_1, \ldots, X_{k-1}), \quad \forall p \in \mathcal{M}, \tag{6.25}$$

where $X_1, \ldots, X_{k-1}$ are tangent vectors to $\mathcal{M}$ at p.

The i_X operator fulfills the following properties:

(1) $i_X(\alpha_1 + \alpha_2) = i_X\alpha_1 + i_X\alpha_2$;
(2) $i_X(\alpha \wedge \beta) = i_X\alpha \wedge \beta + (-1)^k \alpha \wedge i_X\beta$;
(3) if $\alpha \in \Lambda^1(\mathcal{M})$, $(i_X\alpha)(p) = \langle X(p), \alpha_p\rangle = \alpha_p(X(p))$;
(4) if f is a 0-form, then $i_X f = 0$.

Thus, the properties of the interior product i_X on a differential manifold $\mathcal{M}$ are algebraically similar to the ones of the exterior derivative d, namely

$$d^2 = 0, \quad d(\alpha \wedge \beta) = (d\alpha) \wedge \beta + (-1)^r \alpha \wedge d\beta .$$

Of course,

$$i_X : \Lambda^r(\mathcal{M}) \to \Lambda^{r-1}(\mathcal{M})$$
$$d : \Lambda^r(\mathcal{M}) \to \Lambda^{r+1}(\mathcal{M}),$$

and

$$i_X d : \Lambda^r(\mathcal{M}) \to \Lambda^r(\mathcal{M})$$
$$d i_X : \Lambda^r(\mathcal{M}) \to \Lambda^r(\mathcal{M}).$$

The operators $i_X d$ and $d i_X$ do not coincide, as it is easy to verify on the simple example of $\alpha = dx^i$. Indeed, denoting with X^i the components in the basis $\{\partial/\partial x^i\}$ of the vector field X, we get

$$\begin{aligned} (i_X d)dx^i &= 0, \\ (d i_X)dx^i &= d(i_X dx^i) = dX^i = \frac{\partial X^i}{\partial x^j} dx^j . \end{aligned} \tag{6.26}$$

Moreover, the operators $i_X d$ and $d i_X$ are not derivations, since

$$\begin{aligned} i_X d(\alpha \wedge \beta) &= i_X[(d\alpha) \wedge \beta + (-1)^r \alpha \wedge d\beta] \\ &= i_X[(d\alpha) \wedge \beta] + (-1)^r i_X[\alpha \wedge d\beta] \end{aligned}$$

$$= (i_X da) \wedge \beta + (-1)^{r+1}(d\alpha) \wedge i_X \beta$$
$$+ (-1)^r (i_X \alpha) \wedge d\beta + (-1)^{r+r} \alpha \wedge i_X d\beta\,,$$

and

$$d i_X(\alpha \wedge \beta) = d[(i_X da) \wedge \beta + (-1)^r \alpha \wedge i_X \beta]$$
$$= d[(i_X a) \wedge \beta] + (-1)^r d[\alpha \wedge i_X \beta]$$
$$= (d i_X a) \wedge \beta + (-1)^{r+1}(i_X \alpha) \wedge d\beta$$
$$+ (-1)^r (d\alpha) \wedge i_X \beta + (-1)^{r+r} \alpha \wedge d i_X \beta\,.$$

However, by adding $i_X d(\alpha \wedge \beta)$ and $d i_X(\alpha \wedge \beta)$ from the above relations, we obtain the result that the operator $i_X d + d i_X$ is a derivation, since

$$(i_X d + d i_X)(\alpha \wedge \beta) = [(i_X d + d i_X)\alpha] \wedge \beta + \alpha \wedge (i_X d + d i_X)\beta\,. \tag{6.27}$$

Finally, let us remark that the three operators L_X, i_X and d are not independent on $\Lambda^r(\mathcal{M})$. It is easy to see that, on r-forms $\omega \in \Lambda^r(\mathcal{M})$, they satisfy a very useful relation, the so-called *homotopic* or *Cartan identity*:

$$L_X \omega = i_X d\omega + d i_X \omega\,, \tag{6.28}$$

or in operator terms,

$$L_X = i_X \circ d + d \circ i_X\,. \tag{6.29}$$

Proof.

- If $f \in \mathcal{F}(\mathcal{M}) \equiv \Lambda^r(\mathcal{M})$, since $i_X f = 0$, we have

$$i_X df + d i_X f = i_X df = (df)(X) = Xf = L_X f\,,$$

- For a generic 1-form $\alpha = f dg \in \Lambda(\mathcal{M})$:

$$i_X d\alpha = i_X(df \wedge dg) = (Xf)dg - (df)Xg\,,$$
$$d i_X \alpha = d(f Xg) = (df)Xg + f d(Xg)\,. \tag{6.30}$$

Thus,

$$i_X d\alpha + d i_X \alpha = (Xf)dg + fd(Xg) = (L_X f)dg + f L_X dg = L_X \alpha\,, \tag{6.31}$$

where the Cartan identity on functions has been used:

$$fd(Xg) = fd(i_X dg) = f(di_X)dg = f(di_X + i_X d)dg = f L_X dg\,.$$

The proof proceeds now by induction.

A more elegant proof can be found in the Kobayashi–Nomizu book, and it consists in observing that

(1) $i_X d + di_X$ is a derivation of degree 0;
(2) every derivation of degree 0 commuting with d is the Lie derivative with respect to some vector field;
(3) the derivations L_X and $i_X d + di_X$ give the same result on $f \in \mathcal{F}(\mathcal{M})$.

From Eq. (6.29) directly follows the useful formulae

$$[L_X, d] = 0\,, \quad [L_X, i_X] = 0\,.$$

6.2.6 *A different procedure*

The fact that the three operators d, L_X and i_X are not independent on differential forms, suggests the following different procedure to define the exterior derivative in terms of the interior product and the Lie derivative.

Let us observe that, by using the Cartan identity, we have

◇ for a function f:

$$i_X df = \langle df, X\rangle \equiv L_X f\,,$$

◇ for a differential 1-form $\alpha \in \Lambda(\mathcal{M})$:

$$\begin{aligned}(d\alpha)(X,Y) &= i_Y i_X d\alpha \\ &= i_Y(L_X\alpha - di_X\alpha) \\ &= \langle L_X\alpha, Y\rangle - i_Y(di_X\alpha) \\ &= \langle L_X\alpha, Y\rangle - i_Y d(i_X\alpha) \\ &= \langle L_X\alpha, Y\rangle - L_Y\langle\alpha, X\rangle\end{aligned}$$

$$= \langle L_X\alpha, Y\rangle - \langle L_Y\alpha, X\rangle + \langle \alpha, [X,Y]\rangle$$
$$= \langle L_X\alpha, Y\rangle - \langle L_Y\alpha, X\rangle - \langle \alpha, [X,Y]\rangle ,$$

where the property that $f \equiv i_X\alpha = \langle\alpha, X\rangle$ is a function, to which the previous formula can be applied, and the Leibnitz rule has been used,

◇ for a differential 2-form $\omega \in \Lambda^2(\mathcal{M})$:

$$\begin{aligned}
d\omega(X,Y,Z) &= i_Y i_X i_Z d\omega \\
&= i_Y i_X (L_Z - d i_Z)\omega \\
&= i_Y i_X L_Z\omega - i_Y i_X d(i_Z\omega) \\
&= (L_Z\omega)(X,Y) - (d(i_Z\omega))(X,Y) \\
&= (L_Z\omega)(X,Y) - L_X\langle i_Z\omega, Y\rangle + L_Y\langle (i_Z\omega), X\rangle - \langle (i_Z\omega), [X,Y]\rangle \\
&= (L_Z\omega)(X,Y) - L_X(\omega(Z,Y)) + L_Y(\omega(Z,X)) - \omega(Z,[X,Y]) \\
&= (L_Z\omega)(X,Y) - (L_X\omega)(Z,Y) + (L_Y\omega)(Z,X) + \omega(Z,[X,Y]) \\
&\quad + \omega([X,Z],Y) - \omega(Z,[Y,X]) - \omega([Y,Z],X) - \omega(Z,[X,Y]) \\
&= L_X\omega(Y,Z) - L_Y\omega(X,Z) + (L_Z\omega)(X,Y) \\
&\quad + \omega(X,[Y,Z]) - \omega(Y,[X,Z]) + \omega(Z,[X,Y]) ,
\end{aligned}$$

where to $\alpha \equiv i_Z\omega$, which is a differential 1-form, and to $f \equiv \langle i_Z\omega, Y\rangle$, which is a function, previous formulae have been applied. Moreover, Eq. (6.16) has been used.

Thus, the exterior derivative could be defined axiomatically by the following:

- for a function f, as

$$(df)(X) \equiv L_X f ,$$

- for a differential 1-form $\alpha \in \Lambda(\mathcal{M})$, as

$$(d\alpha)(X,Y) \equiv \langle L_X\alpha, Y\rangle - \langle L_Y\alpha, X\rangle + \langle \alpha, [X,Y]\rangle ,$$

- for a differential 2-form $\omega \in \Lambda^2(\mathcal{M})$, as

$$d\omega(X,Y,Z) \equiv L_X\omega(Y,Z) - L_Y\omega(X,Z) + (L_Z\omega)(X,Y) + \omega([X,Y],Z)\,,$$

- for a differential p-form $\omega \in \Lambda^p(\mathcal{M})$, as

$$d\omega(X_1,\ldots,X_{p+1}) \equiv \sum_\sigma (-1)^{|\sigma|} L_{X_i}\omega(X_{i_1},\ldots X_{i_p})$$
$$-\sum_\sigma (-1)^{|\sigma|}\omega([X_i,X_{i_1}],\ldots,X_{i_p})\,, \quad (6.32)$$

where the sum is over all permutation $\sigma = (i, i_1, \ldots, i_p)$ of $(1,\ldots,p+1)$ and $|\sigma| = 0$ or 1, according to the parity (even or odd, respectively) of the permutation.

Exercise 6.2.2 *Prove, by using as definition the one given in the Eq. (6.32), all the properties of the exterior derivative.*

6.2.7 *A dual characterization of holonomic and anholonomic basis*

Let us return to the discussion in Sec. 5.7.1 and consider a generic basis $\{e_i\}$ of vector fields on an n-dimensional manifold $\mathcal{M}$:

$$[e_i, e_j] = c^h_{ij} e_h\,.$$

The dual basis $\{\vartheta^i\}$ has the point-wise property

$$\langle \vartheta^k, e_j\rangle = \delta^k_j\,.$$

By taking the Lie derivative, with respect to the vector field e_i of the previous expression, we obtain

$$\langle L_{e_i}\vartheta^k, e_j\rangle = -\langle \vartheta^k, [e_i,e_j]\rangle = -c^h_{ij}\langle \vartheta^k, e_h\rangle = -c^k_{ij}\,.$$

Then, by using the Cartan identity, we have

$$(d\vartheta^k)(e_i,e_j) = -c^k_{ij}\,.$$

The exterior derivatives $d\vartheta^k$ are differential 2-forms and the above formula allows us to evaluate their coefficients d^k_{ij} in the given basis, in which $d\vartheta^k \equiv d^k_{rs}\vartheta^r \wedge \vartheta^s$.

We obtain

$$d^k_{rs}(\vartheta^r \wedge \vartheta^s)(e_i, e_j) = -c^k_{ij},$$

or

$$2d^k_{ij} = -c^k_{ij}.$$

Therefore, the elements of the dual basis $\{\vartheta^i\}$ have the following property:

$$d\vartheta^k \equiv -\frac{1}{2} c^k_{ij} \vartheta^i \wedge \vartheta^j. \tag{6.33}$$

We can summarize the previous results as follows:

If $\{e_i\}$ is a basis of vector fields and $\{\vartheta^i\}$ its dual basis on an n-dimensional manifold $\mathcal{M}$, then

$$[e_i, e_j] = c^h_{ij} e_h \Leftrightarrow d\vartheta^k \equiv -\frac{1}{2} c^k_{ij} \vartheta^i \wedge \vartheta^j.$$

Therefore, for a holonomic basis, given $c^h_{ij} = 0$, the dual basis consists of closed differential 1-forms ϑ^k, $d\vartheta^k = 0$, and then, locally, coordinates functions $\{x^i\}$ exist such that

$$\vartheta^k = dx^k.$$

As a consequence,

$$e_i = \frac{\partial}{\partial x^i}.$$

Thus, besides the one given in Sec. 5.7.1, a new characterization of a holonomic basis $\{e_i\}$ is given by the closure property of the differential 1-form which composes its dual basis.

6.3 The Metric Tensor Field on a Manifold

A metric tensor field g on a manifold $\mathcal{M}$ is a rule that associates with every point $p \in \mathcal{M}$ a symmetric and not degenerate $(0, 2)$-tensor $g(p)$.

Thus, at every point $p \in \mathcal{M}$, $g(p)$ is a metric tensor for the tangent space $\mathcal{T}_p\mathcal{M}$, and the considerations, already done for a metric tensor on a vector space, can be repeated. In particular, in every tangent space $\mathcal{T}_p\mathcal{M}$, a basis can be chosen such that $g_{ij}(p) = \pm\delta_{ij}$.

Since a metric tensor field is required to be at least continuous and integers do not change continuously, the *canonical form* of g has to be constant everywhere and we speak of *signature* of the field g. The collection of the bases in which g takes on the canonical form, defines a globally orthonormal basis on $\mathcal{M}$, but this global basis is not generally a coordinate basis.

In this sense the space $\Re^n$, considered as a manifold endowed with the Euclidean metric tensor field (δ_{ij} at every point), constitutes just an exceptional case. Even in that case only the Cartesian coordinates generate an orthonormal basis.

6.3.1 *Killing vector fields*

The Killing vector fields play a relevant role in the study of the *isometries* of a metric tensor field; this is why they are usually used in general relativity. They are defined to be the vector fields Δ preserving a metric tensor field g; that is, by the invariance condition

$$L_\Delta g = 0\,.$$

The above equation, given g, admits very few solutions for Δ.

Let the *metric* tensor field

$$g : (X, Y) \to g(X, Y)$$

be locally represented by $g = g_{ij} dx^i \otimes dx^j$.

Its Lie derivative, with respect to the vector field Δ, is given by

$$\begin{aligned} L_\Delta g &= L_\Delta(g_{ij} dx^i \otimes dx^j) \\ &= (L_\Delta g_{ij}) dx^i \otimes dx^j + g_{ij}(L_\Delta dx^i) \otimes dx^j + g_{ij} dx^i \otimes (L_\Delta dx^j) \\ &= \frac{\partial g_{ij}}{\partial x^k} \Delta^k dx^i \otimes dx^j + g_{ij} \frac{\partial \Delta^i}{\partial x^k} dx^k \otimes dx^j + g_{ij} \frac{\partial \Delta^j}{\partial x^k} dx^i \otimes dx^k \\ &= \left(\frac{\partial g_{ij}}{\partial x^k} \Delta^k + g_{kj} \frac{\partial \Delta^k}{\partial x^i} + g_{ik} \frac{\partial \Delta^k}{\partial x^j} \right) dx^i \otimes dx^j\,. \end{aligned} \tag{6.34}$$

Thus, the invariance of g is expressed by

$$\frac{\partial g_{ij}}{\partial x^k} \Delta^k + g_{kj} \frac{\partial \Delta^k}{\partial x^i} + g_{ik} \frac{\partial \Delta^k}{\partial x^j} = 0\,. \tag{6.35}$$

In terms of the matrices

$$\underset{\sim}{g} = (g_{ij})\,, \quad \Delta' = \left(\Delta'^{j}_{k} = \frac{\partial \Delta^j}{\partial x^k}\right)\,,$$

Eq. (6.35) can be written as follows:

$$\frac{d}{d\tau}\,\underset{\sim}{g} = -\left(\Delta'\,\underset{\sim}{g} + \underset{\sim}{g}\,\Delta'^{\mathsf{T}}\right)\,,$$

where the symbol ^{T} denotes matrix transposition.

6.3.2 *Maximally symmetric manifolds*

We may now ask the following question: how many vector fields, leaving a metric tensor field g invariant, exist on an n-dimensional manifold $\mathcal{M}$?

By introducing the differential 1-form ξ by

$$\langle \xi, X\rangle = g(\Delta, X)\,,$$

Eq. (6.35) can also be written in the following form:

$$\frac{\partial \xi_i}{\partial x^j} + \frac{\partial \xi_j}{\partial x^i} = 2\xi_h \Gamma^h_{ij}\,, \tag{6.36}$$

where

$$\Gamma^h_{ij} = \frac{1}{2} g^{hk}\left(\frac{\partial g_{ki}}{\partial x^j} + \frac{\partial g_{kj}}{\partial x^i} - \frac{\partial g_{ij}}{\partial x^k}\right) \tag{6.37}$$

are called the *Christoffel symbols.*

The number of independent differential equations, in the partial differential system (6.36), is $(1/2)n(n+1)$, while the number of unknown functions ξ is n, so that the system (6.36) is overdetermined for $n > 1$, and the number of Killing vectors will be upper-bounded.

By taking the derivative of Eq. (6.36), we obtain

$$\frac{\partial^2 \xi_i}{\partial x^j \partial x^k} + \frac{\partial^2 \xi_j}{\partial x^i \partial x^k} = 2\frac{\partial}{\partial x^k}(\xi_h \Gamma^h_{ij})\,. \tag{6.38}$$

By adding the above equation to itself with the permutation $(i \to j, j \to i, k \to j)$ of the indices, and subtracting the one with the permutation $(i \to j, j \to k, k \to i)$, we finally obtain

$$\frac{\partial^2 \xi_i}{\partial x^j \partial x^k} = A^{rs}_{ijk}\frac{\partial \xi_r}{\partial x^s} + B^r_{ijk}\xi_r\,,$$

where A^{rs}_{ijk} is a function of g_{ij} and its first derivatives, and B^{r}_{ijk} is a function of g_{ij} and its first and second derivatives.

Thus, once the metric tensor field g is given and the functions ξ's and its first derivatives are known at a point $p \in \mathcal{M}$, the above equation allows us also to know the value, at the point p, of the second derivatives of the ξ's. In the same way, by successively differentiating the equation, we can determine all higher derivatives of the ξ's at the point p. This suffices, if the manifold $\mathcal{M}$ is analytic, to determine the differential 1-form ξ everywhere.

Since the assignment of ξ_i at p determine, via Eq. (6.38), the symmetric part of the first derivatives $\partial\xi_i/\partial x^j$, we conclude that every Killing vector on $\mathcal{M}$ is determined by giving the values

$$a_i = \xi_i(p)\,, \quad b_{ij} = \left.\frac{\partial \xi_i}{\partial x^j}\right|_p - \left.\frac{\partial \xi_j}{\partial x^i}\right|_p\,,$$

at any point $p \in \mathcal{M}$.

Therefore, since the number N of the parameters a_i and b_{ij} is given by

$$N = n + \frac{1}{2}n(n-1) = \frac{1}{2}n(n+1)\,,$$

on an n-dimensional manifold $\mathcal{M}$ there exist at most $(1/2)n(n+1)$ Killing vectors, 6 for $n = 3$.

It is worth observing that Eq. (6.36) may not admit solutions.

An n-manifold $\mathcal{M}$, endowed with a metric tensor field, is said to be *maximally symmetric* if, on it, there exist $(1/2)n(n+1)$ Killing vectors.

6.3.3 *The Levi-Civita covariant derivative*

On a differential manifold $\mathcal{M}$ there exist only three natural *derivations*, which are given by the *Lie derivative*, the *interior product* and the *exterior derivative*. Moreover, on differential forms they are not independent, because of the Cartan identity

$$L_X = i_X d + d i_X\,.$$

However, once a metric tensor field g is given on a manifold $\mathcal{M}$, a new derivation, the *Levi-Civita covariant derivative* ∇_X *with respect to the vector field*

X, can be defined by

$$\nabla_X f = L_X f\,,$$
$$(\nabla_X \alpha)(Y) = \frac{1}{2}[(d\alpha)(X,Y) + (L_{X_\alpha} g)(X,Y)]\,, \tag{6.39}$$

where f, α and Y are a differentiable function, a differential 1-form and a vector field, respectively, and where X_α is the vector field associated to α *via* the metric tensor field g; i.e.

$$g(X_\alpha, Y) = \alpha(Y)\,,$$

or, symbolically

$$X_\alpha = g^{-1}(\alpha)\,.$$

We notice that $(\nabla_X \alpha)(Y)$ is the sum of two terms which are antisymmetric and symmetric, respectively, under the interchange $X \leftrightarrow Y$.

The Levi-Civita covariant derivative can be naturally extended to vector fields by the Leibnitz rule; that is, by

$$\langle \alpha, \nabla_X Y\rangle = L_X\langle \alpha, Y\rangle - \langle \nabla_X \alpha, Y\rangle\,.$$

From Eqs. (6.39), it easily follows that the Levi-Civita covariant derivative is $\mathcal{F}$-linear; that is, the following property holds:

$$\nabla_{fX} = f\nabla_X\,, \quad \forall f \in \mathcal{F}(\mathcal{M})\,.$$

It is worth recalling that the same property does not hold for the Lie derivative; i.e.

$$L_{fX} \neq f L_X\,,$$

unless when applied to functions.

Let us evaluate the Levi-Civita covariant derivative in a basis $\{e_i\}$, $\{\vartheta^i\}$. We have

$$X = X^i e_i\,, \quad Y = Y^i e_i\,, \quad \alpha = \alpha_i \vartheta^i\,, \quad g = g_{ij}\vartheta^i \otimes \vartheta^j\,,$$

so that

$$X_\alpha = g^{ij}\alpha_j e_i\,.$$

By repeating, in a generic basis in which $[e_i, e_j] = c^h_{ij} e_h$, the analogous calculations already done to obtain Eq. (6.34), we have

$$L_{X_\alpha} g = [X^k_\alpha (e_k(g_{ij}) + g_{lj} c^l_{ik} + g_{li} c^l_{jk}) + g_{kj} e_i(X^k_\alpha) + g_{ki} e_j(X^k_\alpha)] \vartheta^i \otimes \vartheta^j ,$$

and then

$$(L_{X_\alpha} g)(e_i, e_j) = X^k_\alpha (e_k(g_{ij}) + g_{lj} c^l_{ik} + g_{li} c^l_{jk}) + g_{kj} e_i(X^k_\alpha) + g_{ki} e_j(X^k_\alpha) .$$

On the other hand

$$X_\alpha = g^{kh} \alpha_h e_k ,$$

so that

$$X^k_\alpha = g^{kh} \alpha_h .$$

Therefore, we obtain

$$\begin{aligned}(L_{X_\alpha} g)(e_i, e_j) &= \alpha_h g^{hk} (e_k(g_{ij}) + g_{lj} c^l_{ik} + g_{li} c^l_{jk}) \\ &\quad + g_{kj} e_i(g^{hk} \alpha_h) + g_{ki} e_j(g^{hk} \alpha_h) \\ &= \alpha_h g^{hk} (e_k(g_{ij}) + g_{lj} c^l_{ik} + g_{li} c^l_{jk} - e_i(g_{kj}) - e_j(g_{ki})) \\ &\quad + e_i(\alpha_j) + e_j(\alpha_i) .\end{aligned}$$

Thus, we may write

$$\begin{aligned}(\nabla_X \alpha)(Y) &= X^i Y^j (\nabla_{e_i} \alpha)(e_j) \\ &= \frac{1}{2} X^i Y^j [(d\alpha)(e_i, e_j) + (L_{X_\alpha} g)(e_i, e_j)] \\ &= \frac{1}{2} X^i Y^j [e_i(\alpha_j) - e_j(\alpha_i) - \alpha_h c^h_{ij} + (L_{X_\alpha} g)(e_i, e_j)] \\ &= X^i Y^j e_i(\alpha_j) + \frac{1}{2} X^i Y^j \alpha_h g^{hk} (e_k(g_{ij}) - e_i(g_{kj}) - e_j(g_{ki})) \\ &\quad + \frac{1}{2} X^i Y^j \alpha_h g^{hk} (g_{lj} c^l_{ik} + g_{li} c^l_{jk} - g_{kl} c^l_{ij}) ,\end{aligned}$$

or, shortly

$$(\nabla_X \alpha)(Y) = X^i Y^j [e_i(\alpha_j) - \Gamma^h_{ij} \alpha_h] , \tag{6.40}$$

with

$$\Gamma^h_{ij} = \frac{1}{2} g^{hk}(e_j(g_{ki}) + e_i(g_{kj}) - e_k(g_{ij}) - g_{li}c^l_{jk} - g_{lj}c^l_{ik} + g_{kl}c^l_{ij}) .$$

The above quantities, which are also called the *Levi-Civita connection coefficients*, in a holonomic basis reduce to the *Christoffel symbols* given by Eq. (6.37). They are not the components of any tensor and have the property

$$\Gamma^h_{ij} - \Gamma^h_{ji} - c^h_{ij} = 0 . \qquad (6.41)$$

Exercise 6.3.1 *Show that, in a given basis*

$$\nabla_X Y = X^i[e_i(Y^j) - \Gamma^h_{ji} Y^j] e_h ,$$

so that

$$\nabla_{e^i} e_j = \Gamma^h_{ji} e_h .$$

The covariant derivative can be extended to any tensor field by the Leibnitz rule, so that

$$\nabla_X (S \otimes T) = (\nabla_X S) \otimes T + S \otimes \nabla_X T .$$

Exercise 6.3.2 *Show that the Levi-Civita covariant derivative of the metric tensor field vanishes; i.e.*

$$\nabla_X g = 0 .$$

Equation (6.40), for $\alpha = \vartheta^h$ and $X = e_i$, gives

$$\nabla_{e_i} \vartheta^h = -\Gamma^h_{ji} \vartheta^j ,$$

and can be taken as a starting point to define a more general *covariant derivative*, without use of any metrics, but we will not go on further on this subject. A purely algebraic formulation can be found in Ref. 125.

Equation (6.41) expresses, in a given basis, the vanishing of the $(1, 2)$-tensor field defined by

$$\mathcal{T}(\alpha, X, Y) = \langle \alpha, \nabla_X Y - \nabla_Y X - [X, Y] \rangle ,$$

which is called the *torsion* of the connection ∇.

Exercise 6.3.3 *Show that actually $\mathcal{T}$ is a tensor field and that, in a given basis,*

$$\mathcal{T}^h_{ij} \equiv \mathcal{T}(\vartheta^h, e_i, e_j) = \Gamma^h_{ij} - \Gamma^h_{ji} - c^h_{ij}\,.$$

Thus, the Levi-Civita covariant derivative has vanishing torsion and fulfills the property $\nabla_X g = 0$. It can be shown that, given a metrics on a manifold, the only torsionless connection for which $\nabla_X g = 0$ is the Levi-Civita connection.[43]

6.3.4 *The Riemann tensor field*

The Riemann* $(1,3)$-tensor field $\mathcal{R}$ is defined by

$$\mathcal{R}(\alpha, Z, X, Y) = \langle \alpha, R(X,Y)Z \rangle\,, \tag{6.42}$$

where the $R(X,Y)$ is the *curvature operator of* ∇, defined by

$$R(X,Y) = \nabla_X \nabla_Y - \nabla_Y \nabla_X - \nabla_{[X,Y]}\,.$$

Exercise 6.3.4 *Show that actually $\mathcal{R}$, defined by Eq. (6.42), is a tensor field; that is, it is $\mathcal{F}$-multilinear.*

Exercise 6.3.5 *Show that, in a given basis,*

$$\mathcal{R}^k_{hij} \equiv \mathcal{R}(\vartheta^k, e_i, e_j, e_h) = e_i(\Gamma^k_{hj}) - e_j(\Gamma^k_{hi}) + \Gamma^r_{hj}\Gamma^k_{ri} - \Gamma^r_{hi}\Gamma^k_{rj} - c^r_{ij}\Gamma^k_{hr}\,.$$

Exercise 6.3.6 *Show that the covariant derivative ∇ satisfies the Jacobi identity*

$$[\nabla_X, [\nabla_Y, \nabla_Z]] + [\nabla_Y, [\nabla_Z, \nabla_X]] + [\nabla_Z, [\nabla_X, \nabla_Y]] = 0\,.$$

*Georg Friedrik Bernhard Riemann was born in Breselenz on September 17, 1826 and died in Selasca on July 20, 1866. He studied at Göttingen under Gauss, and subsequently at Berlin under Jacobi, Dirichlet, Steiner and Eisenstein, all of whom were professors there at the same time. In spite of poverty and sickness he struggled to pursue his researches. Riemann was one of the most profound and brilliant mathematicians of his time. In 1857 he was made professor at Göttingen.

Show also that in a coordinates basis the above equation reduces to the so-called Bianchi† *identities*

$$\nabla_{e_l}\mathcal{R}^k_{hij} + \nabla_{e_i}\mathcal{R}^k_{hjl} + \nabla_{e_j}\mathcal{R}^k_{hli} = 0\,.$$

6.3.5 *The Ricci tensor and the scalar curvature*

The *Ricci*‡ *tensor* is the $(0,2)$-tensor field, which is defined in a coordinate basis by

$$\mathcal{R}_{ij} = \mathcal{R}^k_{ikj}\,,$$

and it is a symmetric tensor field

$$\mathcal{R}_{ij} = \mathcal{R}_{ji}\,.$$

The *scalar curvature* is defined by

$$\mathcal{R} = g^{ij}\mathcal{R}_{ij}\,.$$

Exercise 6.3.7 *Show that the Ricci tensor is a symmetric tensor field. Show also that the contracted Bianchi identities*

$$\nabla_{e_l}\mathcal{R}^k_{hkj} + \nabla_{e_l}\mathcal{R}^k_{hjk} + \nabla_{e_k}\mathcal{R}^k_{hlj} = 0$$

imply

$$\nabla_{e_j}\left(\mathcal{R}^{ij} - \frac{1}{2}\mathcal{R}g^{ij}\right) = 0\,.$$

---

†Luigi Bianchi was born in Parma in 1856 and died in Pisa in 1928. He has been a student of E. Betti and U. Dini in Pisa, and of F. Klein in Göttingen. He was a professor at the University and Scuola Normale Superiore in Pisa. He has been one of the most important Italian mathematicians in the last century. His works fill up more than 10 volumes and concern mainly differential geometry and number theory. Both the original papers and the, now classical, books (on differential geometry, transformations groups, elliptic function) are written in a very transparent and elegant form. Bianchi was strongly loved by his students not only for the marked *vis comica* (funny spirit), which was one of his characteristic features.

‡Gregorio Ricci–Curbastro was born in Lugo (Ravenna) in 1853 and died in Bologna in 1925. He studied at Rome, Bologna and Pisa Universities where he obtained his degree in 1875. He has been a student of F. Klein, in Göttingen, and teaching assistant of U. Dini in Pisa. He has been a professor of mathematical physics at Padova University from the year 1880. The main scientific contribution of Ricci has been the invention (together with its student Levi-Civita) of the absolute differential calculus, later an essential tool for the formulation of general relativity.

The previous tensor fields play a basic role in Einstein[§] general relativity in which space-time is represented as a 4-dimensional manifold with a metric tensor field g representing the gravitational field. The empty-space gravitational field g is found by solving Einstein's field equations

$$\mathcal{R}^{ij} - \frac{1}{2}\mathcal{R}g^{ij} = 0 .$$

6.4 Endomorphisms Associated with a Mixed Tensor Field

Let T be a *mixed tensor field*; that is, a tensor field of type $(1,1)$, on the n-dimensional manifold $\mathcal{M}$. In a coordinate basis it can be written as

$$T = \sum_{i,j=1}^{n} T_i^j dx^i \otimes \frac{\partial}{\partial x^j} . \tag{6.43}$$

Let $\mathcal{T}_p\mathcal{M}$ be the tangent space to the manifold at the point p and $\mathcal{T}_p^*\mathcal{M}$ its dual.

The tensor product $\mathcal{T}_p^*\mathcal{M} \otimes \mathcal{T}_p\mathcal{M}$ is isomorphic to $\mathrm{Lin}(\mathcal{T}_p\mathcal{M}, \mathcal{T}_p\mathcal{M})$, the vector space of the linear operators on $\mathcal{T}_p\mathcal{M}$, *via* the *canonical isomorphism* given by

$$\mathcal{I} : \alpha \otimes X \in \mathcal{T}_p^*\mathcal{M} \otimes \mathcal{T}_p\mathcal{M} \to \mathcal{I}(\alpha \otimes X) = L_{\alpha\otimes X} \in \mathrm{Lin}(\mathcal{T}_p\mathcal{M}, \mathcal{T}_p\mathcal{M}) ,$$

where $L_{\alpha\otimes X}$ is the linear map

$$L_{\alpha\otimes X} : Y \in \mathcal{T}_p\mathcal{M} \to L_{\alpha\otimes X}(Y) = \alpha(Y)X \in \mathcal{T}_p\mathcal{M} .$$

[§]Albert Einstein was born in Ulm, Germany in 1879 and died at Princeton in 1955. He spent his youth in Munich and after a period past in Milan, he moved, in 1896, to Swiss. He obtained his Ph.D. at Zurich Polytechnic in 1905. After working at Swiss Patent Office, he was appointed associate professor at Zurich University in 1909 and then at Berlin University in 1913. He wrote in 1905 six papers. The first of them, with the introduction of the *photon*, gave a basic contribution to the rise of quantum theory; the second and the third gave rise to special relativity and, with this, to the new concept of space-time; the others explained the Brownian motion and, with this, introduced new methods to measure the dimensions of atoms. His paper on general relativity is dated 1915. He was appointed a Nobel Prize, for the photoelectric effect, in 1921. In 1933, with the introduction of racial Nazis laws, he moved to Princeton where he was appointed to a chair of professor of physics at *The Institute for Advanced Studies*, where he taught until 1955.

Then, with the tensor (6.43), we can associate the endomorphism $\hat{T}$ on $\mathcal{T}_p\mathcal{M}$

$$\hat{T} : X \in \mathcal{T}_p\mathcal{M} \to \hat{T}X \in \mathcal{T}_p\mathcal{M},$$

defined as follows

$$\hat{T}X = \sum_{i,j=1}^{n} T_i^j X^i \frac{\partial}{\partial x^j}, \quad X = \sum_{k=1}^{n} X^k \frac{\partial}{\partial x^k}. \tag{6.44}$$

Moreover, there is another isomorphism between the tensor product $\mathcal{T}_p^*\mathcal{M} \otimes \mathcal{T}_p\mathcal{M}$ and the space $\mathrm{Lin}(\mathcal{T}_p\mathcal{M}, \mathcal{T}_p\mathcal{M})$ that associates to every $\alpha \otimes X \in \mathcal{T}_p^*\mathcal{M} \otimes \mathcal{T}_p\mathcal{M}$ the linear map

$$L^*_{\alpha\otimes X} : \beta \in \mathcal{T}_p^*\mathcal{M} \to L^*_{\alpha\otimes X}(\beta) = \beta(p)\alpha \in \mathcal{T}_p^*\mathcal{M}.$$

Thus, with the tensor (6.43), we can associate also the endomorphism $\check{T}$ on $\mathcal{T}_p^*\mathcal{M}$

$$\check{T} : \alpha \in \mathcal{T}_p^*\mathcal{M} \to \check{T}\alpha \in \mathcal{T}_p^*\mathcal{M},$$

defined by

$$\check{T}\alpha = \sum_{i,j=1}^{n} \alpha_j T_i^j dx^i, \quad \alpha = \sum_{k=1}^{n} \alpha_k dx^k. \tag{6.45}$$

By using Eqs. (6.44) and (6.45), we find that $\langle \check{T}\alpha, X\rangle = \langle \alpha, \hat{T}X\rangle$.
In fact,

$$\langle \alpha, \hat{T}X\rangle = \alpha(\hat{T}X) = \alpha\left(\sum_{i,j=1}^{n} T_i^j X^i \frac{\partial}{\partial x^j}\right) = \sum_{i,j=1}^{n} \alpha_j T_i^j X^i,$$

$$\langle \check{T}\alpha, X\rangle = (\check{T}\alpha)(X) = \sum_{i,j=1}^{n} \alpha_j T_i^j dx^i(X) = \sum_{i,j=1}^{n} \alpha_j T_i^j X^i.$$

Henceforth, when no confusion possibly arises, we will not distinguish between a tensor T and its associated endomorphisms $\hat{T}$ and $\check{T}$.

6.4.1 *The Nijenhuis bracket of two mixed tensor fields*

If S and T are the endomorphisms associated with two tensor fields of (1.1)-type, and if X, Y are two arbitrary vector fields, the relation

$$2\mathcal{H}_T^S(X,Y) = [SX,TY] + [TX,SY] + ST[X,Y] + TS[X,Y] \\ - S[X,TY] - S[TX,Y] - T[X,SY] - T[SX,Y]$$

is called the *Nijenhuis bracket of S and T*. It defines a vector field $\mathcal{H}_T^S(X,Y)$ which is antisymmetric under the interchange $X \leftrightarrow Y$.

The $(1,2)$-type tensor field defined by

$$\mathcal{N}_T^S(\alpha, X, Y) = \langle \alpha, \mathcal{H}_T^S(X,Y)\rangle$$

is called the *Nijenhuis torsion of S and T*.

Let us observe that $\mathcal{H}_T^S(X,Y)$ can also be written in the form

$$\mathcal{H}_T^S(X,Y) = \frac{1}{2}(L_{SX}T + L_{TX}S - SL_XT - TL_XS)Y\,.$$

The *Nijenhuis torsion $\mathcal{N}_T$ of* a mixed tensor field T with itself is called the *Nijenhuis torsion of T*; it, generally, is not vanishing.

In such a case, the previous relations become

$$\mathcal{N}_T(\alpha, X, Y) = \langle \alpha, \mathcal{H}_T(X,Y)\rangle$$

with

$$\mathcal{H}_T(X,Y) = (L_{TX}T - TL_XT)Y\,.$$

Exercise 6.4.1 *Show that*

- $\mathcal{N}_{S+T} = \mathcal{N}_S + 2\mathcal{N}_T^S + \mathcal{N}_T$
- $\mathcal{N}_T = 0 \Rightarrow \mathcal{N}_{T^2} = 0$.

Exercise 6.4.2 *Let us suppose that the tensor field* (6.43) *has a vanishing Nijenhuis torsion. Thus, it satisfies the condition*

$$(L_{\hat{T}X}T)^\wedge Y = \hat{T}(L_XT)^\wedge Y\,. \tag{6.46}$$

From this relation, it follows that if T is invariant for a vector field Δ, it is invariant for all vector fields $\hat{T}^n\Delta$, generated by repeated application of $\hat{T}$ to Δ. Show that $[T^n\Delta, T^m\Delta] = 0\ \forall n, m$.

The Lie derivative, with respect to Δ, of the tensor (6.43) is given by (see Eq. (6.8))

$$L_\Delta T = \left(\frac{\partial T_i^j}{\partial x^k}\Delta^k + T_k^j\frac{\partial \Delta^k}{\partial x^i} - T_i^k\frac{\partial \Delta^j}{\partial x^k}\right) dx^i \otimes \frac{\partial}{\partial x^j}\,.$$

In local coordinates on $\mathcal{M}$, we have

$$(L_{\hat{T}X}T)^\wedge Y = \left(\frac{\partial T_i^j}{\partial x^k}T_l^k X^l Y^i - \frac{\partial T_i^j}{\partial x^k}T_l^k Y^l X^i + T_i^j\frac{\partial T_l^i}{\partial x^k}Y^k X^l \right.$$
$$\left. + T_k^j T_l^k\frac{\partial X^l}{\partial x^i}Y^i - T_l^j T_i^k\frac{\partial X^l}{\partial x^k}Y^i\right)\frac{\partial}{\partial x^j}$$

and

$$\hat{T}(L_X T)^\wedge Y = \left(T_i^j\frac{\partial T_l^i}{\partial x^k}X^k Y^l + T_k^j T_l^k\frac{\partial X^l}{\partial x^i}Y^i - T_l^j T_i^k\frac{\partial X^l}{\partial x^k}Y^i\right)\frac{\partial}{\partial x^j}\,.$$

Then, the relation (6.46) becomes

$$\frac{\partial T_i^j}{\partial x^k}T_l^k X^l Y^i - \frac{\partial T_i^j}{\partial x^k}T_l^k Y^l X^i = T_i^j\frac{\partial T_l^i}{\partial x^k}X^k Y^l - T_i^j\frac{\partial T_l^i}{\partial x^k}Y^k X^l\,. \qquad (6.47)$$

If $\gamma : \tau \in \Re \to \mathcal{M}$ is a curve on $\mathcal{M}$ such that $\gamma(0) = p$, and

$$\left(\frac{d\gamma(\tau)}{d\tau}\right)_{\tau=0} = X\,,$$

we have

$$\hat{T}'(p)(X,Y) \equiv \left.\frac{d}{d\tau}\hat{T}(\gamma(\tau))Y\right|_{\tau=0}$$
$$= \frac{\partial T_l^j}{\partial x^k}X^k Y^l\frac{\partial}{\partial x^j}\,.$$

Thus, the relation (6.47) can also be written as

$$\hat{T}'(p)(\hat{T}X,Y) - \hat{T}'(p)(\hat{T}Y,X) = \hat{T}[\hat{T}'(p)(X,Y) - \hat{T}'(p)(Y,X)]\,. \qquad (6.48)$$

Chapter 7

Integration Theory

7.1 Orientable Manifolds

A differential manifold $\mathcal{M}$ is said to be *orientable* if a nowhere vanishing continuous differential n-form Ω exists on it. Such a differential n-form is said to be a *volume n-form.*

At each point $p \in \mathcal{M}$, the n-form Ω will define an n-covector, $\Omega_p \in \Lambda^n(\mathcal{T}_p\mathcal{M})$, whose value $\Omega_p(e_1, e_2, \ldots, e_n)$ on a basis $\{e_1, e_2, \ldots, e_n\}$ in $\mathcal{T}_p\mathcal{M}$ will be different from zero. So all the basis in $\mathcal{T}_p\mathcal{M}$ will be divided in two classes according to the sign of $\Omega_p(e_1, e_2, \ldots, e_n)$. The two classes are independent from Ω, because every nowhere vanishing n-form Ω' will be proportional to Ω by a factor $f \neq 0$, and then it will take the same sign (depending on the sign of f) on the elements of each class. The two classes will be called *left-handed* and *right-handed.** Thus, it will be possible to choose, continuously $\forall p \in \mathcal{M}$, a basis $\{e_1, e_2, \ldots, e_n\}_p$ belonging to the same class. If the basis are coordinate basis, the Jacobian determinant in the transition from one basis to another will have, in the neighborhood of each point $p \in \mathcal{M}$, the same sign. The Möbius band is a not an orientable manifold.

---

*Which class has which name is a convention, since the sign of Ω' will depend on f, which is at our disposal.

7.2 Integration on Orientable Manifolds

Let $\mathcal{M}$ be an n-dimensional orientable manifold and ω be a differential n-form on it. In a given chart $(\mathcal{U}, \varphi)$, ω will be locally represented as

$$\omega = f(x^1, \ldots, x^n) dx^1 \wedge \cdots \wedge dx^n .$$

The integral of ω over $\mathcal{U} \subseteq \mathcal{M}$ is defined by

$$\int_{\mathcal{U}} \omega = \int_{\varphi(\mathcal{U})} f(x^1, \ldots, x^n) dx^1 \cdots dx^n , \tag{7.1}$$

where $\varphi(\mathcal{U}) \subseteq \Re^n$ is the *image of* $\mathcal{U}$ and the symbol $dx^1 \cdots dx^n$ denotes the measure for the usual integral of differential calculus.

In order to show that the integral so defined does not depend on the coordinates, let us choose a different coordinate basis $\{\partial/\partial y^i\}$ in which

$$\omega = \tilde{f}(y^1, \ldots, y^n) dy^1 \cdots dy^n .$$

The equality

$$\tilde{f}(y) dy^1 \wedge \cdots \wedge dy^n = f(x) dx^1 \wedge \cdots \wedge dx^n ,$$

evaluated on the basis $\{\partial/\partial y^i\}$

$$\tilde{f}(y)(dy^1 \wedge \cdots \wedge dy^n) \left(\frac{\partial}{\partial y^1}, \ldots, \frac{\partial}{\partial y^n} \right) = f(x)(dx^1 \wedge \cdots \wedge dx^n) \left(\frac{\partial}{\partial y^1}, \ldots, \frac{\partial}{\partial y^n} \right) ,$$

gives

$$\tilde{f}(y) \det \begin{pmatrix} \frac{\partial y^1}{\partial y^1} & \vdots & \frac{\partial y^1}{\partial y^n} \\ \cdot\cdot & \cdot & \cdot\cdot \\ \frac{\partial y^n}{\partial y^1} & \vdots & \frac{\partial y^n}{\partial y^n} \end{pmatrix} = f(x(y)) \det \begin{pmatrix} \frac{\partial x^1}{\partial y^1} & \vdots & \frac{\partial x^1}{\partial y^n} \\ \cdot\cdot & \cdot & \cdot\cdot \\ \frac{\partial x^n}{\partial y^1} & \vdots & \frac{\partial x^n}{\partial y^n} \end{pmatrix} ,$$

or

$$\tilde{f}(y) = J f(x(y)) ,$$

where J is the Jacobian determinant.

Thus, we have

$$\int_{\mathcal{U}} \omega = \int_{\psi(\mathcal{U})} J f(y^1, \ldots, y^n) dy^1 \cdots dy^n . \tag{7.2}$$

Then, our definition of the integral of ω will not depend on the coordinates if

$$\int_{\varphi(\mathcal{U})} f(x^1,\ldots,x^n)dx^1\cdots dx^n = \int_{\psi(\mathcal{U})} Jf(y^1,\ldots,y^n)dy^1\cdots dy^n\,.$$

As we know from differential calculus, the above equality holds only when $J > 0$.

It follows that in the definition (7.1), an orientation for $\mathcal{U}$ must be chosen; that is, a requirement on the handedness of the basis must be added. This explains why, from the very beginning, $\mathcal{M}$ has been supposed to be an orientable manifold; that is, one for which it is possible to choose, continuously at every point $p \in \mathcal{M}$, a coordinate basis $\{\partial/\partial x^i_p\}$ with the same handedness.

However, the integration theory of differential forms has been extended, by de Rham, to nonorientable manifolds[4] by introducing forms of *odd parity*, and this can have interesting physical applications.[173] On the historical side we shall mention that they were introduced by Hermann Weyl[56] and developed by Schouten,[48] and called *Weyl tensors.* Synge and Schild refer to them as *oriented tensors,* while de Rham called them *tensors of odd kind.*[7] Under a change of coordinates ($x \leftrightarrow x'$), a *twisted differential form* transforms as follows

$$\omega'_{ab\cdots c} = \frac{J}{|J|}\frac{\partial x^p}{\partial x'^a}\frac{\partial x^q}{\partial x'^b}\cdots\frac{\partial x^r}{\partial x'^c}\omega_{pq\cdots r}\,,$$

where J is the Jacobian determinant and $|J|$ its absolute value.

7.3 *p*-Vectors and Dual Tensors

Completely antisymmetric tensors of type $(p, 0)$, on a n-dimensional vector space E, are called p-vectors. A Grassmann algebra can be, of course, constructed on them in complete analogy with that of p-covectors. The vector space of p-vectors is denoted with $\mathrm{V}^p(E)$. Its dimension is

$$\dim \mathrm{V}^p(E) = \binom{n}{p} = \binom{n}{n-p} = \dim \mathrm{V}^{n-p}(E)\,,$$

and a basis is given by

$$\underbrace{e_a \wedge e_b \wedge \cdots \wedge e_c}_{p\ \text{times}}\,.$$

Thus,

$$\dim \mathrm{V}^p(E) = \dim \mathrm{V}^{n-p}(E) = \dim \Lambda^p(E) = \dim \Lambda^p(E)\,.$$

If $\{\vartheta^i\}$ is the dual basis of $\{e^i\}$, the n-covector

$$\Omega = \vartheta^1 \wedge \vartheta^2 \wedge \cdots \wedge \vartheta^n = \frac{1}{n!}\varepsilon_{i_1 i_2 \cdots i_n}\vartheta^{i_1} \wedge \vartheta^{i_2} \wedge \cdots \wedge \vartheta^{i_n}$$

is a basis for the vector space Λ^n, and will be called a *volume covector.*

By using the volume covector, we can associate, with any p-vector

$$X = \frac{1}{p!}X^{ab\cdots c}e_a \wedge e_b \wedge \cdots \wedge e_c\,,$$

the $(n-p)$-covector defined by

$$\Omega(X) \equiv i_X\Omega = \frac{1}{p!}X^{i_1 i_2 \cdots i_p}\varepsilon_{i_1 i_2 \cdots i_n}\vartheta^{i_{p+1}} \wedge \vartheta^{i_{p+2}} \wedge \cdots \wedge \vartheta^{i_n}\,.$$

The above $(n-p)$-covector is called the *Ω-dual of X* or also the *Poincaré dual of X.*

A basis independent definition is given by

$$i_X\Omega(Y^{p+1},\ldots,Y^n) = \Omega(X \wedge Y^{p+1} \wedge \cdots \wedge Y^n)\,, \quad \forall Y^{p+1},\ldots,Y^n \in E\,.$$

It is also possible to make the inverse; that is, to associate, with any p-covector α, the $(n-p)$-vector

$$\Omega^{-1}(\alpha) = \frac{1}{p!}\varepsilon^{i_1 i_2 \cdots i_n}\alpha_{i_1 i_2 \cdots i_p}e_{i_{p+1}} \wedge \cdots \wedge e_{i_n}\,,$$

where

$$\varepsilon^{i_1 i_2 \cdots i_n}\varepsilon_{i_1 i_2 \cdots i_n} = n!\,. \tag{7.3}$$

Let us now calculate the "dual of the dual" of a given p-vector X.

We have

$$\begin{aligned}\Omega^{-1}(\Omega(X)) &= \Omega^{-1}\left(\frac{1}{p!}X^{i_1 i_2 \cdots i_p}\varepsilon_{i_1 i_2 \cdots i_n}\vartheta^{i_{p+1}} \wedge \vartheta^{i_{p+2}} \wedge \cdots \wedge \vartheta^{i_n}\right)\\ &= \frac{1}{p!}X^{i_1 i_2 \cdots i_p}\varepsilon_{i_1 i_2 \cdots i_n}\Omega^{-1}(\vartheta^{i_{p+1}} \wedge \vartheta^{i_{p+2}} \wedge \cdots \wedge \vartheta^{i_n})\\ &= \frac{1}{p!(n-p)!}X^{i_1 i_2 \cdots i_p}\varepsilon_{i_1 i_2 \cdots i_n}\varepsilon^{i_{p+1} \cdots i_n j_1 j_2 \cdots j_p}e_{j_1} \wedge \cdots \wedge e_{j_p}\end{aligned}$$

$$= \frac{(-1)^{p(n-p)}}{p!(n-p)!} X^{i_1 i_2 \cdots i_p} \varepsilon_{i_1 i_2 \cdots i_n} \varepsilon^{j_1 j_2 \cdots j_p i_{p+1} \cdots i_n} e_{j_1} \wedge \cdots \wedge e_{j_p} .$$

Then,

$$\begin{aligned}
[\Omega^{-1}(\Omega(X))]^{12\cdots p} &= \frac{(-1)^{p(n-p)}}{p!(n-p)!} X^{i_1 i_2 \cdots i_p} \varepsilon_{i_1 i_2 \cdots i_n} \varepsilon^{1 \cdots p i_{p+1} \cdots i_n} \\
&= \frac{(-1)^{p(n-p)}}{p!} X^{i_1 i_2 \cdots i_p} \varepsilon_{i_1 \cdots i_p (p+1) \cdots n} \varepsilon^{1 \cdots p (p+1) \cdots n} \\
&= (-1)^{p(n-p)} X^{12\cdots p} .
\end{aligned}$$

Let us next calculate the "dual of the dual" of a given p-covector α. We have

$$\begin{aligned}
\Omega(\Omega^{-1}(\alpha)) &= \frac{1}{(n-p)!} \frac{1}{p!} \varepsilon^{i_1 i_2 \cdots i_n} \alpha_{i_1 i_2 \cdots i_p} \Omega(e_{i_{p+1}} \wedge \cdots \wedge e_{i_n}) \\
&= \frac{1}{p!(n-p)!} \varepsilon^{i_1 i_2 \cdots i_n} \alpha_{i_1 i_2 \cdots i_p} \varepsilon_{i_{p+1} \cdots i_n j_1 \cdots j_p} \vartheta^{j_1} \wedge \cdots \wedge \vartheta^{i_p} \\
&= \frac{(-1)^{p(n-p)}}{p!(n-p)!} \varepsilon^{i_{p+1} \cdots i_n i_1 i_2 \cdots i_p} \varepsilon_{i_{p+1} \cdots i_n j_1 \cdots j_p} \alpha_{i_1 i_2 \cdots i_p} \vartheta^{j_1} \wedge \cdots \wedge \vartheta^{i_p} .
\end{aligned}$$

Summing up, we obtain

$$\begin{aligned}
\Omega^{-1}(i_X \Omega) &= (-1)^{p(n-p)} X \quad \forall X \in V^p(E) \\
\Omega(\Omega^{-1}(\alpha)) &= (-1)^{p(n-p)} \alpha \quad \forall \alpha \in \Lambda^p(E) .
\end{aligned}$$

Thus, the volume covector Ω and its inverse Ω^{-1} provide the following mappings:

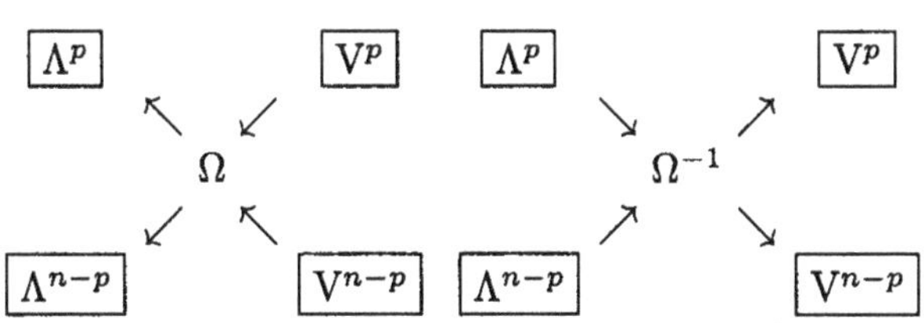

7.4 Metric $\circ$ Volume = Hodge Duality

In Sec. 6.3, it has been shown that a metric tensor g over a n-dimensional vector space provides an isomorphism between vectors and covectors. It is easy to see that it also provides an isomorphism between p-vectors and p-covectors, since

$$\alpha_{ij\cdots k} = g_{il}g_{jm}\cdots g_{kn}X^{lm\cdots n}$$

is completely antisymmetric in the indices $i, j, \ldots, k$.

Then, a metric tensor g allows us to complete the previous picture in the following way:

$$\begin{array}{ccccc} \boxed{\Lambda^p} & & \leftarrow g \leftarrow & & \boxed{V^p} \\ & \nwarrow & & \swarrow & \\ & & \Omega & & \\ & \swarrow & & \nwarrow & \\ \boxed{\Lambda^{n-p}} & & \leftarrow g \leftarrow & & \boxed{V^{n-p}} \end{array} \qquad \begin{array}{ccccc} \boxed{\Lambda^p} & & \leftarrow g \leftarrow & & \boxed{V^p} \\ & \searrow & & \nearrow & \\ & & \Omega^{-1} & & \\ & \nearrow & & \searrow & \\ \boxed{\Lambda^{n-p}} & & \leftarrow g \leftarrow & & \boxed{V^{n-p}} \end{array}$$

The composite map $* \equiv g \circ \Omega^{-1}$

$$* : \Lambda^p \longrightarrow \Lambda^{n-p} ,$$

which provides an isomorphism between p-covectors and $(n-p)$-covectors, is called the *Hodge dual.*

Its transposed $\Omega^{-1} \circ g$ provides an isomorphism between p-vectors and $(n-p)$-vectors, and is denoted by the same symbol.

An example of this isomorphism is given by the so-called *vector product* of two vectors in the 3-dimensional Euclidean space $(\Re^3, g_{ij} = \delta_{ij})$:

- Consider two vectors U, V in $\Re^3$;
- Take the associated covectors $u = g(U,\cdot)$, $v = g(V,\cdot)$ *via* the Euclidean metrics;
- Consider the 2-covector given by their exterior product: $u \wedge v$;
- The volume dual of $u \wedge v$ is a vector which is called the *vector product of* U, V.

Summing up:

$$U \wedge V = \Omega^{-1}(g(U,\cdot) \wedge g(V,\cdot)) .$$

Remark 14 *If a manifold has a metric g, let $\{\vartheta^i\}$ be an orthonormal basis for differential 1-forms, and Ω be the volume-form*

$$\Omega = \vartheta^1 \wedge \vartheta^2 \wedge \cdots \wedge \vartheta^n .$$

If $\{x^k\}$ is an arbitrary coordinate, and Λ is the transformation matrix from $\{dx^k\}$ to $\{\vartheta^i\}$; i.e.

$$\vartheta^i = \Lambda^i_k dx^k ,$$

we have

$$\begin{aligned}\Omega &= \Lambda^1_i \cdot \Lambda^2_j \cdots \Lambda^n_k dx^i \wedge dx^j \wedge \cdots \wedge dx^k \\ &= \Lambda^1_i \cdot \Lambda^2_j \cdots \Lambda^n_k \varepsilon^{ij\cdots k} dx^1 \wedge dx^2 \wedge \cdots \wedge dx^n \\ &= (\det \Lambda) dx^1 \wedge dx^2 \wedge \cdots \wedge dx^n .\end{aligned}$$

On the other hand, we also have

$$g_{ij} = \Lambda^h_i \Lambda^k_j g(e_h, e_k) ,$$

where g_{ij} are the components of the metric tensor g in the coordinate basis $\{\partial/\partial x^i\}$, and $g(e_h, e_k) = \delta_{hk}$, since the original basis was orthonormal.

Therefore,

$$\mathsf{g} \equiv \det(g_{ij}) = (\det \Lambda)^2$$

and

$$\Omega = \sqrt{|\mathsf{g}|} dx^1 \wedge dx^2 \wedge \cdots \wedge dx^n .$$

From Eq. (7.3) it follows that the components $\Omega^{ij\cdots k}$ of Ω^{-1} are given by

$$\Omega^{12\cdots n} = \frac{1}{\Omega_{12\cdots n}} = \frac{1}{\sqrt{|\mathsf{g}|}} .$$

However, in our metric manifold the inverse of Ω could also be defined by

$$\Omega'^{ij\cdots k} = g^{ip} g^{jq} \cdots g^{kr} \Omega_{pq\cdots r} ,$$

so that

$$\Omega'^{12\cdots n} = \frac{\sqrt{|\mathsf{g}|}}{\mathsf{g}} .$$

If g *is negative,* $\Omega^{12\cdots n}$ *and* $\Omega'^{12\cdots n}$ *differ by a sign. In special or general relativity, it is conventional to use* $\Omega'^{12\cdots n}$ *in the dual relations.*

7.5 Stokes Theorem

Let $\mathcal{M}$ be an orientable n-dimensional differential manifold and U be a region of $\mathcal{M}$. We call *boundary* of U an orientable $n-1$ dimensional submanifold of $\mathcal{M}$, namely ∂U, which divides $\mathcal{M} - \partial U$ in two disjoint parts, U and CU, in such a way that any continuous curve joining a point of U with a point in CU contains a point of ∂U. Let us consider the integral of an arbitrary n-form ω over U

$$\int_U \omega \,.$$

Let X be a vector field and $U(\tau)$ the images of U under the flow φ generated by X:

$$U(\tau) = \varphi_\tau(U)\,.$$

From the analysis already performed in Sec. 4.1, it follows that

$$L_X \int_{U(\tau)} \omega = \int_{U(\tau)} L_X \omega \,,$$

or

$$L_X \int_{U(\tau)} \omega = \int_{U(\tau)} d i_X \omega \,, \tag{7.4}$$

since the form ω is obviously closed.

On the other hand, by applying directly the definition of Lie derivative, we obtain

$$\begin{aligned} L_X \int_{U(\tau)} \omega &= \lim_{\tau \to 0} \frac{1}{\tau} \left[\int_{U(\tau)} \omega - \int_{U(0)} \omega \right] \\ &= \lim_{\tau \to 0} \frac{1}{\tau} \int_{\delta U(\tau)} \omega \end{aligned}$$

with $U(0) = U$ and $\delta U(\tau) = U(\tau) - U(0)$.

Let us consider a part $\partial V(0) \subseteq \partial U(0)$, to which the vector field X is not tangent, and a part $\delta V(\tau)$ of $\delta U(\tau)$, representing a region between $\partial U(0)$ and $\partial U(\tau)$ locally given by

$$\delta V(\tau) = \partial V(0) \times]0, \tau[.$$

Then if $(x_2, x_3, \dots, x_n)$ denote the coordinates for $\partial V(0)$, we can introduce, with $x_1 = \tau$, coordinates $(x_1, x_2, \dots, x_n)$ for $\delta V(\tau)$. In these coordinates the differential n-form ω can be expressed as follows:

$$\omega = f(x_1, x_2, \dots, x_n) dx_1 \wedge dx_2 \wedge \cdots \wedge dx_n$$

and

$$\begin{aligned}
\lim_{\tau \to 0} \frac{1}{\tau} \int_{\delta V(\tau)} \omega &= \lim_{\tau \to 0} \frac{1}{\tau} \int_{\partial V(0) \times]0,\tau[} \omega \\
&= \lim_{\tau \to 0} \frac{1}{\tau} \int_{\partial V(0) \times]0,\tau[} f\, dx_1 \wedge dx_2 \wedge \cdots \wedge dx_n \\
&= \lim_{\tau \to 0} \frac{1}{\tau} \int_{\partial V(0)} \left[\int_0^\tau f dx_1 \right] dx_2 \wedge \cdots \wedge dx_n \\
&= \lim_{\tau \to 0} \frac{1}{\tau} \int_{\partial V(0)} [\tau f(0, x_2, \dots, x_n)] dx_2 \wedge \cdots \wedge dx_n \\
&= \int_{\partial V(0)} f|_{\partial U(0)} dx_2 \wedge \cdots \wedge dx_n \\
&= \int_{\partial V(0)} (i_X \omega)|_{\partial U(0)} ,
\end{aligned}$$

where the symbol $|_{\partial U}$ denotes the restriction to ∂U.

The final equation

$$\lim_{\tau \to 0} \frac{1}{\tau} \int_{\delta V(\tau)} \omega = \int_{\partial V(0)} (i_X \omega)|_{\partial U(0)} \tag{7.5}$$

is independent from the constructed coordinates, but it requires that X should not be tangent to ∂U in V. For the whole boundary ∂U, two cases can occur:

- X is tangent to isolated points forming a submanifold of lower dimensionality.

In such a case these points do not contribute to the integral.

- X is tangent to ∂U in an open region of it.
  In this case both sides of the Eq. (7.5) are vanishing and the equation still holds.

So, summing over all parts ∂V of ∂U, we obtain

$$\lim_{\tau \to 0} \frac{1}{\tau} \int_{\delta U(\tau)} \omega = \int_{\partial U(0)} (i_X \omega)|_{\partial U(0)} ,$$

from which

$$L_X \int_U \omega = \int_{\partial U} (i_X \omega)|_{\partial U} . \tag{7.6}$$

By comparing the two expressions, Eqs. (7.4) and (7.6), of the Lie derivative of the integral $\int_U \omega$, we finally obtain

$$\int_U d\, i_X \omega = \int_{\partial U} (i_X \omega)|_{\partial U} ,$$

and since $i_X \omega$, as well as ω, is arbitrary, we conclude that

Theorem 24 (Stokes) *For any $(n-1)$-form α over an n-dimensional manifold U, the following relation*

$$\int_U d\alpha = \int_{\partial U} \alpha|_{\partial U} \tag{7.7}$$

holds.

If $U = [a, b]$ is an interval of the real line, and $f : U \to \Re$ a differentiable function, then the Stokes[†] theorem reduces to

$$\int_a^b f' dx = f(b) - f(a) ,$$

since $\partial U = \{a, b\}$.

[†]George Gabriel Stokes was born in Skreen (Ireland) in 1819 and died in Cambridge in 1903. Physicist and mathematician, he has been a professor of mathematics at Cambridge University and is universally known for the results on the transformations of integrals, on the liquid waves and for his theories on optics, founded on the hypothesis of dragging ether.

7.6 Gradient, Curl and Divergence

On an n-dimensional orientable manifold $\mathcal{M}$, endowed with a metric tensor field g, all properties concerning the volume duality and the Hodge duality can be point-wise carried over directly. This allows us to better understand the meaning of some familiar concepts in $\Re^3$, such as the *gradient*, the *curl* and the *divergence*.

- *The gradient*
  Consider a function f and take its exterior derivative df. The vector field associated to the differential form df, by means of the inverse of the Euclidean metric tensor η, is called the *gradient* of f:

$$\nabla f = \eta^{-1}(df, \cdot) \Longleftrightarrow i_{\nabla f}\eta = df\,.$$

- *The curl*
  Consider a vector field U, take the associated differential form $\alpha = \eta(U, \cdot)$ and its exterior derivative $d\alpha$. The volume-dual of $d\alpha$ is a vector which is called the *curl* of U:

$$\nabla \times U = \Omega^{-1}(d\alpha)\,,$$

  with $\Omega = dx \wedge dy \wedge dz$.
- *The divergence*
  Consider a vector field V, take the associated differential 2-form *via* the volume form $\Omega = dx \wedge dy \wedge dz$. Its exterior derivative is proportional to Ω up to a multiplicative function called the *divergence* of V:

$$(\nabla \cdot V)\Omega = di_V\Omega\,.$$

  Moreover, if $V = \nabla \times U = \Omega^{-1}(d\alpha)$, then

$$\begin{aligned}\nabla \cdot \nabla \times U &= di_{\nabla \times U}\Omega = di_{\Omega^{-1}(d\alpha)}\Omega = d(\Omega(\Omega^{-1}(d\alpha)))\\ &= d(-1)^{3-1}d\alpha = d^2\alpha = 0\,.\end{aligned}$$

Exercise 7.6.1. *Use Stokes' theorem to prove that, for every exact differential 2-form ω on the sphere S^2,*

$$\int_{S^2} \omega = 0\,.$$

Proof.

In order for ω to be exact, a differential 1-form α have to exist such that $\omega = d\alpha$. In this case, Stokes' theorem gives

$$\int_{S^2} \omega = \int_{S^2} d\alpha = \int_{\partial S^2} \alpha = 0\,,$$

since S^2 has no boundary.

Exercise 7.6.2. *Use Stokes' theorem to show that for the differential 2-form $\omega = x^1 dx^2 \wedge dx^3$ on $\Re^3$,*

$$\int_{S^2} \omega|_{S^2} = \frac{4}{3}\pi\,,$$

where S^2 is the unit sphere considered as a submanifold of $\Re^3$.

Answer.

The differential 3-form

$$d\omega = dx^1 \wedge dx^2 \wedge dx^3$$

is the usual volume form, so that when integrated on the volume V of the sphere, gives

$$\int_V \omega = \frac{4}{3}\pi\,.$$

The result then follows by Stokes' theorem.

The above exercise gives an example of a closed differential 2-form on S^2 which is not exact, since it does not satisfy the criterion of the first exercise.

7.7 A Primer for Cohomology

Let $Z^p(\mathcal{M})$ and $B^p(\mathcal{M})$ be the set of all closed differential p-forms and the set of all exact differential p-forms, respectively. Both sets have a natural structure of vector space and, moreover, $B^p(\mathcal{M})$ is a subspace of $Z^p(\mathcal{M})$. Then, we can introduce in $Z^p(\mathcal{M})$ an *equivalent relation*, namely $\approx$ by declaring

$$\alpha \approx \beta \Leftrightarrow (\alpha - \beta) \in B^p(\mathcal{M})\,;$$

i.e.,

$$\alpha \approx \beta \Leftrightarrow \alpha - \beta = d\gamma\,.$$

The set of all equivalence classes is denoted with $H^p(\mathcal{M})$ and is called the *pth de Rham cohomology vector space of* $\mathcal{M}$.

It is easy to show that, if $\mathcal{M}$ is any connected manifold, then

$$H^0(\mathcal{M}) = Z^0(\mathcal{M}) = \Re\,.$$

Indeed, a zero-differential form is just a function, so that $Z^0(\mathcal{M})$ is the space of functions f for which $df = 0$; i.e. $Z^0(\mathcal{M}) = \Re$. Moreover, $B^0(\mathcal{M}) = \{0\}$; i.e. the zero function, so that constants f and g are equivalent if they coincide.

If $\mathcal{M}$ is not connected, then an element in $Z^0(\mathcal{M})$ will be constant on each connected component of $\mathcal{M}$, but the value of the constant can be different on different components, so that

$$H^0(\mathcal{M}) = Z^0(\mathcal{M}) = \Re^m\,,$$

where m is the number of components of $\mathcal{M}$.

Exercise 7.7.1. *Show that for the n-dimensional open ball or any region* $\mathcal{U}$ *diffeomorphic to it,*

$$H^p(\mathcal{U}) = 0\,, \quad p \geq 1\,.$$

(Hint: All closed differential p-forms are equivalent to one another, and hence to the zero differential p-form).

Exercise 7.7.2. *Show that*

$$H^n(S^n) \neq 0\,, \quad H^{n-1}(S^n) = 0\,.$$

It can be proven that[51]

$$\begin{aligned} &H^n(S^n) = \Re\,,\\ &H^p(S^n) = 0\,, \quad 0 < p < n\,,\\ &H^0(S^n) = \Re\,. \end{aligned}$$

Remark 15 *The dimension of $H^p(\mathcal{M})$ is called the pth-Betti[‡] number.*

Remark 16 *The given definition of $H^p(\mathcal{M})$ relied on the differential structure of $\mathcal{M}$. However, it can be proven (see, for instance Ref. 55) that the cohomology groups depend, only on the topology of $\mathcal{M}$ and not its differentiability.*

7.8 Scalar Product of Differential p-Forms

Let $\mathcal{M}$ be an orientable n-dimensional compact differential manifold and let α, β be differential p-forms $\Lambda^p(\mathcal{M})$. The Hodge dual $*\beta$ of β is a differential $(n-p)$-form: $*\beta \in \Lambda^{n-p}(\mathcal{M})$, so that $\alpha \wedge *\beta$ is a differential n-form. This allows us to define the scalar product (α, β) of α and β, by

$$(\alpha, \beta) = \int_{\mathcal{M}} \alpha \wedge *\beta$$

Exercise 7.8.1. *Show that the previous formula defines a scalar product on $\Lambda^p(\mathcal{M})$.*

7.8.1 *Exterior codifferential*

By using the above scalar product, we can define a new operator δ, the adjoint of d, by

$$(\gamma, \delta\beta) := (d\gamma, \beta) \quad \gamma \in \Lambda^{p-1}(\mathcal{M}), \beta \in \Lambda^p(\mathcal{M})\,.$$

Clearly,

$$\delta \,:\, \Lambda^p(\mathcal{M}) \longrightarrow \Lambda^{p-1}(\mathcal{M})\,, \quad 1 \leq p \leq n\,,$$
$$\delta f = 0 \quad for\ every\ function\ f\,.$$

The operator δ is called the *codifferential.* It is worth to observe that it can be introduced only by using a metric tensor field defined on $\mathcal{M}$.

[‡]Enrico Betti, born in Pistoia in 1823 and died in Pisa in 1892, has been a professor of mathematical physics at Pisa University and Director of the Scuola Normale Superiore in Pisa. He gave deep contributions to algebra, topology, elasticity theory, and potential theory. An excellent teacher, and among his students were Luigi Bianchi and Vito Volterra.

Exercise 7.8.2. *Show that for every differential p-form* α

$$\delta = (-1)^{np+n+1} * \ d \ * \tag{7.8}$$

and that δ *is not a derivation.*

Let us finally observe that, from the Eq. (7.8), it follows that $\delta^2 = 0$.

The Laplace–Beltrami operator

The Laplace–Beltrami[§] operator Δ is defined by the relation

$$\Delta = d \circ \delta + \delta \circ d = (d + \delta)^2 ,$$

and it is a self-adjoint operator, since

$$(\Delta\alpha, \beta) = (\alpha, \Delta\beta) .$$

Exercise 7.8.3. *Give the expression of* Δ *in* $\Re^3$ *by using the Cartesian coordinates and the spherical-polar ones.*

A differential form ω that satisfies the differential equation

$$\Delta\omega = 0$$

is called *harmonic*.

Clearly,

$$\Delta\omega = 0 \Longleftrightarrow d\omega = 0, \delta\omega = 0 .$$

Indeed

$$\begin{aligned}(\Delta\omega, \omega) &= (d\delta\omega, \omega) + (\delta d\omega, \omega) \\ &= (\delta\omega, \delta\omega) + (d\omega, d\omega) ,\end{aligned}$$

with $(\delta\omega, \delta\omega) \geq 0$, $(d\omega, d\omega) \geq 0$.

[§] Eugenio Beltrami, born in Cremona in 1835 and died in Rome in 1900, has been a professor of algebra and analytical geometry at Bologna University and, after, at Pisa, Pavia and Rome universities. His research activity on the Newtonian potential and on the differential parameters can be considered of basic importance, and his *Saggio sulla interpretazione della Geometria non Euclidea* is now considered as classical work.

7.8.2 *Hodge theorem*

Hodge theorem is an important *decomposition theorem* that we quote without proof.

Theorem 25 (Hodge) *Every differential p-form ω can be written as*

$$\omega = d\alpha + \delta\beta + \gamma,$$

where α is a differential $(p-1)$-form, β is a differential $(p+1)$-form, and γ a harmonic form. Moreover, the differential forms $d\alpha$, $\delta\beta$ and γ are unique.

Chapter 8

Lie Groups and Lie Algebras

8.1 Lie Groups

A finite-dimensional Lie group is a C^∞ manifold G of dimension n, endowed with a group structure, such that the product

$$(g, h) \in G \times G \longmapsto gh \in G \tag{8.1}$$

and the inverse

$$g \in G \longmapsto g^{-1} \in G \tag{8.2}$$

are C^∞ maps. The diffeomorphisms

$$L_g : h \in G \to gh \in G \,, \quad R_g : h \in G \to hg \in G$$

are called the *left translation by* g and the *right translation by* h, respectively.

Let $(\mathcal{U}, \varphi)$ be a chart in G such that $e \in \mathcal{U}$ and $\varphi^i(e) = 0$, where e is the identity of the group. For every open set $\mathcal{U}$ containing e, there exists an open set $\mathcal{V} \subset \mathcal{U}$, to which e belongs, such that $\mathcal{V} \cdot \mathcal{V} \subset \mathcal{U}$, where $\mathcal{V} \cdot \mathcal{V} = \{gh : g, h \in \mathcal{V}\}$.

Then, the product $\varphi(\mathcal{V}) \times \varphi(\mathcal{V})$ is an open set in $\Re^n \times \Re^n$ containing the point $(0, 0)$. Since G is a Lie group, the map (8.1) is differentiable, so that, if g and h are two elements in $\mathcal{V}$, the coordinates $\varphi^i(gh) = (gh)^i$ of their product are differentiable functions of the coordinates $x^i = \varphi^i(g)$ of g and $y^i = \varphi^i(h)$ of h, and we can set

$$(gh)^i = f^i(x^1, \dots, x^n, y^1, \dots, y^n)$$

or shortly,

$$(gh)^i = f^i(x, y) .$$

The group structure implies that the functions f^i must satisfy the following properties:

$$f^i(f(x,y), z) = f^i(x, f(y,z)) \quad f^i(x,0) = f^i(0,x) .$$

The first property follows from the associativity of the group product; i.e.

$$((gh)u)^i = (g(hu))^i , \quad \forall g, h, u \in G \Longrightarrow f^i(f(x,y),z) = f^i(x, f(y,z)) ,$$

for every choice of x, y, z in $\varphi(\mathcal{V})$, with $z^i = \varphi^i(u)$. The second follows from

$$(ge)^i = (eg)^i = x^i \Longrightarrow f^i(x,0) = f^i(0,x) = x^i .$$

Moreover, for the f^i's the following expansion can be performed:

$$f^i(x,y) = x^i + y^i + \sum_{\alpha \geq 1, \beta \geq 1} \lambda^i_{\alpha\beta} x^\alpha y^\beta . \tag{8.3}$$

To build the inverse g^{-1} of an element g it is sufficient to solve the following system of equations with respect to y^i:

$$f^i(x^1, \dots, x^n, y^1, \dots, y^n) = 0 . \tag{8.4}$$

From Eq. (8.3), we have

$$\left(\frac{\partial f^i}{\partial y^j}\right)_{(0,0)} = \delta^i_j ,$$

so that the Jacobian determinant $\partial(f^1, \dots, f^n)/\partial(y^1, \dots, y^n)$ at the point $(0,0)$ in $\Re^n \times \Re^n$ is 1. Therefore, by continuity, the Jacobian does not vanish in a neighborhood of the origin and, by the implicit functions theorem, there exists an open set $\mathcal{V}' \subset \mathcal{V}$ in e, such that for every $g \in \mathcal{V}'$ the system (8.4) has a unique solution $(y^1, \dots, y^n)$.

8.1.1 *Local Lie groups*

A *local Lie group* is a local version of a Lie Group. Then, it is a pair (A, f), where A is an open set in $\Re^n$ containing the origin of the coordinates, and f a differentiable map

$$f : A \times A \to \Re^n$$

satisfying, $\forall\, x, y, z \in A$, the following conditions:

(a) $f(x, f(y, z)) = f(f(x, y), z)$;
(b) $f(0, x) = f(x, 0) = x$;
(c) there exists a differentiable map $\varepsilon : A \to \Re^n$, such that

$$f(x, \varepsilon(x)) = f(\varepsilon(x), x) = 0\,.$$

Thus, given a Lie group G it is always possible to build a local Lie group, with the identification $A \equiv \varphi(\mathcal{V}') \subset \Re^n$.

Two local Lie groups, (A_1, f_1) and (A_2, f_2), are isomorphic if there exists any two neighborhoods, $A_1' \subset A_1$ and $A_2' \subset A_2$, of the origin of $\Re^n$ and a diffeomorphism $\psi : A_1' \to A_2'$ such that the diagram

$$\begin{array}{rccc} & A_1' \times A_1' & \xrightarrow{\alpha_1} & \Re^n \supset A_1' \\ \psi \times \psi & \big\downarrow & & \psi \ \big\downarrow \\ & A_2' \times A_2' & \xrightarrow[\alpha_2]{} & \Re^n \supset A_2' \end{array}$$

is commutative, as to say

$$\psi(f_1(x, y)) = f_2((\psi \times \psi)(x, y)) = f_2(\psi(x), \psi(y))\,, \quad \forall\,(x, y) \in A_1' \times A_1'\,.$$

Of course, all local Lie groups obtained from the same Lie group G, with the previous procedure, are isomorphic among themselves.

8.2 Building of a Lie Algebra from a Lie Group

8.2.1 *Lie algebras*

The algebraic definition of Lie algebra has been already given in Part I. Here we are going to give just a mention about three types of algebras with great relevance; they are *Abelian Lie algebras*, *simple Lie algebras*, and *semi-simple Lie algebras.*

Vector spaces endowed with an identically vanishing commutator constitute the so-called commutative or Abelian Lie algebras. The definitions of simple and semi-simple Lie algebra need the introduction of a further concept, that of *ideal* in a Lie algebra.

A subspace $\mathcal{I}$ of a Lie algebra $\mathcal{A}$ is said to be an *ideal* if

$$[\mathcal{A},\mathcal{I}] \subset \mathcal{I},$$

that is $x \in \mathcal{I}$, if $[y,x] \in \mathcal{I}$ for every $y \in \mathcal{A}$.

Of course, since $[y,x] = -[x,y]$ and $\mathcal{I}$ is a vector space, if $[x,y] \in \mathcal{I}$ for every $y \in \mathcal{A}$, then $x \in \mathcal{I}$. Notice that this implies $\mathcal{I}$ is a subalgebra.

Trivial ideals in $\mathcal{A}$ are $\{0\}$ and $\mathcal{A}$. A Lie algebra containing just trivial ideals is called *simple.* A Lie algebra containing nontrivial ideals, but none of them Abelian, is called a *semi-simple Lie algebra.*

There exist different methods to build a Lie algebra from a Lie group. Here we are going to give account of the two most significant methods.

The first of them is based on the use of differential operators on the group.

8.2.2 *Left invariant vector fields*

Let G be a finite dimensional Lie group. For every $g \in G$, the left translation

$$L_g : h \in G \to L_g(h) = gh \in G$$

is a diffeomorphism from G to itself. Any neighborhood of e is mapped by left translation along a particular g onto a neighborhood of g, so that the map carries curves through e into curves through g, and curves through h into curves through gh. Then, the derivative of the map at point h, namely $(L_g)_{*h}$, is a linear map from the tangent space $\mathcal{T}_hG$ to the tangent space $\mathcal{T}_{gh}G$,

$$(L_g)_{*h} : \mathcal{T}_hG \to \mathcal{T}_{gh}G\,.$$

If V is a vector field, its value $V(h)$ at point h belongs to $\mathcal{T}_hG$. Its image by $(L_g)_{*h}$, which belongs to $\mathcal{T}_{gh}G$, will be denoted by $(gV)(gh)$; i.e.

$$(gV)(gh) \equiv (L_g)_{*h}V(h)\,,$$

so that we have

$$(gV)(h) \equiv (L_g)_{*g^{-1}h} V(g^{-1}h)\,. \tag{8.5}$$

A vector field V on a Lie group G is said to be *left invariant* if

$$(gV)(h) = V(h)\,, \quad \forall g, h \in G\,,$$

or equivalently, if

$$V(gh) = (L_g)_{*h} V(h)\,.$$

The addition of vector fields on G and their product with real numbers can be naturally defined as follows:

$$\left.\begin{aligned} (V+W)(g) &= V(g) + W(g)\,, \\ (\alpha V)(g) &= \alpha V(g)\,, \end{aligned}\right. \quad \forall g \in G\,,$$

so that, by the linearity of the operator $(L_g)_{*g}$, it follows that

- *The set of left invariant vector fields on G is a vector space over $\Re$.*

Moreover,

- *A left invariant vector field on G is uniquely determined by its valuc at the identity element e of the group G.*

Indeed, if $V(h)$ and $W(h)$ are left invariant vector fields on G

$$\left.\begin{aligned} (gV)(h) &= V(h)\,, \\ (gW)(h) &= W(h)\,, \end{aligned}\right. \quad \forall g, h \in G\,,$$

such that $V(e) = W(e)$, we have (with $h = g$)

$$\begin{aligned} V(g) &= (gV)(g) = (L_g)_{*g^{-1}g}(V(g^{-1}g)) = (L_g)_{*e}(V(e)) \\ &= (L_g)_{*e}(W(e)) = (L_g)_{*g^{-1}g}(W(g^{-1}g)) = (gW)(g) = W(g)\,. \end{aligned}$$

- *The vector space of the left invariant vector fields is isomorphic to T_eG, the tangent space to G at e.*

Indeed, with every vector $V_e \in T_eG$, we can associate a vector field $V(h)$ on G by means of the operator $(L_h)_{*e}$

$$V(h) = (L_h)_{*e}(V_e)\,, \quad \forall h \in G\,, \tag{8.6}$$

the left invariance of $V(g)$ following from

$$\begin{aligned}(gV)(h) &= (L_g)_{*g^{-1}h}(V(g^{-1}h)) \\ &= (L_g)_{*g^{-1}h}((L_{g^{-1}h})_{*e}V_e) \\ &= (L_g \circ L_{g^{-1}h})_{*e}V_e \\ &= (L_h)_{*e}V_e = V(h)\,, \quad \forall g,h \in G\,.\end{aligned}$$

Let us consider now the set of differential 1-forms α on G that constitutes a vector space on $\Re$ if the sum and the product with real number r are defined as

$$\begin{aligned}(\alpha+\alpha')(g) &= \alpha(g)+\alpha'(g)\,, \\ (r\alpha)(g) &= r\alpha(g)\,,\end{aligned} \qquad \forall g \in G\,.$$

A useful notation for the operator $(L_g)_*$ is given by the symbol dL_g. Then, relations (8.5) and (8.6) can be rewritten as follows:

$$(gV)(h) = dL_g(V(g^{-1}h))\,, \tag{8.7}$$

$$V(h) = dL_h(V_e)\,. \tag{8.8}$$

Let us introduce the transposed operator dL_g^* of dL_g, which acts on the differential 1-forms α, by

$$\langle V, dL_g^*(\alpha)\rangle = \langle dL_g(V)\,,\alpha\rangle\,, \tag{8.9}$$

where V and α are a vector field and a differential 1-form on G, respectively, and the brackets $\langle\cdot,\cdot\rangle$ denotes, as it is usual, the interior product.

Thus, dL_s and dL_s^* are the following operators:

$$dL_g : T_hG \to T_{gh}G\,, \quad dL_g^* : T_{gh}^*G \to T_h^*G\,. \tag{8.10}$$

A differential 1-form α on G, is transformed by means of the translation dL_s^* in a differential 1-form $g\alpha$ on G, according to the relation

$$(g\alpha)(h) = dL_s^*(\alpha(gh))\,.$$

A differential 1-form α is said to be *left invariant* if

$$(g\alpha)(h) = \alpha(h)\,, \quad \forall g,h \in G\,.$$

Since the linearity of dL_s implies the linearity of dL_s^*, we have that

- *The set of left invariant differential* 1*-forms on* G *is a vector space on* $\Re$.

Moreover,

- *A left invariant differential* 1*-form is uniquely determined by its value in* e.

Indeed, if α and α' are two left invariant differential 1-forms, such that $\alpha(e) = \alpha'(e)$, we have

$$\begin{aligned}\alpha(g) &= (g^{-1}\alpha)(g) = dL^*_{g^{-1}}(\alpha(g^{-1}g)) = dL^*_{g^{-1}}(\alpha(e)) \\ &= dL^*_{g^{-1}}(\alpha'(e)) = dL^*_{g^{-1}}(\alpha'(g^{-1}g)) = \alpha'(g)\,.\end{aligned}$$

As the vector space of the left invariant vector fields is isomorphic to $\mathcal{T}_eG$, so the vector space of left invariant differential 1-forms is isomorphic to $\mathcal{T}_e^*G$.

If α_e denotes a covector on $\mathcal{T}_eG$, the differential 1-form $\alpha(g)$, defined by

$$\alpha(g) = dL^*_{g^{-1}}(\alpha_e)\,, \quad \forall\, g \in G\,,$$

is a left invariant differential 1-form.

Indeed, since

$$dL^*_{gh} = dL^*_g \circ dL^*_h\,,$$

we have

$$\begin{aligned}(g\alpha)(h) &= dL^*_g(\alpha(gh)) = (dL^*_g \circ dL^*_{(gh)^{-1}})(\alpha_e) \\ &= (dL^*_g \circ dL^*_{h^{-1}g^{-1}})(\alpha_e) = dL^*_{h^{-1}}(\alpha_e) \\ &= \alpha(h)\,, \quad \forall\, gh \in G\,.\end{aligned}$$

Thus, with every covector on $\mathcal{T}_eG$ we can associate, in a unique way, a left invariant differential form on G.

An interesting and useful result is the following:

- *The contraction* $\langle\alpha, V\rangle$, *between a left invariant differential* 1*-form* α and *a left invariant vector field* V, *is constant on* G.

Indeed,

$$\begin{aligned}\langle\alpha, V\rangle(g) &= \langle\alpha(g), V(g)\rangle \\ &= \langle dL^*_{g^{-1}}(\alpha(e)), dL_g(V(e))\rangle\end{aligned}$$

$$= (dL_{g^{-1}} \circ dL_g)(\langle \alpha(e), V(e) \rangle)$$
$$= \langle \alpha(e), V(e) \rangle, \quad \forall g \in G.$$

There is a converse to this, namely

- *A vector field V on G, for which $\langle \alpha, V \rangle$ is constant on G for every left invariant differential 1-form, is a left invariant vector field.*

Indeed,

$$\langle \alpha(h), gV(h) \rangle = \langle \alpha(h), dL_g(V(g^{-1}h)) \rangle = \langle dL_g^*(\alpha(h)), V(g^{-1}h) \rangle$$
$$= \langle \alpha(g^{-1}h), V(g^{-1}h) \rangle = \langle \alpha(h), V(h) \rangle .$$

Since the left invariant form α is arbitrary, then

$$(gV)(h) = V(h), \quad \forall g, h \in G.$$

These two properties, together with the useful relation*

$$d\alpha(X,Y) = L_X(\langle \alpha, Y \rangle) - L_Y(\langle \alpha, X \rangle) + \langle \alpha, [X,Y] \rangle, \tag{8.11}$$

allow us to prove the following statement:

- *If X and Y are two left invariant vector fields on a Lie group G, their Lie brackets $[X,Y]$ is a left invariant vector field.*

To this purpose, we just have to prove that $\langle \alpha, [X,Y] \rangle$ is constant on G for every left invariant differential 1-form $\alpha(g)$. Indeed, if α is left invariant, then $\langle \alpha, X \rangle$ and $\langle \alpha, Y \rangle$ are constant on G, because $X(g)$ and $Y(g)$ are, by hypothesis, left invariant vector fields, so that the Lie derivatives $L_X(\langle \alpha, Y \rangle)$ and $L_Y(\langle \alpha, X \rangle)$ vanish identically.

Therefore, we have

$$d\alpha(X,Y) = \langle \alpha, [X,Y] \rangle . \tag{8.12}$$

Since $d\alpha$ is an exact 2-form, for which $dd\alpha = 0$, and the right hand of Eq. (8.12) is a 0-form; that is, a function on G, then

$$\langle \alpha, [X,Y] \rangle = \text{constant}.$$

*In this chapter the Lie derivative, with respect to a vector field X, has been denoted with the symbol $\mathcal{L}_X$, instead of L_X, to avoid confusion with the left translation L_x.

Thus, by using the isomorphism between $\mathcal{T}_e G$ and the vector space of the left invariant vector fields, it is possible to introduce, in the tangent space $\mathcal{T}_e G$, a commutation relation which, being bilinear, antisymmetric, and satisfying the Jacobi identity, endows it with a Lie algebra structure.

To be specific, if X_e and Y_e denote two vectors belonging to $\mathcal{T}_e G$, the Lie brackets of the two left invariant vector field on G corresponding to them, still is a left invariant vector field which also is uniquely determined by its value at the identity element of the group.

Thus, given X_e and Y_e belonging to $\mathcal{T}_e G$, we define the *Lie commutator* of X and Y as the value, at the identity element e of the group G, of the Lie bracket of the corresponding left invariant vector fields

$$[X_e, Y_e] = [dL_g(X_e), dL_g(Y_e)]_{g=e} .$$

This Lie algebra is called the Lie algebra of the Lie group G.

8.2.3 *The adjoint representation of a Lie group*

There exists a second method, as well, which allows us to introduce a Lie algebra structure in the tangent space $\mathcal{T}_e G$.

Let us observe that the map

$$A_g : h \in G \to A_g(h) = ghg^{-1} \in G ,$$

composed of the left translation by g and the right translation by g^{-1}

$$A_g = R_{g^{-1}} L_g : h \in G \to (R_{g^{-1}} L_g)(h) = ghg^{-1} \in G ,$$

is a one-to-one, differentiable map. Actually, since $A_g^{-1} = A_{g^{-1}}$, the map is a diffeomorphism of G into itself.

Since

$$A_g(h_1 h_2) = g h_1 h_2 g^{-1} = g h_1 g^{-1} g h_2 g^{-1} = A_g(h_1) A_g(h_2) ,$$

A_g is a homomorphism of G into itself. Actually, A_g is an isomorphism of G into itself, as to say an *inner automorphism of* G, since $A_{g^{-1}} = A_g^{-1}$.

Notice that each A_g maps the identity element e into itself, so that every curve through e is mapped into a, possibly different, curve through e.

Therefore, the derivative $(A_g)_{*e}$ at the unit e, usually denoted with Ad_g,

$$Ad_g : \mathcal{T}_e G \to \mathcal{T}_e G\,,$$

is an invertible linear map of any tangent vector of $\mathcal{T}_e G$ to another one in $\mathcal{T}_e G$.

For the automorphism A_g, we have

$$A_f \circ A_g = A_{fg}\,, \quad \forall f, g \in G\,,$$

and for derivatives

$$(A_f \circ A_g)_{*e} = (A_f)_{*e} \circ (A_g)_{*e}\,,$$

so that

$$Ad_{fg} = Ad_f \circ Ad_g\,.$$

The set of all invertible linear maps of $\mathcal{T}_e G$ into itself is a group whose internal composition law is the usual composition of maps. This group is denoted by Aut $\mathcal{T}_e G$.

Thus, the map

$$Ad : g \in G \to Ad(g) = Ad_g \in \text{Aut}\,\mathcal{T}_e G \tag{8.13}$$

is a homomorphism of G into the group $Aut\,\mathcal{T}_e G$ of the invertible linear maps of the vector space $\mathcal{T}_e G$.

Once a basis in $\mathcal{T}_e G$ is chosen, the map Ad becomes a homomorphism of G into the group $GL(n, \Re)$, where n is the dimension of $\mathcal{T}_e G$. The group $GL(n, \Re)$ is the group of nonsingular real matrices $n \times n$ and can be endowed with a differential manifold structure.

The compatibility of group and differential manifold structures promotes the group $GL(n, \Re)$ to a Lie group. Obviously, the dimension of $GL(n, \Re)$ is n^2.

Thus, the map Ad is a representation of G on $\mathcal{T}_e G$ and is called the *adjoint representation* of the Lie group G.

The tangent space to $GL(n, \Re)$ at the identity I (the unit matrix) is the space, denoted with $Mat_n(\Re)$, of the not necessarily invertible real matrixes $n \times n$.

The map Ad is differentiable and its derivative $(Ad)_{*e}$ at the unit e is a linear map of $\mathcal{T}_e G$ in $End\,\mathcal{T}_e G$, the vector space of the (not necessarily invertible) linear maps of $\mathcal{T}_eG$ into itself, that are endomorphisms of $\mathcal{T}_eG$. In other terms,

$$\text{End}\,\mathcal{T}_eG \equiv T(\text{Aut}\,\mathcal{T}_eG)\,.$$

The derivative $(Ad)_{*e}$ is denoted with the symbol *ad*,

$$ad : V \in \mathcal{T}_eG \to ad(V) = ad_V \in \text{End}\,\mathcal{T}_eG\,.$$

A *one parameter subgroup* of a Lie group G is a representation of $\Re$ in G, as to say a homeomorphism of $\Re$ in G; that is, a differentiable map

$$\rho : t \in \Re \to \rho(t) \in G\,,$$

such that

$$\rho(0) = e\,, \quad \rho(t+t') = \rho(t)\rho(t')\,, \quad \forall t, t' \in \Re\,.$$

Let V_e be an element in $\mathcal{T}_eG$ and let

$$\rho_{V_e} : t \in \Re \to \rho_{V_e}(t) = e^{tV_e} \in G \tag{8.14}$$

be the integral curve of the left invariant vector field $(L_g)_{*e}(V_e)$ on G.

Let us also fix $s \in \Re$ and define the map

$$\chi_1 : t \in \Re \to \chi(t) = \rho_{V_e}(s)\rho_{V_e}(t) = L_{\rho_{V_e}(\sigma)}\rho_{V_e}(t) \in G\,,$$

where $L_{\rho_{V_e}(s)}$ is the left translation by $\rho_{V_e}(s)$.

Since the vector field $(L_g)_{*e}(V)$ on G is left invariant, we have

$$\left(\frac{d\chi_1(t)}{dt}\right)_{t=0} = \frac{d}{dt}\rho_{V_e}(s)\rho_{V_e}(t)\bigg|_{t=0} = \frac{d}{dt}L_{\rho_{V_e}(s)}\rho_{V_e}(t)\bigg|_{t=0} = (L_{\rho_{V_e}(s)})_{*e}(V_e)\,,$$

so that $\chi_1(t)$ is an integral curve of $V = (L_g)_{*e}(V_e)$ through $\rho_{V_e}(s)$ at $t = 0$.

On the other hand, the map

$$\chi_2 : t \in \Re \to \chi_2\,(t) = \rho_{V_e}\,(s+t) \in G$$

is also an integral curve of $V = (L_g)_{*e}(V_e)$ through $\rho_{V_e}(s)$ at $t = 0$.

Thus, $\chi_1(t) \equiv \chi_2(t)$, since the integral curve of $V = (L_g)_{*e}(V_e)$ through $\rho_{V_e}(s)$ at $t = 0$ is unique. As a consequence, we have

$$\rho_{V_e}(s+t) = \rho_{V_e}(s)\rho_{V_e}(t) \tag{8.15}$$

and

$$e^{(s+t)V_e} = e^{sV_e} e^{tV_e} .$$

From Eq. (8.15), it follows that the map (8.14) is a homeomorphism of $\Re$ in G, and then a one parameter subgroup of G. This subgroup is unique.

Indeed, if

$$\sigma : t \in \Re \to \sigma(t) \in G$$

is another one parameter subgroup of G such that

$$\sigma(0) = e , \quad \left(\frac{d\sigma(t)}{dt} \right)_{t=0} = V_e ,$$

then

$$\sigma(t+s) = \sigma(t)\sigma(s) = L_{\sigma(t)}\sigma(s) .$$

Thus,

$$\left(\frac{d\sigma(t)}{dt} \right)_{t=t'} = \frac{d}{ds}\sigma(t'+s)\bigg|_{s=0} = \frac{d}{ds}L_{\sigma(t')}\sigma(s)\bigg|_{s=0} = (L_{\sigma(t')})_{*e}(V_e) ;$$

that is, $\sigma(t)$ is an integral curve of $V = (L_g)_{*e}(V_e)$ through e at $t = 0$. Since, Eq. (8.14) shows that ρ_{V_e} is an integral curve of $V = (L_g)_{*e}(V_e)$ through e, then $\rho_V \equiv \sigma$.

We can conclude that, with every vector $V_e \in \mathcal{T}_e G$, there is associated a unique one-parameter subgroup $\rho_{V_e}(t)$ of G.

By using the notation $\rho_{V_e}(t) = e^{tV_e}$, we can write

$$V_e = \frac{d}{dt} e^{tV_e} \bigg|_{t=0} .$$

The explicit expression of the operator ad_V can be easily found. Indeed,

$$ad_{V_e} = (Ad)_{*e}(V_e) = \frac{d}{dt} Ad(e^{tV_e})\bigg|_{t=0} = \frac{d}{dt} Ad_{e^{tV_e}}\bigg|_{t=0} ,$$

so that the value of the operator ad_{V_e} on a vector $W_e \in \mathcal{T}_e G$ will be given by

$$ad_{V_e}(W_e) = \frac{d}{dt} Ad_{e^{tV_e}}(W_e)\bigg|_{t=0}$$

$$= \frac{d}{dt}(R_{e^{-tV_e}} L_{e^{tV_e}})_{*e}(W_e)\Big|_{t=0}$$

$$= \frac{d}{dt}(R_{e^{-tV_e}})_{*e^{tV_e}}(L_{e^{tV_e}})_{*e}(W_e)\Big|_{t=0}$$

$$= \frac{d}{dt}(R_{e^{-tV_e}})_{*e^{tV_e}}(W(e^{tV_e}))\Big|_{t=0}.$$

On the other hand, e^{tV_e} is just the value at e of the flow of the vector field $V(g) = (L_g)_{*e}(V_e)$, as it is easily followed by

$$\sigma_V^t(g) = ge^{tV_e} = R_{e^{tV_e}}g,$$

so that

$$e^{tV_e} = \sigma_V^t(e).$$

Thus, we have

$$ad_{V_e}(W_e) = \frac{d}{dt}(\sigma_V^{-t})_{*\sigma_V^t(e)}(W(\sigma_V^t(e)))\Big|_{t=0} = (L_V W)(e). \tag{8.16}$$

By Eq. (8.16) and by the well-known properties of the Lie derivative, we can define the following bracket in $\mathcal{T}_eG$:

$$[\cdot,\cdot] = (V_e, W_e) \in \mathcal{T}_eG \times \mathcal{T}_eG \to [V_e, W_e] = ad_{V_e}(W_e) \in \mathcal{T}_eG, \tag{8.17}$$

which can be easily seen to be

- bilinear because

$$ad_{c_1X_1+c_2X_2}(Y) = c_1 ad_{X_1}(Y) + c_2 ad_{X_2}(Y),$$
$$ad_X(c_1Y_1 + c_2Y_2) = c_1 ad_X(Y_1) + c_2 ad_X(Y_2),$$

  whatever $X_1, X_2, V, Y_1, Y_2, Y \in \mathcal{T}_eG$ and $c_1, c_2 \in \Re$ are chosen;
- antisymmetric because

$$ad_X(Y) = -ad_Y(X), \quad \forall X, Y \in \mathcal{T}_eG; \quad \text{and}$$

- satisfying the Jacobi identity

$$ad_X(ad_Y(Z)) + ad_Y(ad_Z(X)) + ad_Z(ad_X(Y)) = 0.$$

The composition law, defined by Eq. (8.17), provides the vector space $\mathcal{T}_e G$ with a Lie algebra structure.

In conclusion, given a Lie group G, it is always possible to build from it, a Lie algebra. Now we can ask whether, given a Lie algebra $\mathcal{A}$, there exists a Lie group of which, *vice versa*, $\mathcal{A}$ is the algebra. The answer to this question is given by the following theorem:

Theorem 26 (Cartan) *Every Lie algebra is the Lie algebra of some Lie group.*

In the previous section we spoke about local Lie groups. They are also related to the Lie algebras because every Lie algebra is a Lie algebra of some local Lie group. Moreover, the local Lie groups are isomorphic if and only if the corresponding algebras are isomorphic as well.

Let $\mathcal{A}_1$ and $\mathcal{A}_2$ be two Lie algebras, and G_1 and G_2 the corresponding Lie groups. By the last we can build two local Lie groups G_1' and G_2', which will be isomorphic if $\mathcal{A}_1$ and $\mathcal{A}_2$ are isomorphic. However, the fact that G_1' and G_2' are isomorphic does not imply that G_1 and G_2 are so. In this case, we speak of *local isomorphism* between G_1 and G_2.

For a simply connected Lie group G, the following theorem holds:

Theorem 27 (Monodromy) *If G is a simply connected Lie group and F any Lie group, every local homomorphism*[†] *of G in F is uniquely prolonged in a global homomorphism of G in F.*

Let us denote with $GL(n, \Re)$ the Lie group of $n \times n$ invertible real matrices and with $\mathcal{G}L(n, \Re)$ the corresponding Lie algebra, which is given by the vector space of $n \times n$ real matrices with the commutator as Lie bracket. A very important result is the following:

Theorem 28 (Ado) *Every Lie algebra of a Lie group is a subalgebra of $\mathcal{G}L(n, \Re)$ for some value of n.*

For Lie groups, the analogous statement holds only locally; i.e.

Every Lie group is locally isomorphic to a subgroup of $GL(n, \Re)$ for some value of n.

By this theorem the local isomorphism between G_1 and G_2 is prolonged to a global isomorphism.

[†] A local homeomorphism is a homeomorphism of the correspondent local groups.

Thus, we can conclude that just one simply connected Lie group G corresponds to a Lie algebra $\mathcal{A}$.

8.2.4 *The coadjoint representation of a Lie group*

We can introduce translation operators also in the dual space $\mathcal{T}_e^*G$ of $\mathcal{T}_eG$. As for the left translation, we can use relations (8.9) and (8.10); the right translations are defined in a perfectly analogous way.

$$\langle V, dR_g^*(\alpha)\rangle = \langle dR_g(V), \alpha\rangle$$
$$dR_g : \mathcal{T}_hG \to \mathcal{T}_{hg}G\,, \quad dR_g^* : \mathcal{T}_{hg}^*G \to \mathcal{T}_h^*G\,.$$

It is also possible to define the operator Ad_g^*, dual of the operator Ad_g, as follows:

$$\langle V, Ad_g^*(\alpha)\rangle = \langle Ad_g(V), \alpha\rangle\,. \tag{8.18}$$

By Eq. (8.18) and by the properties of Ad_g, we argue that

$$Ad_g^* : \mathcal{T}_e^*G \to \mathcal{T}_e^*G$$

is an invertible linear map of $\mathcal{T}_c^*G$ into itself.

The map

$$Ad^* : g \in G \to Ad^*(g) = Ad_g^* \in Aut\,\mathcal{T}_e^*G\,, \tag{8.19}$$

as the one in Eq. (8.13), is also a representation of the Lie group G. It is called *coadjoint representation of the Lie group G*

The map (8.19) is differentiable. Its derivative $(Ad^*)_{*e}$ at the unit, denoted by ad^*, is the map

$$ad^* : V \in \mathcal{T}_eG \to ad^*(V) = ad_V^* \in \mathrm{End}\,\mathcal{T}_e^*G\,.$$

The operator ad_V^* is the conjugate of ad_V, and

$$\langle W, ad_V^*(\alpha)\rangle = \langle ad_V(W), \alpha\rangle\,, \quad \forall\,\alpha \in \mathcal{T}_e^*G\,, \quad \forall\,W \in \mathcal{T}_eG\,.$$

The vector spaces $\mathcal{T}_eG$ and $\mathcal{T}_e^*G$, endowed with a bracket giving them a Lie algebra structure, are usually denoted by the symbols $\mathcal{G}$ and $\mathcal{G}^*$.

The coadjoint representation of a Lie group has an important role in classical mechanics. As we will see, the orbits of the group under the coadjoint representation are symplectic manifolds.

This will be shown in Part III, in the chapter *Orbits method*, after the introduction of some preliminary concepts. An exhaustive discussion on this subject can be found in Ref. 41. The second part of this book is in fact completely devoted to *Reduction, Actions of Group and Algebras.*

Further Readings

- R. Abraham, J. E. Marsden, and T. Ratiu, *Manifolds, Tensor Analysis, and Applications* (Addison-Wesley, 1983).
- B. Dubrovin, S. Novikov, A. Fomenko, *Géométrie Contemporaine, Éditions Mir* (Moscow 1979, 1982).
- C. J. Isham, *Modern Differential Geometry for Physicists* (World Scientific, 1989).
- J. L. Koszul, *Lectures on Fibre Bundles and Differential Geometry* (Tata Institute of Fundamental Research, Bombay, 1960).
- A. Trautman, *Differential Geometry for Physicists* (Bibliopolis, Naples, 1984).

Part III
Geometry and Physics

Part III is devoted to a revisiting of analytical mechanics in terms of geometrical structures. Chapter 9 is devoted to the intrinsic formulation of Maxwell's differential equations in terms of differential forms, so that it can be considered as an introduction for *Gauge Theories*.

Chapter 9

Symplectic Manifolds and Hamiltonian Systems

9.1 Symplectic Structures on a Manifold

If $\mathcal{M}$ is a $2n$-dimensional differentiable manifold, a *symplectic structure* on $\mathcal{M}$ is a differential 2-form ω, required to be

- closed

$$d\omega = 0,$$

- and *not degenerate*

$$(\omega_p(X,Y) = 0 \quad \forall Y \in \mathcal{T}_p\mathcal{M}) \Rightarrow (X = 0) \quad \forall p \in \mathcal{M}. \tag{9.1}$$

A pair $(\mathcal{M}, \omega)$, with $\mathcal{M}$ a $2n$-dimensional differentiable manifold and ω a *symplectic structure*, is called a *symplectic manifold*

In a given basis $\{e_i\}$ for vector fields on $\mathcal{M}$, we may write

$$X = X^i e_i, \quad Y = Y^i e_i,$$

so that, with $\omega_{ij}(p) \equiv \omega_p(e_i, e_j)$, the relation (9.1) becomes

$$(X^i Y^i \omega_{ij}(p) = 0 \quad \forall\, Y^i) \quad \Rightarrow \quad (X^i = 0) \quad \forall p \in \mathcal{M},$$

or equivalently,

$$(X^i \omega_{ij} = 0) \Rightarrow (X^i = 0).$$

Thus, a differential 2-form is not degenerate iff

$$\det(\omega_{ij}(p)) \neq 0 \quad \forall p \in \mathcal{M}.$$

A generic differential 2-form ω on a manifold defines a homomorphism

$$\omega : \mathcal{T}_p\mathcal{M} \to \mathcal{T}_p^*\mathcal{M}$$

of the vector space $\mathcal{T}_p\mathcal{M}$, of tangent vectors at the point $p \in \mathcal{M}$, into $\mathcal{T}_p^*\mathcal{M}$, the vector space of differential 1-forms to the manifold $\mathcal{M}$ at the point $p \in \mathcal{M}$, since with the vector $X_p \in \mathcal{T}_p\mathcal{M}$, ω associates the differential 1-form α_p, defined as

$$\alpha_p = i_{X_p}\omega(p).$$

As a consequence, with the vector field X, ω associates the differential 1-form α, defined point-wise as

$$\alpha = i_X\omega.$$

When the differential 2-form is not degenerate, the above relation can be point-wise solved with respect to the vector field X.

Then, a not degenerate differential 2-form ω defines an isomorphism, between the vector spaces $\mathcal{T}_p\mathcal{M}$ and $\mathcal{T}_p^*\mathcal{M}$, given by

$$X = \Lambda(\alpha, \cdot),$$

where the 2-vector field

$$\Lambda : \mathcal{T}_p^*\mathcal{M} \to \mathcal{T}_p\mathcal{M}, \tag{9.2}$$

is the inverse of ω,

$$\Lambda \circ \omega = \omega \circ \Lambda = 1.$$

The above relation in a given coordinate basis, in which

$$\omega = \frac{1}{2}\omega_{ij}dx^i \wedge dx^i, \quad \Lambda = \frac{1}{2}\Lambda^{ij}\frac{\partial}{\partial x^i} \wedge \frac{\partial}{\partial x^j}, \tag{9.3}$$

is simply written as follows:

$$\Lambda^{ih}\omega_{hj} = \delta^i_j.$$

9.2 Locally and Globally Hamiltonian Vector Fields

A vector field X on symplectic manifold $(\mathcal{M}, \omega)$ is called a (*locally*) *Hamiltonian vector field* if

$$L_X \omega = 0 ,$$

that is, if the symplectic structure is invariant under the flow generated by X.

Since a symplectic form is closed, the above relation can also be written, by using the Cartan identity, in the following form:

$$d i_X \omega = 0 .$$

Thus, a locally Hamiltonian vector field on $\mathcal{M}$ is a vector field satisfying the requirement that the differential 1-form α, defined by

$$\alpha = i_X \omega ,$$

is closed.

If the differential 1-form $\alpha = i_X \omega$ is also exact; that is, a function H on $\mathcal{M}$ exists such that

$$i_X \omega = -dH , \tag{9.4}$$

the vector field is called a *globally Hamiltonian vector field*, or simply a *Hamiltonian vector field*, and the function H is called the *Hamiltonian function corresponding* to X. The minus sign in Eq. (9.4) is introduced just for historical reasons.

Vice versa, any differentiable function on a symplectic manifold $\mathcal{M}$,

$$f : \mathcal{M} \to \Re ,$$

defines a Hamiltonian vector field X_f by the relation

$$i_{X_f} \omega = df .$$

9.2.1 *Integral curves of a Hamiltonian vector field*

In a coordinate basis, we may write

$$X = X^i \frac{\partial}{\partial x^i} , \quad \omega = \frac{1}{2} \omega_{ij} dx^i \wedge dx^j , \quad dH = \frac{\partial H}{\partial x^i} dx^i ,$$

so that

$$i_X\omega = \frac{1}{2}\omega_{ij}X^i dx^j - \frac{1}{2}\omega_{ij}dx^i X^j = \omega_{ij}X^i dx^j\,,$$

and Eq. (9.4) becomes

$$\omega_{ij}X^i dx^j = -\frac{\partial H}{\partial x^j}dx^j\,,$$

or

$$\omega_{ji}X^i = -\frac{\partial H}{\partial x^j}\,.$$

Since $\det(\omega_{ij}(x)) \neq 0$, the last relation gives

$$X^i = \Lambda^{ij}\frac{\partial H}{\partial x^j}\,.$$

Thus, the first order differential equations for the integral curves of the Hamiltonian vector fields X have the following form:

$$\frac{dx^i}{dt} = \Lambda^{ij}\frac{\partial H}{\partial x^j}\,, \tag{9.5}$$

and they are very similar to the Eqs. (2.24) of Sec. 2.4.1.

Equations (2.24) and (9.5) coincide, provided that the antisymmetric matrix, whose elements are Λ^{ij}, satisfies the Jacobi identity

$$\Lambda^{ij}\frac{\partial \Lambda^{hk}}{\partial x^j} + \Lambda^{hj}\frac{\partial \Lambda^{ki}}{\partial x^j} + \Lambda^{kj}\frac{\partial \Lambda^{ih}}{\partial x^j} = 0\,. \tag{9.6}$$

Actually, the Jacobi identity is satisfied because of the closure of the symplectic form ω. Indeed, in a coordinate basis, we may write

$$d\omega = \left(\frac{\partial \omega_{ij}}{\partial x^k} + \frac{\partial \omega_{jk}}{\partial x^i} + \frac{\partial \omega_{ki}}{\partial x^j}\right) dx^i \wedge dx^j \wedge dx^k\,,$$

so that

$$d\omega = 0 \Leftrightarrow \frac{\partial \omega_{ij}}{\partial x^k} + \frac{\partial \omega_{jk}}{\partial x^i} + \frac{\partial \omega_{ki}}{\partial x^j} = 0\,.$$

The reader can easily check that, if $\Lambda^{ih}\omega_{hj} = \delta^i_j$, Eqs. (9.6) are equivalent to

$$\frac{\partial \omega_{ij}}{\partial x^k} + \frac{\partial \omega_{jk}}{\partial x^i} + \frac{\partial \omega_{ki}}{\partial x^j} = 0\,.$$

9.3 Hamiltonian Flows

What has been said in the previous section can be repeated, more geometrically, as follows.

Let us consider a function f defined on the differentiable symplectic manifold $(\mathcal{M}, \omega)$. Its differential df_p at the point $p \in \mathcal{M}$ belongs to $\mathcal{T}_p^*\mathcal{M}$

$$df_p : \mathcal{T}_p\mathcal{M} \to \Re, \quad \forall p \in \mathcal{M}.$$

The bi-vector field Λ associates to df_p a tangent vector to $\mathcal{M}$ at the point $p \in \mathcal{M}$ as follows:

$$X_f(p) \equiv \Lambda(df(p), \cdot).$$

With the vector field $X_f(p)$, a one-parameter group of diffeomorphisms is associated (Eq. (5.13)) as

$$\sigma^t : \mathcal{M} \to \mathcal{M},$$

such that

$$\left.\frac{d}{dt}\sigma^t(p)\right|_{t=0} = X_f(p).$$

The group σ^t, which is called *Hamiltonian flow with Hamilton function* H, preserves the symplectic structure, that is

$$\sigma^{t*}\omega = \omega, \tag{9.7}$$

where σ^{t*} is the derivative of σ^t.

More explicitly, Eq. (9.7) can be written in the following form:

$$(\sigma^{t*}\omega)_p(X, Y) = \omega_{\sigma^t(p)}(\sigma^t_{*p}(X), \sigma^t_{*p}(Y)) = \omega_p(X, Y), \tag{9.8}$$

where $X, Y \in \mathcal{T}_p\mathcal{M}$ and

$$\sigma^t_{*p} : \mathcal{T}_p\mathcal{M} \to \mathcal{T}_{\sigma^t(p)}\mathcal{M}$$

is the derivative of σ^t at the point p.

The Lie derivative of the 2-form ω along X_f is given by

$$(L_{X_f}\omega)_p(X, Y) = \frac{d}{dt}\omega_{\sigma^t(p)}(\sigma^t_{*p}(X), \sigma^t_{*p}(Y)) = 0,$$

where the relation (9.8) has been used.

Since ω is closed, we may write

$$L_{X_f}\omega = 0 \quad \Leftrightarrow \quad di_{X_f}\omega = 0\,. \tag{9.9}$$

Of course, $i_{X_f}\omega$ is an exact differential 1-form, since

$$i_{X_f}\omega = \omega(X_f, \cdot) = \omega(\Lambda(df, \cdot),) = df\,.$$

9.3.1 *Lie algebras of Hamiltonian vector fields and of Hamilton functions*

It is worth recalling that a Lie algebra is a vector space $\mathcal{A}$ supplied with a bracket

$$[\cdot\,,\,\cdot] : (x, y) \in \mathcal{A} \times \mathcal{A} \to [x, y] \in \mathcal{A}\,,$$

which is

- bilinear

$$\begin{aligned} [\alpha x + \beta y, z] &= \alpha[x, z] + \beta[y, z]\,, \\ [x, \alpha y + \beta z] &= \alpha[x, y] + \beta[x, z]\,, \end{aligned} \quad \forall\, x, y, z \in \mathcal{A}, \quad \forall\, \alpha, \beta \in \Re(\text{or } \mathbb{C})\,; \tag{9.10}$$

- antisymmetric

$$[x, y] = -[y, x] \quad \forall\, x, y \in \mathcal{A}\,; \tag{9.11}$$

- and satisfying the Jacobi identity

$$[[x, y], z] + [[y, z], x] + [[z, x], y] = 0\,, \quad \forall\, x, y, z \in \mathcal{A}\,. \tag{9.12}$$

The Lie bracket

$$[X, Y] = L_X Y\,, \tag{9.13}$$

which satisfies the relations (9.10), (9.11) and (9.12), provides the infinite-dimensional vector space of differentiable vector fields, on a manifold $\mathcal{M}$, with a Lie algebra structure.

Let X and Y be two vector fields and σ_X^t and σ_Y^s be the corresponding flows, respectively. As already said, such flows are diffeomorphisms defined over all $\mathcal{M}$, if the manifold is compact. Otherwise, σ_X^t and σ_Y^s are defined only in open sets in $\mathcal{M}$ and for small values of the parameters t and s. However, this suffices for our purposes.

An important property of the Lie bracket, of two vector fields, is that its vanishing is a necessary and sufficient condition for the corresponding flows to commute[2]:

$$[X,Y]=0 \quad \Leftrightarrow \quad \sigma_X^t \sigma_Y^s = \sigma_Y^s \sigma_X^t \,. \tag{9.14}$$

Let $(\mathcal{M},\omega)$ be a symplectic manifold, and f and g two differentiable functions on $\mathcal{M}$.

The bracket $\{f,g\}$, defined by

$$\{f,g\}(p) = \frac{d}{dt} g(\sigma_f^t(p))\bigg|_{t=0} , \tag{9.15}$$

where σ_f^t denotes the Hamiltonian flow corresponding to the Hamiltonian vector field X_f defined by $i_{X_f}\omega = df$, is called the *Poisson bracket of the functions* f and g.

From definition (9.15), we have that the vanishing of the Poisson bracket $\{f,g\}$ is a necessary and sufficient condition for the function g to be a first integral of the flow σ_f^t with Hamilton function f.

Because of the isomorphism (9.2) between vector fields and differential forms, the Poisson bracket (9.15) can be written in the following form:

$$\{f,g\}(p) = \Lambda_p(df,dg) = \omega_p(X_f,X_g)\,. \tag{9.16}$$

Indeed, $\forall p \in \mathcal{M}$

$$\begin{aligned}\{g,f\}(p) &= \frac{d}{dt} g(\sigma_f^t(p))\bigg|_{t=0} \\ &= dg_p(X_f) = (X_f g)_p = i_{X_f} dg_p = i_{X_f} i_{X_g} \omega_p \\ &= \omega_p(X_g,X_f)\,.\end{aligned}$$

Exercise 9.3.1. *Prove, by using (9.16), that the Poisson Bracket is bilinear, antisymmetric, and satisfies the Jacobi identity*

$$\{\{f,g\},h\} + \{\{g,h\},f\} + \{\{h,f\},g\} = 0\,. \tag{9.17}$$

Thus, the Poisson bracket provides the set $\mathcal{F}(\mathcal{M})$, of differentiable functions on $\mathcal{M}$, with a Lie algebra structure. This Lie algebra, as it has already been shown in Part I, modulo the constants, is isomorphic to the Lie algebra of differentiable vector fields on $\mathcal{M}$.

Exercise 9.3.2. *Let ω be a closed differential 2-form and X and Y be any two vector fields on a manifold $\mathcal{M}$, locally represented by*

$$\omega = \omega_{ij} dx^i \wedge dx^j , \quad X = X^j \frac{\partial}{\partial x^j} , \quad Y = Y^k \frac{\partial}{\partial x^k} ,$$

We have

$$\begin{aligned} L_X i_Y \omega - i_Y L_X \omega &= d i_X i_Y \omega + i_X d i_Y \omega - i_Y d i_X \omega \\ &= d(\omega(Y, X)) + i_X d i_Y \omega - i_Y d i_X \omega \\ &= d(-\omega_{ij} X^i Y^j) + i_X d(\omega_{ij} Y^i dx^j - \omega_{ij} Y^j dx^i) \\ &\quad - i_Y d(\omega_{ij} X^i dx^j - \omega_{ij} X^j dx^i) \\ &= \omega_{ij} [X, Y]^i dx^j - \omega_{ij} [X, Y]^j dx^i \\ &= i_{[X,Y]} \omega . \end{aligned}$$

Prove the relation

$$i_{[X,Y]} \omega = L_X i_Y \omega - i_Y L_X \omega , \tag{9.18}$$

without use of the coordinates.

Let X_f and X_g be the Hamiltonian vector fields associated with the functions f and g, respectively; i.e.

$$i_{X_f} \omega = df , \quad i_{X_g} \omega = dg .$$

By using Eq. (9.18) for the Hamiltonian vector fields X_f and Y_f, we have

$$i_{[X_f, X_g]} \omega = L_{X_f} i_{X_g} \omega = d i_{X_f} i_{X_g} \omega + i_{X_f} d i_{X_g} \omega = d i_{X_f} i_{X_g} \omega = d\{f, g\} , \tag{9.19}$$

so that

$$L_{[X_f, X_g]} \omega = 0 . \tag{9.20}$$

Therefore, $[X_f, X_g]$ is a globally Hamiltonian vector field with Hamilton function given by

$$H(p) = \omega_p(X_f, X_g) = \{f, g\}(p) .$$

Thus, the set of globally Hamiltonian vector fields on a symplectic manifold close on a Lie subalgebra of all vector fields.

Exercise 9.3.3. *Prove that the set of first integrals of a Hamiltonian flow constitute a subalgebra of the Lie algebra of all differentiable functions.*

Exercise 9.3.4. *Prove, by using Eq. (9.18), that the Lie bracket of two locally Hamiltonian vector fields, X and Y, is a globally Hamiltonian vector field, with Hamiltonian function given by $H(p) = \omega_p(Y, X)$.*

It follows that the set of locally Hamiltonian vector fields constitute a subalgebra of the Lie algebra of all vector fields too.

The considerations developed in Sec. 2.4.4 (*Further generalizations of the Jacobi–Poisson dynamics*), can be repeated, of course, also in this new context. A useful reading on the theory of ordinary Jacobi–Poisson manifolds is given by Vaisman's book.[54]

9.4 The Cotangent Bundle and Its Symplectic Structure

An example of symplectic manifold is given by the cotangent bundle $\mathcal{T}^*Q$ of an n-dimensional manifold Q. An element ϑ of $\mathcal{T}^*Q$ is a differential 1-form on $\mathcal{T}_pQ$, the tangent space to Q at a point p. In a coordinates basis $(q^1, \ldots, q^n)$, a differential 1-form ϑ has components $p_1, \ldots, p_n$ and the $2n$ numbers $(p_1, \ldots, p_n, q^1, \ldots, q^n)$ can be taken as local coordinates of a point in $\mathcal{T}^*Q$.

Thus, the cotangent bundle $\mathcal{M} = \mathcal{T}^*Q$ has a natural structure of a $2n$-dimensional differential manifold.[2,1]

Moreover, it can be proven (see Appendix E) that $\mathcal{T}^*Q$ has a natural symplectic structure ω_c which, in local coordinates, can be written as follows:

$$\omega_c = dp_1 \wedge dq^1 + \cdots + dp_n \wedge dq^n \,, \tag{9.21}$$

or

$$\omega_c = d\vartheta_c \,,$$

with

$$\vartheta_c = p_i dq^i \,.$$

The differential forms ϑ_c and ω_c are called the *canonical differential 1-form* and the *canonical symplectic structure*, respectively.

But there is much more, in the sense that any symplectic manifold can be locally considered as a cotangent bundle. This is guaranteed by the Darboux* theorem,[2,1,7] according to which:

Theorem 29 (Darboux) *At every point p_0 of a 2n-dimensional symplectic manifold $\mathcal{M}$, there exists a chart $(\mathcal{U}, \varphi_D)$ in which the symplectic structure ω assumes the form*

$$\omega = dx^i \wedge dx^{i+n}, \quad i = 1, \ldots, n.$$

Such a chart $(\mathcal{U}, \varphi_D)$ is called a *Darboux chart.* In a Darboux chart, setting

$$(p_1 \equiv x^1, \ldots, p_n \equiv x^n, q^1 \equiv x^{n+1}, \ldots, q^n \equiv x^{2n}),$$

the symplectic structure ω and the bivector field Λ, given by Eqs. (9.3), assume the canonical forms

$$\omega_c = dp_i \wedge dq^i$$

and

$$\Lambda_c = \frac{\partial}{\partial p_i} \wedge \frac{\partial}{\partial q^i},$$

respectively. Moreover, the Eq. (9.3),

$$\frac{dx^i}{dt} = \Lambda^{ij} \frac{\partial H}{\partial x^j},$$

become the familiar Hamilton equations

$$\begin{cases} \dot{p}_i = -\dfrac{\partial H}{\partial q^i}, \\ \dot{q}^i = \dfrac{\partial H}{\partial p_i}, \end{cases} \forall i = 1, \ldots, n. \tag{9.22}$$

An atlas for $\mathcal{M}$, composed by Darboux charts, is called a *Darboux atlas* or a *symplectic atlas.*

*Gaston Darboux, born in Nimes in 1842 and died in Paris in 1917, has been a professor at the Sorbonne University for about 40 years. His work in four volumes on *Théorie des Surfaces* is considered a classic. Besides giving new and remarkable contributions to differential geometry, he deeply influenced the development of the theory of differential equations and, thanks to a deep geometrical insight and a sagacious use of algorithms, gave solutions to relevant problems in calculus and mechanics.

At this point it is clear that the *Hamiltonian formulation of the dynamics*, described in Part I (Analytical Mechanics) is, at least for systems which do not depend explicitly on time, the local version (i.e. in a Darboux chart) of the theory of Hamiltonian vector fields on a symplectic manifold $\mathcal{M}$.

9.5 Revisited Analytical Mechanics

The reader can discover by himself the global version of many results obtained in Part I.

Indeed,

- A system of particles has n-degrees of freedom if its configurations define an n-dimensional differential manifold Q. The *state space* of the system is the tangent bundle $\mathcal{T}Q$, while the *phase space* is the cotangent bundle $\mathcal{T}^*Q$.
- A Lagrangian function $\mathcal{L}$ is a differentiable map

$$\mathcal{L} : \mathcal{T}Q \to \Re ,$$

from $\mathcal{T}Q$ into $\Re$.
- Lagrange's equations

$$\begin{cases} \dfrac{dq^h}{dt} = v^h , \\ \dfrac{dv^h}{dt} = L^{hk} F_k , \end{cases}$$

can be written in the intrinsic form,

$$L_\Delta \vartheta_\mathcal{L} = d\mathcal{L} , \tag{9.23}$$

or

$$i_\Delta d\vartheta_\mathcal{L} = -dE_\mathcal{L} , \tag{9.24}$$

where

— Δ is the vector field given by $\Delta = v^h \partial/\partial q^h + L^{hk} F_k(q/v) \partial/\partial v^h$;
— L^{hk} are the elements of the matrix L^{-1}, with $L = (\partial^2 \mathcal{L}/\partial v^h v^k)$;
— $\vartheta_\mathcal{L}$ the differential 1-form on $\mathcal{T}Q$ defined by $\vartheta_\mathcal{L} = (\partial \mathcal{L}/\partial v^h) dq^h$;

— $E_{\mathcal{L}}$ is the energy

$$E_{\mathcal{L}} = i_{\Delta}\vartheta_{\mathcal{L}} - \mathcal{L}\,.$$

We notice that, if the Hessian determinant of the Lagrangian is not vanishing, then $\omega_{\mathcal{L}} = d\vartheta_{\mathcal{L}}$ is a symplectic structure on $\mathcal{T}Q$.

The intrinsic form of Lagrange's equation allows us to introduce the Nöther theorem as follows.

Consider a complete vector field X on $\mathcal{T}Q$; i.e. the generator of a one-parameter group φ_τ of diffeomorphisms on $\mathcal{T}Q$. Let us calculate the *infinitesimal transformation*, $\delta\mathcal{L} \equiv L_X\mathcal{L}$, which X induces on the Lagrangian function $\mathcal{L}$. From Eq. (9.23), we have

$$\begin{aligned}\delta\mathcal{L} \equiv L_X\mathcal{L} = i_X d\mathcal{L} = i_X L_{\Delta}\vartheta_{\mathcal{L}} &= \langle L_{\Delta}\vartheta_{\mathcal{L}}, X\rangle \\ &= L_{\Delta}\langle\vartheta_{\mathcal{L}}, X\rangle - \langle\vartheta_{\mathcal{L}}, L_{\Delta}X\rangle \\ &= i_{[X,\Delta]}\vartheta_{\mathcal{L}} + L_{\Delta}i_X\vartheta_{\mathcal{L}}\,.\end{aligned}$$

It follows that

$$L_X\mathcal{L} = 0, \text{and } [X,\Delta] = 0 \Rightarrow L_{\Delta}i_X\vartheta_{\mathcal{L}} = 0\,,$$

i.e.

Theorem 30 (Nöther) *A symmetry X of both the Lagrangian $\mathcal{L}$ and the dynamics Δ gives rise to a first integral given by $L_{\Delta}(i_X\vartheta_{\mathcal{L}})$.*

The translation of the previous geometrical formulation in coordinate language gives back the original formulation by Emmy Nöther.

Remark 17 *In order for $i_X\vartheta_{\mathcal{L}}$ to be a first integral, it suffices that $i_{[X,\Delta]}\vartheta_{\mathcal{L}} - L_X\mathcal{L}$ vanishes, which is less stringent than the separate vanishing of each term.*

- The Legendre transformation defines a vector bundle isomorphism between $\mathcal{T}Q$ and $\mathcal{T}^*Q$.
  Indeed, the map

$$f : (q,v) \in \mathcal{T}Q \longrightarrow (q,p) \in \mathcal{T}^*Q\,,$$

with $p_h = (\partial\mathcal{L}/\partial v^h)(q,v)$, induces the derivative map

$$f_* : X \in \mathcal{T}_{(q,v)}(\mathcal{T}Q) \longmapsto X_* = f_*X \in \mathcal{T}_{(q,p)}(\mathcal{T}^*Q)\,.$$

The Legendre transformation is then defined by

$$(q, v/Q, V) \longrightarrow (q, p/\tilde{Q}, P),$$

where Q, V denote the sets of "q" and the "v" components of the vector field X, respectively, and $\tilde{Q}, P$ the ones of the "q" and the "p" components of the vector field X_*.

$$X = Q^h \frac{\partial}{\partial q^h} + V^h \frac{\partial}{\partial v^h},$$

$$X_* = \tilde{Q}^h \frac{\partial}{\partial q^h} + P^h \frac{\partial}{\partial p^h}.$$

In matrix notation, setting

$$f_* = \begin{pmatrix} I & 0 \\ M & L \end{pmatrix},$$

with I the $n \times n$ identity matrix, and

$$M = \left(\frac{\partial^2 \mathcal{L}}{\partial v^h q^k} \right),$$

we have

$$\begin{pmatrix} \tilde{Q} \\ P \end{pmatrix} = \widetilde{\begin{pmatrix} I & 0 \\ M & L \end{pmatrix}} \cdot \widetilde{\begin{pmatrix} Q \\ V \end{pmatrix}} = \widetilde{\begin{pmatrix} I & 0 \\ M & L \end{pmatrix}} \cdot \widetilde{\begin{pmatrix} v \\ L^{-1}F \end{pmatrix}} = \begin{pmatrix} \alpha(q/p/t) \\ \widetilde{\left(\frac{\partial \mathcal{L}}{\partial q}\right)} \end{pmatrix},$$

where the *tilde* $\sim$ indicates that the velocities v's must be expressed in terms of the q's and p's by inverting the relations $p_h = (\partial \mathcal{L}/\partial v^h)(q, v)$. If the Lagrangian is degenerate; i.e. the Hessian determinant vanishes, the Legendre map defines only a vector bundle homomorphism from $\mathcal{T}Q$ into $\mathcal{T}^*Q$.

The theory of constraints by Dirac and Bergman[85,17,18,26] just starts from this observation. A geometrical analysis can be found in Refs. 41, 108, 177, 146 and 136.

- An algebraic formulation of Lagrangian dynamics, suitable to be used in a general context, including situations with no global Lagrangian and/or fermionic variables, can be found in Ref. 76.

- A symplectic transformation, as defined in Part I, is just a map between two Darboux chart $(\mathcal{U}, \varphi \equiv (p/q))$ and $(\mathcal{V}, \psi \equiv (\pi/\chi))$; that is, it is a map such that

$$dp_i \wedge dq^i = d\pi_i \wedge d\chi^i \,.$$

- The above relation is the exterior derivative of

$$p_i dq^i = \pi_i d\chi^i + dF \,,$$

which is just the Lie condition for a transformation to be symplectic.
- A completely canonical transformation is a map between two *almost* Darboux charts, in the sense that

$$dp_i \wedge dq^i = c d\pi_i \wedge d\chi^i \,.$$

- The Lagrange bracket

$$[p_i, q^j] = \frac{\partial \pi_h}{\partial p_i}\frac{\partial \chi^h}{\partial q^j} - \frac{\partial \pi_h}{\partial q^j}\frac{\partial \chi^h}{\partial p_i} \,,$$

in which the inversion of the position of covariant and contravariant indices is just caused by the old notations, can be then obtained much more easily by expanding the previous equality to the form

$$dp_i \wedge dq^i = c d\pi_h \wedge d\chi^h = c[p_i, q^j] dp_i \wedge dq^j \,,$$

which gives the familiar conditions for a transformation to be completely canonical,

$$c[p_i, q^j] = \delta^i_j \,.$$

- The Poisson bracket $\{\pi_h, \chi^k\}$ as defined in Part I is, *vice versa*, obtained expanding the inverse equality

$$\frac{\partial}{\partial \pi_h} \wedge \frac{\partial}{\partial \chi^h} = c \frac{\partial}{\partial p_i} \wedge \frac{\partial}{\partial q^i} \,,$$

in the inverted direction, to obtain

$$\frac{\partial}{\partial \pi_h} \wedge \frac{\partial}{\partial \chi^h} = c \left(\frac{\partial \pi_h}{\partial p_i}\frac{\partial \chi^k}{\partial q^i} - \frac{\partial \pi_h}{\partial q^i}\frac{\partial \chi^k}{\partial p_i} \right) \frac{\partial}{\partial \pi_h} \wedge \frac{\partial}{\partial \chi^k} \,,$$

which gives the old conditions for a transformation to be completely canonical,

$$c\{\pi_h, \chi^k\} = \delta_h^k \,,$$

this time with the right covariance of indices!

- The operator

$$X_f = \{f, \cdot\} = \frac{\partial f}{\partial p_i}\frac{\partial}{\partial q^i} - \frac{\partial f}{\partial q^i}\frac{\partial}{\partial p_i} \,,$$

that we called *Hamiltonian vector field* in Part I, is just the local expression of the Hamiltonian vector field

$$i_{X_f}\omega = df \,,$$

here introduced.

- A complete analogy exists between the intrinsic Lagrange equations and the Hamilton equations,

$$i_\Delta \omega_{\mathcal{L}} = -dE_{\mathcal{L}} \,, \quad i_X \omega = -dH \,.$$

The main difference between them, consists in the fact that, in the Hamilton equations, the "interaction" is present only in the Hamiltonian function, while in the Lagrange equations, the "interaction," *via* the Lagrangian function, is also present in the symplectic structure $\omega_{\mathcal{L}}$. In other words, the symplectic structure ω, in the Hamilton equations, is *universal*, in the sense that it does not depend on the considered dynamical system. This is not true for Lagrange's equations. This feature is a consequence of the fact that the cotangent bundle $\mathcal{T}^*Q$, of a manifold Q, carries a *natural* symplectic structure, while the tangent bundle $\mathcal{T}Q$ has not such a structure.

- A Nöther-type theorem, connecting a symmetry to a first integral, can be stated in the Hamiltonian formalism as well as in the Lagrangian, even more easily. Indeed, let Δ and X_f be globally Hamiltonian vector fields, with Hamiltonian functions given by H and f, respectively; i.e.

$$i_\Delta \omega = -dH \,, \quad i_{X_f}\omega = -df \,.$$

We thus have

$$\begin{aligned} L_{X_f}H = 0 \Leftrightarrow i_{X_f}dH = 0 \Leftrightarrow i_{X_f}i_\Delta\omega = 0 \Leftrightarrow \omega(X_f, \Delta) \\ = 0 \Leftrightarrow \{H, f\} = 0 \,, \end{aligned}$$

so that

$$L_{X_f}H = 0 \Leftrightarrow L_\Delta f = 0\,,$$

that is, to any symmetry of the Hamiltonian corresponds a constant of the motion and, *vice versa*, any constant of the motion is the *infinitesimal generator* of a symmetry transformation.
In other words,
A first integral, for a Hamiltonian dynamics, generates a one-parameter group of symplectomorphisms, which leaves the Hamiltonian function H invariant and, vice versa, with any one-parameter group of symplec-to-morphisms, leaving H invariant, we can associate a first integral.

- The Sec. (4.1), on the *Integral invariants*, can be revisited as follows. Let us observe that the Lie derivative, with respect to the vector field X, of the differential n-form

$$\alpha = \rho(x)dx^1 \wedge dx^2 \wedge \cdots \wedge dx^n\,,$$

on an n-dimensional manifold $\mathcal{M}$, is given by

$$L_X\alpha = di_X\alpha = \mathrm{div}(\rho\vec{X})dx^1 \wedge dx^2 \wedge \cdots \wedge dx^n\,,$$

so that the relation (4.5), for a function ρ not depending explicitly on time, simply says that

$$L_X \int_U \alpha = \int_U L_X\alpha\,.$$

It follows that a necessary and sufficient condition for $\int_U \alpha$ to be invariant is $L_X\alpha = 0$.
What has been said can be generalized as follows.
A differential k-form $\alpha \in \Lambda^k(\mathcal{M})$, on an n-dimensional manifold $\mathcal{M}$, is said to be an *absolute integral invariant* of the complete vector field X, if

$$L_X\alpha = 0\,.$$

The latter is equivalent to

$$\varphi_\tau^*(\alpha(\varphi_\tau(p))) = \alpha(p)\,, \tag{9.25}$$

where φ_τ denote the flow of the vector field X.

If U is a k-dimensional submanifold of $\mathcal{M}$ and i the immersion map

$$i : U \hookrightarrow \mathcal{M},$$

$(\varphi_\tau \circ i)(U)$ is a new k-dimensional submanifold of $\mathcal{M}$, and

$$\int_{(\varphi_\tau \circ i)(U)} \alpha = \int_U (\varphi_\tau \circ i)^* \alpha = \int_U (i^* \circ \varphi_\tau^*)\alpha .$$

It follows, from Eq. (9.25), that if α is invariant, then

$$\int_{(\varphi_\tau \circ i)(U)} \alpha = \int_U i^* \alpha = \int_{i(U)} \alpha .$$

Vice versa, if the relation

$$\int_{(\varphi_\tau \circ i)(U)} \alpha = \int_{i(U)} \alpha$$

holds, for any choice of U, i and τ, then $(i^* \circ \varphi_\tau^*)\alpha = i^*\alpha$ or, equivalently, $\varphi_\tau^* \alpha = \alpha$.

We can conclude that a necessary and sufficient condition for a differential k-form to be an absolute integral invariant is that

$$\int_{(\varphi_\tau \circ i)(U)} \alpha = \int_{i(U)} \alpha ,$$

for any choice of U, i and τ.

A differential $(k-1)$-form $\beta \in \Lambda^{k-1}(\mathcal{M})$, on an n-dimensional manifold $\mathcal{M}$, is said to be a *relative integral invariant* of the complete vector field X, if $d\beta$ is an absolute integral invariant; that is, if

$$dL_X \beta = L_X d\beta = 0 .$$

- A revisiting of the Hamilton–Jacobi theory can be found in Ref. 149.

Another difficult task is to globalize the Liouville theorem. Undertaking this task would also be useless, since as we shall see in the next section, it has already been accomplished,[2,185,1,3].

9.6 The Liouville Theorem

Let $(\mathcal{M}, \omega)$ be a $2n$-dimensional symplectic manifold on which n differentiable functions are defined

$$f_i : \mathcal{M} \to \Re, \quad \forall i = 1, \ldots, n.$$

Let us suppose that the functions $f_1, \ldots, f_n$ are in involution; i.e.

$$\{f_i, f_j\} = 0, \quad \forall i, j = 1, \ldots, n, \tag{9.26}$$

and that the n differential 1-forms $df_1, \ldots, df_n$ are linearly independent at every point p of the *level set* $\mathcal{M}_{f(\pi)}$ defined by

$$\mathcal{M}_{f(\pi)} = \{p \in \mathcal{M} : f_i(p) = \pi_i, \quad i = 1, \ldots, n\}.$$

From the implicit functions theorem, the level set $\mathcal{M}_{f(\pi)}$ is an n-dimensional submanifold of $\mathcal{M}$, which is called the *level manifold.*

Because of the isomorphism (9.2), with each differential 1-form df_i, we can associate a vector field X_{f_i} on $\mathcal{M}$ such that

$$i_{X_{f_i}} \omega = df_i .$$

These vector fields X_{f_i}, which are supposed to be complete, are linearly independent at every point of $\mathcal{M}_{f(\pi)}$ since the differentials $df_1, \ldots, df_n$ are linearly independent and the symplectic form ω is not degenerate.

In addition, by Eq. (9.26), the vector fields X_{f_i} commute each other,

$$[X_{f_i}, X_{f_j}] = 0, \quad \forall i, j = 1, \ldots, n.$$

Moreover, since

$$(L_{X_{f_j}(p)} f_i)(p) = (i_{X_{f_j}} df_i)(p) = df_i|_p(X_{f_j}(p)) = \{f_i, f_j\}(p) = 0,$$

the fields $X_{f_1}, \ldots, X_{f_n}$ are tangent to $\mathcal{M}_{f(\pi)}$.

Thus, there exist n commuting tangent vector fields on $\mathcal{M}_{f(\pi)}$ that are linearly independent at every point.

These vector fields form a local basis of an involutive distribution which, by Frobenius' theorem, is completely integrable.

Moreover, $\mathcal{M}_{f(\pi)}$ is invariant with respect to each one of the n commuting flows σ_i^t associated with the functions f_i.

It can be proven that the differential manifold $\mathcal{M}_{f(\pi)}$, if compact and connected, is diffeomorphic to an n-dimensional torus T^n, which admits the angles

$\varphi^1, \ldots, \varphi^n$, as local coordinates, being T^n the product of n circles. Indeed, let us observe that, by hypothesis, on $\mathcal{M}_{f(\pi)}$ there exist n functions f_i, which define an n-dimensional Abelian Lie algebra with the Poisson bracket as a Lie bracket. They generate, at each point, n independent flows under which $\mathcal{M}_{f(\pi)}$ is invariant. It follows that, *a priori*, $\mathcal{M}_{f(\pi)} \simeq \Re^{n-k} \times T^k$, but if $\mathcal{M}_{f(\pi)}$ is compact, we can only have $k = n$.

Under the action of the Hamiltonian flow, generated by $H = f_1$, the angular coordinates φ^i will change according to

$$\frac{d\varphi^i}{dt} = \omega^i, \quad \forall i = 1, \ldots, n,$$

where $\omega^i = \omega^i(f_1, \ldots, f_n)$, so that the motion on $\mathcal{M}_{f(\pi)}$

$$\varphi^i(t) = \varphi^i(0) + \omega^i t, \quad \forall i = 1, \ldots, n \tag{9.27}$$

is *almost periodic.*

Let us consider a neighborhood $\mathcal{U} \subseteq \mathcal{M}$ of $\mathcal{M}_{f(\pi)}$. If we use the functions $f_1, \ldots, f_n$ as coordinates in $\mathcal{U}$, we can find a neighborhood $\mathcal{U}' \subset \mathcal{U} \subset \mathcal{M}$ of $\mathcal{M}_{f(\pi)}$, which is diffeomorphic to the direct product $T^n \times S^n$, where S^n is a sphere of an n-dimensional Euclidean space; i.e. a neighborhood of π in $\Re^n$.

The Hamiltonian flow, generated by $H = f_1$, expressed in terms of coordinates $(\varphi^1, \ldots, \varphi^n, f_1, \ldots, f_n)$ becomes

$$\frac{df_i}{dt} = 0, \quad \frac{d\varphi^i}{dt} = \omega^i(f_1, \ldots, f_n), \quad \forall i = 1, \ldots, n. \tag{9.28}$$

The system (9.28) can be directly integrated to

$$f_i(t) = f_i(0), \quad \varphi^i(t) = \varphi^i(0) + \omega^i(f_1(0), \ldots, f_n(0))t, \quad \forall i = 1, \ldots, n.$$

The integration of the original canonical system is, then, equivalent to finding the angular variables $\varphi^1, \ldots, \varphi^n$. This can be done by only using quadratures.

What has been previously said concerning the compact case, can be summarized by the following theorem.[2]

Theorem 31 (Liouville) *If on the 2n-dimensional symplectic manifold $\mathcal{M}$ are defined n functions $f_1, \ldots, f_n$ in involution*

$$\{f_i, f_j\} = 0, \quad \forall i, j = 1, \ldots, n,$$

and the n differential 1-forms $df_1, \ldots, df_n$ are linearly independent at every point in the level manifold

$$\mathcal{M}_{f(\pi)} = \{p \in \mathcal{M} : f_i(p) = \pi_i\,, \quad i = 1, \ldots, n\}\,,$$

then

(a) *$\mathcal{M}_{f(\pi)}$ is an n-dimensional submanifold of $\mathcal{M}$, invariant with respect to the Hamiltonian flow generated by $H = f_1$;*
(b) *if compact and connected, $\mathcal{M}_{f(\pi)}$ is diffeomorphic to the n-dimensional torus T^n, with angular coordinates $(\varphi^1, \ldots, \varphi^n)$;*
(c) *the motion on $\mathcal{M}_{f(\pi)}$, determined by the Hamiltonian flow generated by H, is almost-periodic*

$$\frac{d\varphi^i}{dt} = \omega^i\,;$$

(d) *the canonical equations with Hamilton function H can be integrated by pure quadratures.*

Let us now observe that, in general, the coordinates $(f_1, \ldots, f_n, \varphi^1, \ldots, \varphi^n)$ do not form a symplectic coordinates system. However, there exist functions

$$J_h = J_h(f_1, \ldots, f_n)\,, \quad \forall i = 1, \ldots, n\,, \tag{9.29}$$

such that the coordinates $(J_1, \ldots, J_n, \varphi^1, \ldots, \varphi^n)$ are symplectic; that is, such that the original symplectic form ω can be expressed as

$$\omega = dJ_h \wedge d\varphi^h\,.$$

The variables (9.29), which conjugate with the angles, are called *action variables*; they are first integrals of the Hamiltonian flow generated by H.

In terms of these coordinates, the system (9.28) takes the form

$$\frac{dJ_i}{dt} = 0\,, \quad \frac{d\varphi^i}{dt} = \nu^i(J_1, \ldots, J_n)\,, \quad \forall i = 1, \ldots, n\,. \tag{9.30}$$

9.6.1 *The construction of the action-angle coordinates*

An analysis for the construction of global action coordinates can be found in Refs. 19 and 13 and a general analysis on the possibility to introduce "action-angle type" coordinates can be found in Refs. 158 and 41.

Let us consider the case in which the manifold $\mathcal{M}$ is a cotangent bundle, so that $\omega = d\vartheta_c = d(p_h dq^h)$.

Let us then consider the immersion

$$i : \mathcal{M}_{f(\pi)} \hookrightarrow \mathcal{M}$$

of the level manifold $\mathcal{M}_{f(\pi)}$ into $\mathcal{M}$ and the pull-back $i^*\omega$ to $\mathcal{M}_{f(\pi)}$ of the symplectic structure. Since $di^* = i^*d$, we have

$$di^*\omega = i^*d\omega = 0\,.$$

Thus, $i^*\omega$ is a closed differential 2-form on the torus. It is not an exact differential form since the torus is not simply connected; that is, there exist curves on the torus which cannot be contracted to a point. We have

$$i^*\omega = i^*d\vartheta_c = di^*\vartheta_c\,.$$

On the other hand, the vector fields $e_{f_i} \equiv X_{f_i}$ are a basis for vector field, which are tangent to $\mathcal{M}_{f(\pi)}$, so that, for any two such fields X and Y, we may write

$$X = X^i e_{f_i}\,, \quad Y = Y^i e_{f_i}\,.$$

It thus follows that

$$(i^*\omega)(X,Y) = X^iY^j(i^*\omega)(e_{f_i},e_{f_i}) = X^iY^j\{f_i,f_j\} = 0\,.$$

Therefore, over any bidimensional region Σ on the torus, we have

$$\int_\Sigma i^*\omega = 0\,.$$

Since two homotopic curves γ_1 and γ_2 on the torus will be the boundary of a two dimensional region Σ, we obtain

$$0 = \int_\Sigma i^*\omega = \int_{\gamma_1\cup\{-\gamma_2\}} i^*\vartheta_c\,,$$

by Stokes' theorem.

As a consequence, we have

$$\int_{\gamma_1} i^*\vartheta_c = \int_{\gamma_2} i^*\vartheta_c\,.$$

The curves on the torus T^n can be divided in n equivalence classes $[\gamma_h]$, each class containing noncontractible homotopic curves. The action variables are then defined as follows:

$$J_h = \frac{1}{2\pi}\int_{[\gamma_h]} i^*\vartheta_c = \frac{1}{2\pi}\int_{[\gamma_h]} p_h(q/\pi)dq^h ,$$

where $h = 1, 2, \ldots, n$.

The construction can be finally completed as in Sec. 4.6.4.

9.7 A New Characterization of Complete Integrability

In general, the peculiarities of a given dynamics Δ can be characterized by the invariance of some geometric structure. For instance, the symplectic character of a dynamics is characterized by the invariance of a symplectic structure. This is the case of both Lagrangian and Hamiltonian dynamics.

It is then interesting to note the question whether the integrability properties of a dynamics can be characterized from this point of view. As we shall see, this can be done.

Let us start with the following considerations.

Meaning of the vanishing Nijenhuis torsion of a mixed tensor field. A consequence of the vanishing Nijenhuis torsion $\mathcal{N}_T$, of a mixed tensor field T, is that, given a vector field Δ_1, the vector fields of the sequence

$$\Delta_{n+1} = \hat{T}\Delta_n , \quad n \geq 1 ,$$

close on an Abelian Lie algebra

$$[\Delta_n, \Delta_m] = 0 , \quad n, m \geq 1 ,$$

and that the transposed endomorphism $\check{T}$ generates sequences of exact differential 1-forms,[78] in the sense that

$$(d\alpha = 0 , \quad d\check{T}\alpha = 0 , \quad \mathcal{N}_T =.0) \Rightarrow d(\check{T}^n\alpha) = 0 , \quad n \geq 1 , \quad \alpha \in T^*_u\mathcal{M} .$$

Moreover, the invariance of T, under the flow generated by a vector field Δ, implies the invariance of the vector fields $\hat{T}\Delta_n$ and of the differential 1-forms $\check{T}^n\alpha$.

Properties of eigenvectors. It is also interesting to analyze the properties of vector fields which are eigenvectors of a torsionless diagonalizable mixed tensor field.

Let $\mathcal{M}$ be a differentiable manifold and T a diagonalizable mixed tensor field; i.e.

$$\hat{T}e_k = \lambda_k e_k\,, \quad \check{T}\vartheta^k = \lambda_k \vartheta^k\,,$$

where $\{e_k\}$ is a generic basis of $\mathcal{T}_p\mathcal{M}$,

$$[e_i, e_j] = c^k_{ij} e_k\,,$$

and $\{\vartheta^k\}$ its dual basis of $\mathcal{T}^*_p\mathcal{M}$,

$$d\vartheta^k = -\frac{1}{2} c^k_{rs} \vartheta^r \wedge \vartheta^s\,.$$

We recall that the Nijenhuis torsion of T is defined by

$$\mathcal{N}_T(\alpha, X, Y) = \langle \alpha, \mathcal{H}_T(X, Y)\rangle\,,$$

with

$$\begin{aligned}\mathcal{H}_T(X,Y) &= (L_{\hat{T}X}T)^\wedge Y - \hat{T}(L_X T)^\wedge Y \\ &= [\hat{T}X, \hat{T}Y] + \hat{T}^2[X,Y] - \hat{T}[\hat{T}X, Y] - \hat{T}[X, \hat{T}Y]\,.\end{aligned}$$

Let us evaluate $\mathcal{H}_T$ on the basis $\{e_k\}$:

$$\mathcal{H}_T(e_i, e_j) = [\hat{T}e_i, \hat{T}e_j] + \hat{T}^2[e_i, e_j] - \hat{T}[\hat{T}e_i, e_j] - \hat{T}[e_i, \hat{T}e_j]\,. \tag{9.31}$$

Since, for any two differentiable functions f and g, and any two vector fields X and Y on $\mathcal{M}$, we may write

$$[fX, gY] = fg[X,Y] + g(L_Y f)X - f(L_X g)Y\,,$$

and have

$$\begin{aligned}[\hat{T}e_i, \hat{T}e_j] &= [\lambda_i e_i, \lambda_j e_j] = \lambda_i\lambda_j[e_i, e_j] + \lambda_j(L_{e_j}\lambda_i)e_i - \lambda_i(L_{e_i}\lambda_j)e_j\,,\\ \hat{T}[\hat{T}e_i, e_j] &= \hat{T}(\lambda_i[e_i, e_j] + (L_{e_j}\lambda_i)e_i) = \lambda_i\hat{T}[e_i, e_j] + \lambda_i(L_{e_j}\lambda_i)e_i\,,\\ \hat{T}[e_i, \hat{T}e_j] &= \hat{T}(\lambda_j[e_i, e_j] - (L_{e_i}\lambda_j)e_j) = \lambda_j\hat{T}[e_i, e_j] - \lambda_j(L_{e_i}\lambda_j)e_j\,.\end{aligned}$$

Thus, the relation (9.31) becomes

$$\mathcal{H}_T(e_i, e_j) = (\hat{T} - \lambda_i)(\hat{T} - \lambda_j)[e_i, e_j] + (\lambda_j - \lambda_i)[(\mathcal{L}_{e_i}\lambda_j)e_j + (\mathcal{L}_{e_j}\lambda_i)e_i]$$

and the vanishing of Nijenhuis torsion $\mathcal{H}_T(e_i, e_j) = 0$ implies the following:

$$(\hat{T} - \lambda_i)(\hat{T} - \lambda_j)[e_i, e_j] = 0\,, \tag{9.32}$$

$$(\lambda_i - \lambda_j)\mathcal{L}_{e_i}\lambda_j = 0\,. \tag{9.33}$$

It follows that, if the eigenvalues λ_k of T are supposed to have nowhere vanishing differentials

$$(d\lambda^j)_p \neq 0\,, \quad \forall p \in \mathcal{M}\,,$$

and to be doubly degenerate, then the two vector fields e_i and e_j, belonging to the same eigenvalue $\lambda_i = \lambda_j$, satisfy the relation

$$[e_i, e_j] = ae_i + be_j\,. \tag{9.34}$$

Therefore, the vector fields e_i, e_j are a local basis of a 2-dimensional involutive distribution and, by Frobenius' theorem, define a 2-dimensional submanifold of $\mathcal{M}$.

A dual point of view is that, by contracting the relation (9.32) with the elements of the dual basis, we also find

$$\begin{aligned}
0 &= (\lambda_k - \lambda_i)(\lambda_k - \lambda_j)\langle [e_i, e_j], \vartheta^k\rangle \\
&= -(\lambda_k - \lambda_i)(\lambda_k - \lambda_j)\langle e_j, L_{e_i}\vartheta^k\rangle \\
&= -(\lambda_k - \lambda_i)(\lambda_k - \lambda_j)\langle e_j, i_{e_i} d\vartheta^k\rangle \\
&= \frac{1}{2}(\lambda_k - \lambda_i)(\lambda_k - \lambda_j)\langle e_j, i_{e_i} c^k_{rs}\vartheta^r \wedge \vartheta^s\rangle \\
&= c^k_{rs}\frac{1}{2}(\lambda_k - \lambda_i)(\lambda_k - \lambda_j)\langle e_j, \delta^r_i\vartheta^s - \delta^s_i\vartheta^r\rangle \\
&= c^k_{rs}\frac{1}{2}(\lambda_k - \lambda_i)(\lambda_k - \lambda_j)(\delta^r_i\delta^s_j - \delta^r_j\delta^s_i) \\
&= c^k_{ij}(\lambda_k - \lambda_i)(\lambda_k - \lambda_j)\,,
\end{aligned} \tag{9.35}$$

where the relation $[e_i, e_j] = c^k_{ij}e_k$ has been used.

In this way, the relation (9.35) simply says that

$$c^k_{ij} = 0\,, \quad \forall k \neq i \text{ and } k \neq j\,, \tag{9.36}$$

so that we discover again Eq. (9.34). Moreover, we also have

$$d\vartheta^k = c^k_{ks}\vartheta^s \wedge \vartheta^k \quad (\textit{no sum over } k)\,.$$

The last relation implies that

$$\vartheta^k \wedge d\vartheta^k = 0\,,$$

which again, by Frobenius' theorem (in the dual form), ensures the holonomicity of the basis.

In conclusion, the relations (9.32) or (9.35), which directly follows from the Nijenhuis condition, ensure the holonomicity of the basis $\{e_k\}$, in which the tensor field T is diagonal,

$$T = \sum_i \lambda_i e_i \otimes \vartheta^i\,. \tag{9.37}$$

Invariance of the eigenvalues of an invariant mixed tensor field. It is easy to check that the invariance of T, under the flow generated by a vector field Δ, implies the invariance of its eigenvalues λ.

Indeed, let $V \in \mathcal{T}_p\mathcal{M}$ and $\alpha \in \mathcal{T}_p^*\mathcal{M}$ be eigenvectors of $\hat{T}$ and $\check{T}$, respectively,

$$\hat{T}V = \lambda V\,, \quad \check{T}\alpha = \lambda\alpha\,,$$

belonging to the same eigenvalue λ, such that $i_V\alpha \neq 0$.

If T is supposed to be Δ-invariant, we have

$$L_\Delta(\hat{T}V) = (L_\Delta T)^\wedge V + \hat{T}(L_\Delta V) = \hat{T}(L_\Delta V)\,, \quad V \in \mathcal{T}_p\mathcal{M}\,, \tag{9.38}$$

so that

$$\hat{T}V = \lambda V \Rightarrow \hat{T}(L_\Delta V) = L_\Delta(\hat{T}V) = (L_\Delta\lambda)V + \lambda(L_\Delta V)\,,$$

and

$$\langle L_\Delta V, \check{T}\alpha\rangle = \langle \hat{T}L_\Delta V, \alpha\rangle = (L_\Delta\lambda)\langle V, \alpha\rangle + \langle\lambda L_\Delta V, \alpha\rangle\,.$$

Then, from

$$\langle L_\Delta V, \check{T}\alpha\rangle = \langle L_\Delta V, \lambda\alpha\rangle = \langle\lambda L_\Delta V, \alpha\rangle\,,$$

we finally have

$$(L_\Delta\lambda)\langle V, \alpha\rangle = 0 \Rightarrow L_\Delta\lambda = 0\,;$$

that is, we obtain the invariance of λ under the flow generated by Δ.

If a tensor field T is invariant under the flow generated by a vector field Δ, the vector field Δ is said to be an *automorphism* of the tensor field T.

Peculiarities of automorphisms of a torsionless mixed tensor field. The Δ-invariance implies Eq. (9.38), so that

$$\begin{aligned}(\lambda_i - \lambda_j)L_{e_i}\langle\Delta, \vartheta^j\rangle &= \lambda_i L_{e_i}\langle\Delta, \vartheta^j\rangle - \lambda_j L_{e_i}\langle\Delta, \vartheta^j\rangle \\ &= \lambda_i\langle L_{e_i}\Delta, \vartheta^j\rangle - \lambda_j\langle L_{e_i}\Delta, \vartheta^j\rangle \\ &= -\lambda_i\langle L_\Delta e_i, \vartheta^j\rangle + \langle L_\Delta e_i, \lambda_j\vartheta^j\rangle \\ &= -\lambda_i\langle L_\Delta e_i, \vartheta^j\rangle + \langle L_\Delta e_i, \check{T}\vartheta^j\rangle \\ &= -\lambda_i\langle L_\Delta e_i, \vartheta^j\rangle + \langle \hat{T} L_\Delta e_i, \vartheta^j\rangle \\ &= -\lambda_i\langle L_\Delta e_i, \vartheta^j\rangle + \langle L_\Delta \hat{T} e_i, \vartheta^j\rangle \\ &= -\lambda_i\langle L_\Delta e_i, \vartheta^j\rangle + \lambda_i\langle L_\Delta e_i, \vartheta^j\rangle \\ &= 0\,.\end{aligned}$$

At this point, it is worth recalling that a dynamical vector field Δ is said to be *separable*, in dynamics with smaller dimensions, in an open set $O \subseteq \mathcal{M}$, if a frame $\{e_i\}$ exists such that

$$L_{e_i}\langle\Delta, \vartheta^j\rangle \neq 0 \Rightarrow i = j\,,$$

where $\{\vartheta^j\}$ is the dual basis of $\{e_i\}$. If O coincides with $\mathcal{M}$, we'll say that Δ is separable.

Since, as it has been shown,

$$L_\Delta T = 0 \Rightarrow (\lambda_i - \lambda_j)L_{e_i}\langle\Delta, \vartheta^j\rangle = 0\,,$$

the Δ-invariance of T implies the separability of the dynamics.

Remark 18 *This notion of separability is different from the one (see Ref. 65) in the Hamilton–Jacobi theory.*

Equation (9.33) can also be written in the form

$$\lambda_i L_{e_i}\lambda_j = \lambda_j L_{e_i}\lambda_j\,,$$

so that

$$\check{T} d\lambda_j = \check{T}\vartheta^i L_{e_i}\lambda_j = \lambda_i\vartheta^i L_{e_i}\lambda_j = \lambda_j\vartheta^i L_{e_i}\lambda_j = \lambda_j d\lambda_j\,. \tag{9.39}$$

Since the eigenvalues of T are doubly degenerate, the decomposition (9.37) can also be written in the form

$$T = \sum_{j=1}^{n} \lambda_j (e_j \otimes \vartheta^j + e_{n+j} \otimes \vartheta^{j+n}) .$$

By means of Eq. (9.39), which implies the functional independence of the λ_j's, and, as a consequence, the linear independence of $d\lambda_j$'s, it is now possible to choose the basis in such a way as T has the following expression:

$$T = \sum_{j=1}^{n} \lambda_j (e_j \otimes \vartheta^j + e_{n+j} \otimes d\lambda^j) , \tag{9.40}$$

that is, as if $d\lambda_j \equiv d\lambda^j$ were part of such basis.

Equation (9.40) explicitly shows the integrability of the projected dynamics.

The equation $\dot{x} = \Delta(x)$ can be decomposed in the following decoupled systems:

$$\begin{cases} i_{\dot{x}} \vartheta^j = i_\Delta \vartheta^j , \\ i_{\dot{x}} d\lambda^j = 0 , \end{cases} \qquad j = 1, \ldots, n . \tag{9.41}$$

Equation (9.40) can be rewritten in terms of the coordinates (φ^j, λ^j) in the form

$$T = \sum_{j=1}^{n} \lambda_j \left(\frac{\partial}{\partial \varphi^j} \otimes d\varphi^j + \frac{\partial}{\partial \lambda^j} \otimes d\lambda^j \right) ,$$

where the λ's are defined globally on $\mathcal{M}$, while the φ's, such that $d\varphi = \vartheta^j$, can be defined only locally on $\mathcal{M}$; in this way, all fields satisfying the equation $L_\Delta T = 0$ can be expressed as follows:

$$\Delta = \Delta^j (\lambda^j, \varphi^j) \frac{\partial}{\partial \varphi^j} ,$$

and the systems (9.41) become

$$\begin{cases} \dot{\varphi}^j = \Delta^j (\lambda^j, \varphi^j) , \\ \dot{\lambda}^j = 0 , \end{cases} \qquad j = 1, \ldots, n . \tag{9.42}$$

It is easy to check that the separable and integrable vector field Δ is also a Hamiltonian vector field.

In fact, given Δ, we can build many invariant symplectic structures ω

$$\omega = \sum_k f_k(\lambda^k, \varphi^k) d\varphi^k \wedge d\lambda^k, \tag{9.43}$$

where f_k are arbitrary functions required to ensure the invariance of the differential 2-form (Eq. (9.43)). If we suppose that the field Δ has not got singular points, the generic symplectic structure ω will have the form

$$\omega = \sum_k \frac{g_k(\lambda^k)}{\Delta^k} d\varphi^k \wedge d\lambda^k .$$

Choosing, as a basis, the one associated to the action-angle variables (J^k, φ^k), the tensor field T becomes

$$T = \sum_k \lambda^k(J^k) \left(\frac{\partial}{\partial J^k} \otimes dJ^k + \frac{\partial}{\partial \varphi^k} \otimes d\varphi^k \right),$$

and ω takes the following form:

$$\omega = \sum_k dJ^k \wedge d\varphi^k .$$

What has been said in the present section can be summarized as follows.[78,80]

Theorem 32 (DMSV) *Let Δ be a dynamical vector field on a manifold $\mathcal{M}$ which admits a diagonalizable mixed tensor field T which*

- *is invariant*

$$L_\Delta T = 0 ,$$

- *has a vanishing Nijenhuis torsion*

$$\mathcal{N}_T = 0 ,$$

- *has doubly degenerate eigenvalues λ^j with nowhere vanishing differentials*

$$\deg \lambda^j = 2 , \quad (d\lambda^j)_p \neq 0 , \quad \forall p \in \mathcal{M} .$$

Then, the vector field Δ is separable, completely integrable and Hamiltonian.

Remark 19 *The conditions $L_\Delta T = 0$ and $\mathcal{N}_T = 0$ and the bidimensionality of the eigenspaces of T was extracted from the existence of dynamics with infinitely many degrees of freedom, admitting a Lax representation (see Part IV). The fact that nonlinear field theories, integrable with the inverse scattering method show an endomorphism, invariant under the dynamics, with vanishing Nijenhuis torsion and bidimensional invariant eigenspaces, suggested that the analysis of the integrability of dynamical systems could be realized, instead that in terms of a mixed tensor field T, rather than symplectic structure ω.*

The integrability conditions in terms of symplectic structures ω strictly depend on the finite dimensionality of the space and cannot easily be extended to the infinite-dimensional case. On the contrary, the integrability in terms of T is expressed by conditions which do not depend on the finite number of degrees of freedom of the dynamical system Δ.

Remark 20 *It is worth remarking that the vector field Δ is not taken to be a priori a Hamiltonian vector field. As we shall see in Part IV, integrability of dissipative dynamics can be put in the same setting by assuming different spectral hypothesis for the tensor field T.*

9.7.1 *From the Liouville integrability to invariant mixed tensor fields*

Let us now study the problem of constructing invariant mixed tensor fields, with the appropriate properties (also called a *recursion operator*), for a given Liouville's integrable Hamiltonian dynamics Δ. If H is the Hamiltonian function and $\{\cdot,\cdot\}$ is the Poisson bracket, we have

$$\Delta f = \{H, f\}\,.$$

Let us introduce in some neighborhood of a Liouville's torus T^n action-angle variables $(J_1,\ldots,J_n,\varphi^1,\ldots,\varphi^n)$.

We have

$$dJ_1 \wedge dJ_2 \wedge \cdots \wedge dJ_n \neq 0\,, \quad \{H, J_h\} = 0\,,$$

or equivalently,

$$dH \wedge dJ_1 \wedge dJ_2 \wedge \cdots \wedge dJ_n = 0\,,$$

$$\omega = \sum_h dJ_h \wedge d\varphi^h ,$$

$$\Delta = \frac{\partial H}{\partial J_h} \cdot \frac{\partial}{\partial \varphi^h} .$$

Let us distinguish the two following cases:

- The Hamiltonian H is a separable one

$$H = \sum_k H_k(J_k) .$$

  In this case a class of recursion tensor fields can be easily defined as

$$T = \sum_h \lambda_h(J_h) \left(dJ_h \otimes \frac{\partial}{\partial J_h} + d\varphi^h \otimes \frac{\partial}{\partial \varphi^h} \right) ,$$

  with the λ's arbitrary functions required to have nowhere vanishing differentials. Indeed, the tensor field T is invariant and has vanishing Nijenhuis torsion and doubly degenerate eigenvalues.
- The Hamiltonian has a nonvanishing Hessian

$$\det \left(\frac{\partial^2 H}{\partial J_h \partial J_k} \right) \neq 0 .$$

  In this case new coordinates

$$\nu^h(J) = \frac{\partial H}{\partial J_h} ,$$

  which satisfy the condition

$$d\nu^1 \wedge d\nu^2 \wedge \cdots \wedge d\nu^h \neq 0 ,$$

  can be introduced.
  A new symplectic structure in this neighborhood can be then defined as

$$\omega_1 = \sum_h d\nu^h \wedge d\varphi^h = \sum_{hk} \frac{\partial^2 H}{\partial J_h \partial J_k} dJ_k \wedge d\varphi^h ,$$

  with respect to which the Hamiltonian becomes a separable one

$$H = \frac{1}{2} \sum_h (\nu^h)^2 .$$

The class of recursion tensor fields is then given by

$$T = \sum_h \lambda_h(\nu^h) \left(d\nu^h \otimes \frac{\partial}{\partial \nu^h} + d\varphi^h \otimes \frac{\partial}{\partial \varphi^h} \right) .$$

By means of this construction, it is possible to find the second symplectic structure for a completely integrable Hamiltonian system.

9.8 Applications

9.8.1 *A Recursion operator for the rigid body dynamics*

An invariant mixed tensor field, with vanishing Nijenhuis tensor and doubly degenerate eigenvalues, can be easily constructed,[86] for the Lagrange–Poisson gyroscope dynamics, without gravity for the sake of simplicity, by using the constants of the motion found by Mishenko, Dikii, Manakov, and Ratiu.[154,84,141,165]

The Hamiltonian function for the rigid body is locally given by

$$H = \frac{1}{2} \left(\frac{(p_\vartheta \cos\varphi + \sigma \sin\varphi)^2}{\mathcal{A}} + \frac{(p_\vartheta \sin\varphi - \sigma \cos\varphi)^2}{\mathcal{B}} + \frac{p_\varphi^2}{\mathcal{C}} \right) ,$$

where ϑ, φ and ψ are the Euler angles (of the body principal axes frame $Oxyz$ with respect to a generic fixed frame $O\xi\eta\varsigma$), p_ϑ, p_φ and p_ψ their conjugate variables, and $\mathcal{A}$, $\mathcal{B}$ and $\mathcal{C}$ the components of the inertial tensor with respect to $Oxyz$, and

$$\sigma = \frac{p_\psi - p_\varphi \sin\vartheta}{\sin\vartheta} .$$

When $\mathcal{A} = \mathcal{B}$ the Hamiltonian H reduces to

$$H_S = \frac{1}{2} \left(\frac{p_\vartheta^2 + \sigma^2}{\mathcal{A}} + \frac{p_\varphi^2}{\mathcal{C}} \right) ,$$

and the rigid body is said to possess a gyroscopic structure (*Lagrange–Poisson gyroscope*). Its complete integrability is obviously granted by the Liouville theorem in the open submanifold where H, p_φ and p_ψ are independent.

The tensor field defined by

$$T = T^i_j \frac{\partial}{\partial u^i} \otimes du^j \,,$$

with $u \equiv (\vartheta, \varphi, \psi, p_\vartheta, p_\varphi, p_\psi)$ and the matrix $\hat{T} = (T^i_j)$ given by

$$\hat{T} = \begin{pmatrix} L & 0 & 0 & 0 & \dfrac{\sigma}{L+p_\varphi} & \dfrac{\tau}{L+p_\psi} \\ \dfrac{p_\vartheta \tau}{(L+p_\varphi)\sin\vartheta} & p_\varphi & 0 & -\dfrac{\sigma}{L+p_\varphi} & 0 & -N \\ \dfrac{p_\vartheta \sigma}{(L+p_\psi)\sin\vartheta} & 0 & p_\psi & -\dfrac{\tau}{L+p_\psi} & N & 0 \\ 0 & 0 & 0 & L & \dfrac{p_\vartheta \tau}{(L+p_\varphi)\sin\vartheta} & \dfrac{p_\vartheta \sigma}{(L+p_\psi)\sin\vartheta} \\ 0 & 0 & 0 & 0 & p_\varphi & 0 \\ 0 & 0 & 0 & 0 & 0 & p_\psi \end{pmatrix},$$

fulfills the following properties:

- $L_\Delta T = 0$
- $\mathcal{H}_T = 0$
- deg (eigenvalues of T) = 2,

where

$$\Delta = -\frac{p_\vartheta}{\mathcal{A}} \frac{\partial}{\partial\vartheta} + \left(\frac{\sigma}{\mathcal{A}} \cot\vartheta - \frac{p_\varphi}{\mathcal{C}}\right) \frac{\partial}{\partial\varphi} - \frac{\sigma}{\mathcal{A}\sin\vartheta} \frac{\partial}{\partial\psi} + \frac{\sigma\tau}{\mathcal{A}\sin\vartheta} \frac{\partial}{\partial p_\vartheta}$$

denotes the Hamiltonian vector field corresponding to H_S

$$i_\Delta \omega_c = -dH_S \,,$$

by means of the canonical symplectic structure ω_c.

The above properties can be easily verified by using the action-angle coordinates $v \equiv (\varphi^1, \varphi^2, \varphi^3, J_1, J_2, J_3)$ linked to $u \equiv (\vartheta, \varphi, \psi, p_\vartheta, p_\varphi, p_\psi)$ by the following symplectic map:

$$\begin{cases} \vartheta = \arccos \dfrac{\eta + \cos\varphi^1}{\xi}\,, \\[2ex] \varphi = \varphi^2 - \arctan\left(J_1 \dfrac{\xi - \eta + 1}{\xi(J_3 - J_2)} \tan\dfrac{\varphi^1}{2}\right) + \arctan\left(J_1 \dfrac{\xi + \eta - 1}{\xi(J_3 + J_2)} \tan\dfrac{\varphi^1}{2}\right), \\[2ex] \psi = \varphi^3 - \arctan\left(J_1 \dfrac{\xi - \eta + 1}{\xi(J_3 - J_2)} \tan\dfrac{\varphi^1}{2}\right) + \arctan\left(J_1 \dfrac{\xi + \eta - 1}{\xi(J_3 + J_2)} \tan\dfrac{\varphi^1}{2}\right), \\[2ex] p_\vartheta = \dfrac{J_1 \sin\varphi^1}{[\xi^2 - (\eta + \cos\varphi^1)^2]^{\frac{1}{2}}}\,, \\[2ex] p_\varphi = J_2\,, \\[1ex] p_\psi = J_3\,, \end{cases}$$

where

$$\begin{cases} \xi = \dfrac{J_1^2}{[(J_3^2 - J_1^2)(J_2^2 - J_1^2)]^{\frac{1}{2}}}\,, \\[2ex] \eta = \dfrac{J_2 J_3}{[(J_3^2 - J_1^2)(J_2^2 - J_1^2)]^{\frac{1}{2}}}\,. \end{cases}$$

In these coordinates the tensor field T has the form

$$T = \tilde{T}^i_j \frac{\partial}{\partial v^i} \otimes dv^j$$

with

$$\tilde{T} = \mathrm{diag}\,(J_1, J_2, J_3, J_1, J_2, J_3)\,.$$

On the other hand, the complete integrability can be explained in terms of coadjoint orbits of Lie groups[62] so that the previous invariant tensor field can be useful to establish a connection with completely integrable systems on coadjoint orbits of a Lie group.[61,178,60]

9.8.2 A Recursion operator for the Kepler dynamics

The vector field for the Kepler problem, in spherical-polar coordinates, for $\Re^3 - \{0\}$, is given by

$$\Delta = \frac{1}{m}\left(p_r\frac{\partial}{\partial r} + \frac{p_\vartheta}{r^2}\frac{\partial}{\partial\vartheta} + \frac{p_\varphi}{r^2\sin^2\vartheta}\frac{\partial}{\partial\varphi} - \frac{1}{r^2}\left[mk + \frac{(p_\vartheta^2\sin^2\vartheta + p_\varphi^2)}{r\sin^2\vartheta}\right]\frac{\partial}{\partial p_r} - \frac{p_\varphi^2\cos\vartheta}{r^2\sin^3\vartheta}\frac{\partial}{\partial p_\vartheta}\right).$$

It is globally Hamiltonian with respect to the following symplectic form:

$$\omega = \sum_i dp_i \wedge dq^i\,, \quad i = r, \vartheta, \varphi\,, \tag{9.44}$$

with the Hamiltonian H given by

$$H = \frac{1}{2m}\left(p_r^2 + \frac{p_\vartheta^2}{r^2} + \frac{p_\varphi^2}{r^2\sin^2\vartheta}\right) + U(r)\,, \quad U(r) = -\frac{k}{r}\,.$$

In action-angle coordinates (J, φ), the Kepler Hamiltonian H, the symplectic form ω and the vector field Δ become

$$H = -\frac{mk^2}{(J_r + J_\vartheta + J_\varphi)^2}\,,$$

$$\omega = \sum_h dJ_h \wedge d\varphi^h\,,$$

$$\Delta = \frac{2mk^2}{(J_r + J_\vartheta + J_\varphi)^3}\left(\frac{\partial}{\partial\varphi^1} + \frac{\partial}{\partial\varphi^2} + \frac{\partial}{\partial\varphi^3}\right).$$

Unfortunately, the Hessian of the Hamiltonian identically vanishes and we cannot apply the previously described methods for the construction of the recursion operator. Nevertheless, starting from the observation that the Hamiltonian depends only upon the sum of action variables, it is possible to define a new coordinates system in which the Hamiltonian appears to be separated. In these new coordinates we can easily apply the previous methods and then, using the covariance of our formulation, construct a recursion operator in the original coordinates.

The results can be summarized as follows:[150]

The vector field Δ is globally Hamiltonian also with respect to the symplectic form ω_1, where

$$\omega_1 = \sum_{hk} S_k^h dJ_h \wedge d\varphi^k , \tag{9.45}$$

with the Hamiltonian H_1 given by

$$H_1 = -\frac{2mk^2}{J_r + J_\vartheta + J_\varphi} ,$$

or equivalently,

$$\Delta = \{H_1, \cdot\}_1 ,$$

where

$$\{f, g\}_1 = \sum_{hk} (S^{-1})_h^k \left(\frac{\partial f}{\partial J_h} \frac{\partial g}{\partial \varphi^k} - \frac{\partial f}{\partial \varphi^k} \frac{\partial g}{\partial J_h} \right) ,$$

and the matrix S is defined by

$$S = \frac{1}{2} \begin{pmatrix} J_1 & J_2 & J_3 \\ J_2 - J_3 & J_1 + J_3 & J_3 \\ J_3 - J_2 & J_2 & J_1 + J_2 \end{pmatrix} .$$

Remark 21 *The matrix S cannot be identified as a transformation Jacobian as it is clear from the fact that the $Sd\phi^h$'s are not closed 1-forms.*

In the original coordinates (p, q), the symplectic form ω_1 is simply written as follows:

$$\omega_1 = \sum_i dK_i \wedge d\alpha^i , \tag{9.46}$$

where the functions $K_i(p, q)$ and $\alpha^i(p, q)$, defined by

$$\begin{cases} K_1 = \frac{1}{4}[J_1^2 + (J_2 - J_3)^2](p, q) , \\ K_2 = \frac{1}{2} J_2 [J_1 + J_3](p, q) , \\ K_3 = \frac{1}{2} J_3 [J_1 + J_2](p, q) , \\ \alpha^i = \varphi^i(p, q) , \end{cases}$$

are considered as functions of p, q by means of the map $J_i = J_i(p,q), \varphi^i = \varphi^i(p,q)$.

As a consequence, a mixed invariant tensor field T, defined for nondegenerate ω by

$$\omega(\hat{T}X, Y) = \omega_1(X, Y),$$

can be constructed.

The vanishing of the Nijenhuis torsion and the double degeneracy of the eigenvalues of T is more easily checked, however, in the angle-action coordinates, where the tensor field T is simply written as

$$T = \sum_{hk} \left(S_k^h dJ_h \otimes \frac{\partial}{\partial J_k} + (S^+)_h^k d\varphi^h \otimes \frac{\partial}{\partial \varphi^k} \right).$$

Moreover, we have

$$\check{T} dH = k \left(-\frac{m}{H} \right)^{\frac{1}{2}} dH.$$

Thus, the iterated application of T does not produce new functionally independent constants of the motion. It has been shown that this particular situation prevails for periodic systems when the period P is a smooth function of the initial condition.[150]

It is now clear that all various alternative Hamiltonian descriptions that we may build, *via* a recursion operator T, will satisfy

$$dP \wedge (\check{T})^k dH = 0,$$

i.e.

$$dP \neq 0 \to (\check{T})^k dH \wedge (\check{T})^{k+r} dH = 0.$$

However, in this finite dimensional setting, $\{\mathrm{Tr}(\hat{T})^k, \mathrm{Tr}(\hat{T})^h\} = 0$, and $\mathrm{Tr}\hat{T}$, $\mathrm{Tr}(\hat{T})^2$, $\mathrm{Tr}(\hat{T})^3$ are functionally independent. On the other hand, in the infinite dimensional case, it is not easy to give a meaning to the trace of an endomorphism.

The Γ scheme

Let us observe that the symplectic form ω_1, given by Eq. (9.46), can be considered[181] as the Lie derivative of the symplectic form ω, given by Eq. (9.44), with respect to the vector field

$$\Gamma = K_h \frac{\partial}{\partial J_h},$$

so that

$$\omega_1 = L_\Gamma \omega .$$

The vector field Γ generates a sequence of finitely many (Abelian) symmetries according to the following scheme:

$$\Delta_{h+1} = [\Delta_h, \Gamma],$$

where $\Delta_0 = \Delta$ and where the bracket $[\cdot, \cdot]$ denotes the usual commutator between differential operators. Such vector fields turn out to be Hamiltonian with respect to both the symplectic structures, so that

$$\Delta_k = \{H_k, \cdot\} = \{H_{k+1}, \cdot\}_1,$$

and commute between them

$$[\Delta_h, \Delta_k] = 0 .$$

9.9 Poisson–Nijenhuis Structures

We may also mention a somewhat different approach to the same problem when the manifold $\mathcal{M}$ is supposed, from the very beginning, to be equipped with a Poisson structure Λ, so that $(\mathcal{M}, \Lambda)$ is a Poisson manifold.

9.9.1 *Compatible Poisson pairs*

Following Ref. 139, we shall say that a *Poisson–Nijenhuis structure* is defined on the manifold $\mathcal{M}$; if on $\mathcal{M}$ are defined, simultaneously a Poisson tensor field Λ and a Nijenhuis torsionless tensor field T that satisfy the following coupling conditions:

$$\begin{aligned} &(a)\ \hat{T}\check{\Lambda} = \check{\Lambda}\check{T} \\ &(b)\ \check{\Lambda}L_{\hat{T}X}\alpha - \check{\Lambda}L_X(\check{T}\alpha) + (L_{\check{\Lambda}\alpha}T)^\vee(X) = 0 , \end{aligned} \tag{9.47}$$

for arbitrary choices of the vector field X and the differential 1-form α.

As a matter of fact, we shall see that, on the same manifold, there are infinitely Poisson–Nijenhuis structures, because it turns out that all the tensors $T^k\Lambda$, for $k = 1, 2, \ldots$, are Poisson tensors too and satisfy the coupling condition.

The structure we have introduced seems very specific, but it is interesting to note that it is very natural for soliton dynamics. In fact, almost in every approach to the theory of completely integrable systems, one can notice that a crucial role is played by the so-called compatible Poisson tensors,[88,22,139,140,121] or as they are also called, Hamiltonian pairs.[103]

Two Poisson tensors P and Q are said to be *compatible*, if the tensor $P+Q$ is a Poisson tensor too.

We shall quote now the following theorem[139]:

Theorem 33 (Magri I) *Let P and Q be Poisson tensors on $\mathcal{M}$. Assume that Q^{-1} exists and is a smooth field of continuous linear mappings $p \in \mathcal{M} \to Q_p^{-1}$. Then, the tensor fields $T = P \circ Q^{-1}$ and Q endow the manifold with Poisson–Nijenhuis structure. Conversely, if T is Nijenhuis torsionless tensor field, satisfying the coupling conditions with the Poisson tensor Q, then Q and $T \circ Q \equiv TQ$ are compatible Poisson tensors on $\mathcal{M}$.*

A construction, similar to the one used in the above theorem, can be also applied in the following situation. Suppose we have, on the manifold $\mathcal{M}$, simultaneously a Poisson tensor Λ and a closed 2-form ω (not necessarily nondegenerate), or as it is often referred, a presymplectic form. Then, the following theorem[139] holds:

Theorem 34 (Magri II) *If the form $\omega \circ \Lambda \circ \omega$ is closed, then the tensor fields Λ and $T = \Lambda \circ \omega$ define a Poisson–Nijenhuis structure on the manifold $\mathcal{M}$.*

It is worth noting that here we consider the 2-form ω as a field of mappings

$$p \in \mathcal{M} \to \omega_p : \mathcal{T}_p\mathcal{M} \to \mathcal{T}_p^*\mathcal{M} \,. \tag{9.48}$$

An interesting situation arises on a symplectic manifold $(\mathcal{M}, \omega)$, if in addition, there is a nondegenerate Nijenhuis tensor T for which the following condition is satisfied:

$$\omega \circ T = T \circ \omega \,. \tag{9.49}$$

This condition is obviously an analogous of the coupling condition (a) for the Poisson–Nijenhuis structure. In this case, it can be shown that, if the eigenvalues of T are smooth functions on $\mathcal{M}$, they generate a system of integrable vector fields, without the additional requirements which are usually imposed on ω and T (see for example Ref. 139). More precisely, we have the following (see Refs. 100 and 139):

Theorem 35 (Florko–Magri–Yanovski) *Let $(\mathcal{M}, \omega)$ be a 2n-dimensional symplectic manifold on which there exists a Nijenhuis torsionless tensor field T, such that $T^* \circ \omega = \omega \circ T$. Let, for every point $p \in \mathcal{M}$, T_p be a semisimple operator and the dimension of its eigenspaces be a constant on $\mathcal{M}$. Then*

- *The eigenspaces S_i, corresponding to the eigenvalues λ_i, are orthogonal with respect to ω and have even dimension.*
- *If none of the functions λ_i is nowhere constant; that is, there is no open subset $V \subset \mathcal{M}$ such that $\lambda_i|_V =$ constant, then the forms $d\lambda_i$ are independent and are in involution. The corresponding vector fields, $i_{X_j}\omega = -d\lambda_j$, belong to subspaces S_j, pointwise.*
- *If, for every $p \in \mathcal{M}$, $\dim S_p = 2$; that is, if every eigenvalue is doubly degenerate, and if these eigenvalues are nowhere constants, then*
  (a) *The set $\{\lambda_i, i = 1, 2, \ldots, n\}$ is a complete set of functions in involution and each vector field X_j is a completely integrable Hamiltonian system.*
  (b) *The differential 2-form ω can be expressed in the following way: $\omega = \sum_{i=1}^{n} \omega_i, \omega_i \equiv \omega|_{S_i}, \omega_i = d\lambda_i \wedge \gamma_i$, where γ_i are some differential 1-forms on $\mathcal{M}$. If we denote by Y_i the vector fields corresponding to $\gamma_i(-\gamma_i = i_{Y_i}\omega)$, then X_i, Y_i span the subspaces S_i. If X_i, Y_i are chosen in such a way that $[X_i, Y_i]^L = 0$, then $L_{X_i}T = 0$.*
  (c) *If the eigenvalues λ_i have no zeroes on $\mathcal{M}$, then the differential 2-forms $\omega_n = \omega \circ T^n$, for $n = 0, 1, 2, \ldots$, are again symplectic structures on $\mathcal{M}$.*

Chapter 10

The Orbits Method

10.1 Reduced Phase Space

An *action* of a Lie group G on a symplectic manifold $(\mathcal{M}, \omega)$ is a differentiable map

$$\Phi : (g, p) \in G \times \mathcal{M} \to \Phi(g, p) \in \mathcal{M}\,, \tag{10.1}$$

satisfying the following requirements:

$$\Phi(e, p) = p\,, \quad \Phi(f, \Phi(g, p)) = \Phi(fg, p)\,, \quad \forall\, f, g \in G,\ \forall\, p \in \mathcal{M}\,.$$

The action (10.1) is said to be *symplectic* if the diffeomorphisms

$$\Phi_g : p \in \mathcal{M} \to \Phi_g(p) = \Phi(g, p) \in \mathcal{M}$$

are symplectic; i.e.

$$\Phi_g^* \omega = \omega\,, \quad \forall\, g \in G\,.$$

For every $\xi \in \mathcal{G}$, the map

$$\Phi^\xi : (t, p) \in \Re \times \mathcal{M} \to \Phi^\xi(t, p) = \Phi(e^{t\xi}, p) \in \mathcal{M} \tag{10.2}$$

is an action of the additive group $(\Re, +)$ on the manifold $\mathcal{M}$. Thus, with every element ξ in $\mathcal{G}$ we can associate a vector field on $\mathcal{M}$ defined by

$$\xi_{\mathcal{M}}(p) = \frac{d}{dt}\Phi(e^{t\xi}, p)\bigg|_{t=0} . \tag{10.3}$$

Then, every one parameter subgroup $e^{t\xi}$ of the group G operates as a locally Hamiltonian flow on $\mathcal{M}$. In fact, since the action (10.1) is symplectic*

$$\mathcal{L}_{\xi_{\mathcal{M}}}\omega = 0$$

and

$$\mathcal{L}_{\xi_{\mathcal{M}}}\omega = di_{\xi_M}\omega + i_{\xi_M}d\omega = di_{\xi_M}\omega ,$$

we have

$$di_{\xi_{\mathcal{M}}}\omega = 0 ,$$

that is, the differential 1-form $i_{\xi_M}\omega$ is closed.

Therefore, for every $\xi \in \mathcal{G}$, a function J_ξ can be locally found on $\mathcal{M}$ such that

$$i_{\xi_{\mathcal{M}}}\omega = dJ_\xi . \tag{10.4}$$

Furthermore, since

$$i_{\xi_{\mathcal{M}}}\omega + i_{\eta_{\mathcal{M}}}\omega = i_{(\xi+\eta)_{\mathcal{M}}}\omega ,$$

we have

$$dJ_{\xi+\eta} = dJ_\xi + dJ_\eta ,$$

or

$$J_{\xi+\eta} = J_\xi + J_\eta + Q ,$$

where Q is a constant. It is possible to choose this constant to be zero, so that

$$J_{\xi+\eta} = J_\xi + J_\eta . \tag{10.5}$$

*In this chapter the Lie derivative, with respect to a vector field X, has been denoted with the symbol $\mathcal{L}_X$, instead of L_X, to avoid confusion with the left translation L_g.

If, for every $\xi \in \mathcal{G}$, $\xi_{\mathcal{M}}$ is a globally Hamiltonian vector field, then the J_ξ's are globally defined on $\mathcal{M}$ and Eq. (10.5) allows us to define a map

$$J : p \in \mathcal{M} \to J(p) \in \mathcal{G}^* \,, \tag{10.6}$$

where $J(p)$ is the element of $\mathcal{G}^*$ such that

$$\langle \xi, J(p) \rangle = J_\xi(p) \,, \quad \forall \xi \in \mathcal{G} \,.$$

Following Souriau,[52] the map (10.6) is called a *momentum map.*

For every $\xi \in \mathcal{G}$ and $p \in \mathcal{M}$, let $\gamma_{\xi,p}$ be the map

$$\gamma_{\xi,p} : g \in G \to \gamma_{\xi,p}(g) = J_\xi(\Phi_g(p)) - J_{Ad_{g^{-1}}\xi}(p) \,,$$

whose derivative at the identity $(\gamma_{\xi,p})_{*e}$, once evaluated on a vector $\eta \in \mathcal{G}$, gives

$$\begin{aligned}
(\gamma_{\xi,p})_{*e}(\eta) &= \frac{d}{dt}\gamma_{\xi,p}(e^{t\eta})\bigg|_{t=0} \\
&= \frac{d}{dt} J_\xi(\Phi(e^{t\eta}, p))\bigg|_{t=0} - \frac{d}{dt} J_{Ad_{e^{-t\eta}}\xi}(p)\bigg|_{t=0} \\
&= \{J_\xi, J_\eta\}(p) - \frac{d}{dt}\langle Ad_{e^{-t\eta}}\xi, J(p)\rangle\bigg|_{t=0} \\
&= \{J_\xi, J_\eta\}(p) - \langle [\xi, \eta], J(p)\rangle \\
&= (\{J_\xi, J_\eta\} - J_{[\xi,\eta]})(p) \,.
\end{aligned}$$

If $\gamma_{\xi,p}$ is constant on G, then $(\gamma_{\xi,p})_{*e}(\eta) = 0$, so that

$$(\{J_\xi, J_\eta\} - J_{[\xi,\eta]})(p) = 0 \,, \quad \forall \eta \in \mathcal{G} \,.$$

Moreover, since $\gamma_{\xi,p}(e) = 0$, we have

$$J_\xi(\Phi_g(p)) = J_{Ad_{g^{-1}}\xi}(p) \,, \quad \forall g \in G \,,$$

or equivalently,

$$\langle \xi, J(\Phi_g(p)) \rangle = \langle \xi, Ad^*_{g^{-1}}(J(p)) \rangle \,, \quad \forall g \in G \,.$$

We can summarize by saying that, if $\gamma_{\xi,p}$ is constant on G for every $\xi \in \mathcal{G}$ and for every $p \in \mathcal{M}$; that is, if

$$J(\Phi_g(p)) = Ad^*_{g^{-1}} J(p) \,, \quad \forall p \in \mathcal{M} \,, \quad \forall g \in G \,, \tag{10.7}$$

or equivalently, if the diagram

$$\begin{array}{ccc} \mathcal{M} & \overset{\Phi_g}{\longrightarrow} & \mathcal{M} \\ J\downarrow & & \downarrow J \\ \mathcal{G}^* & \underset{Ad^*_{g^{-1}}}{\longrightarrow} & \mathcal{G}^* \end{array}$$

is commutative, then

$$(\{J_\xi, J_\eta\} - J_{[\xi,\eta]})(p) = 0\,, \quad \forall \xi, \eta \in \mathcal{G},\ \forall p \in \mathcal{M}\,.$$

Therefore, the condition (10.7) assures that the linear map

$$\xi \longmapsto J_\xi$$

is a homomorphism of the Lie algebra of G in the Lie algebra of the Hamilton functions on $\mathcal{M}$. In this case, the action (10.1) is called a *Poissonian action.* Of course, not all the symplectic actions Φ on symplectic manifolds $\mathcal{M}$ are Poissonian.

If the symplectic form ω defined on the manifold is also exact; i.e.

$$\omega = -d\theta\,, \tag{10.8}$$

and the action Φ leaves θ invariant, as to say

$$\Phi_g^*\theta = \theta\,, \quad \forall g \in G\,, \tag{10.9}$$

then the map $J : \mathcal{M} \to \mathcal{G}^*$, defined as below

$$\langle \xi, J(p)\rangle = (i_{\xi_\mathcal{M}}\theta)(p)\,, \quad \forall p \in \mathcal{M},\ \forall \xi \in \mathcal{G}\,, \tag{10.10}$$

is a momentum map and Φ is a Poissonian action.

Indeed, since Φ leaves θ invariant, then

$$L_{\xi_\mathcal{M}}\theta = 0\,,$$

i.e.

$$d i_{\xi_\mathcal{M}}\theta + i_{\xi_\mathcal{M}} d\theta = 0\,.$$

In addition, from Eq. (10.8), we have

$$d i_{\xi_\mathcal{M}}\theta - i_{\xi_\mathcal{M}}\omega = 0\,,$$

as to say

$$dJ_\xi = i_{\xi_{\mathcal{M}}}\omega\,,$$

that is, J is a momentum map. Now, because (see Appendix D)

$$(Ad_{g^{-1}}\xi)_{\mathcal{M}}(p) = (\Phi_{g^{-1}})_{*\Phi_g(p)}(\xi_{\mathcal{M}}(\Phi_g(p)))\,,$$

we also have

$$\begin{aligned}(i(Ad_{g^{-1}}\xi)_{\mathcal{M}}\theta)(p) &= \theta_p((Ad_{g^{-1}}\xi)_{\mathcal{M}})\\ &= \theta_p((\Phi_{g^{-1}})_{*\Phi_g(p)}(\xi_{\mathcal{M}}(\Phi_g(p))))\\ &= \theta_{\Phi_g(p)}(\xi_{\mathcal{M}}(\Phi_g(p)))\,,\end{aligned} \tag{10.11}$$

since Φ leaves θ invariant. Finally, the above equation is equivalent to the condition

$$J_\xi(\Phi_g(p)) = J_{Ad_{g^{-1}}\xi}(p)\,, \quad \forall\xi\in\mathcal{G},\ \forall p\in\mathcal{M},\ \forall g\in G\,,$$

i.e.

$$J(\Phi_g(p)) = Ad^*_{g^{-1}}(J(p))\,, \quad \forall p\in\mathcal{M},\ \forall g\in G\,.$$

If μ is an element of $\mathcal{G}^*$, the set

$$G_\mu = \{g\in G : Ad^*_{g^{-1}}\mu = \mu\}$$

is a subgroup of G, acting on $\mathcal{M}$.

The group G usually permutes the sets of the type

$$J^{-1}(\mu) = \{p\in\mathcal{M} : J(p) = \mu\}\,.$$

In fact, if $p\in J^{-1}(\mu)$, then $J(p)=\mu$ but $\Phi_g(p)$, with $g\in G$, may not belong to $J^{-1}(\mu)$.

Let $p\in J^{-1}(\mu)$ and g be an element of G_μ, then

$$Ad^*_{g^{-1}}(J(p)) = Ad^*_{g^{-1}}\mu = \mu\,,$$

which, by using Eq. (10.7), can be written as

$$Ad^*_{g^{-1}}(J(p)) = J(\Phi_g(p))\,,$$

so that

$$J(\Phi_g(p)) = \mu\,, \quad \forall g \in G_\mu,\ \forall p \in J^{-1}(\mu)\,,$$

that is, G_μ leaves $J^{-1}(\mu)$ fixed.

Then, by Eq. (10.1), we can define an action Φ of G_μ on $J^{-1}(\mu)$,

$$\Phi : (g,p) \in G_\mu \times J^{-1}(\mu) \to \Phi(g,p) \in J^{-1}(\mu)\,. \tag{10.12}$$

The orbit of the point $p \in J^{-1}(\mu)$ under the action of the group G_μ is given by

$$G_\mu \cdot p = \{\Phi_g(p) : g \in G_\mu\}\,. \tag{10.13}$$

We can introduce an *equivalence relation* on $J^{-1}(\mu)$ by defining two points of $J^{-1}(\mu)$ to be *equivalent* if they belong to the same orbit (Eq. (10.13)).

The set of equivalence classes, denoted as usual by $J^{-1}(\mu)/G_\mu$, is the set of the orbits of the points of $J^{-1}(\mu)$ under the action of G_μ.

The point $\mu \in \mathcal{G}^*$ is said to be a *regular value of* J if, for every $p \in J^{-1}(\mu)$, the derivative

$$J_{*p} : \mathcal{T}_p\mathcal{M} \to \mathcal{T}_\mu\mathcal{G}^*$$

is a surjective map; in such case it can be proven[1] that the set $J^{-1}(\mu)$ is a differential manifold.

The action (Eq. (10.12)),

$$\Phi : (g,p) \in G_\mu \times J^{-1}(\mu) \to \Phi(g,p) \in J^{-1}(\mu)\,,$$

is called a *proper action* if it satisfies the following condition:

If $(p_n)_{n\in\aleph}$ and $(\Phi_{g_n}(p_n))_{n\in\aleph}$ are sequences of points of $J^{-1}(\mu)$, which converge in $J^{-1}(\mu)$, then $(g_n)_{n\in\aleph}$ admits a subsequence which converges in G_μ.

The hypotheses that $\mu \in \mathcal{G}^*$ is a regular value of J and that the action (10.12) is a proper action are sufficient conditions to assure that $J^{-1}(\mu)/G_\mu$ is a differential manifold and that the map

$$\pi_\mu : J^{-1}(\mu) \to J^{-1}(\mu)/G_\mu\,,$$

which associates with every point of $J^{-1}(\mu)$, the orbit to which it belongs, and its derivative

$$(\pi_\mu)_{*p} : \mathcal{T}_p J^{-1}(\mu) \to \mathcal{T}_{G_\mu\cdot p}(J^{-1}(\mu)/G_\mu)\,, \tag{10.14}$$

are surjective maps.[1]

The manifold $J^{-1}(\mu)/C_\mu$ is called the *reduced phase space* and can be endowed with a natural symplectic structure.

Indeed, let us consider two vectors $\tilde{V}$ and $\tilde{W}$ tangent to $J^{-1}(\mu)/G_\mu$ at the point y, which is the orbit of a point $J^{-1}(\mu)$ under the action of G_μ. Let us choose a point p in this orbit: the vectors $\tilde{V}$ and $\tilde{W}$ tangent to the orbit at y can be obtained from some vectors V and W, tangent to $J^{-1}(\mu)$ at the point p, by using the map (10.14).

We can thus define on $J^{-1}(\mu)/G_\mu$ a bilinear form Ω_μ expressible in terms of the symplectic form ω on $\mathcal{M}$

$$\pi_\mu^* \Omega_\mu = \omega \,,$$

that is,

$$\begin{aligned}
(\pi_\mu^* \Omega_\mu)_p(V, W) &= \omega_p(V, W)\,, \quad V, W \in \mathcal{T}_p J^{-1}(\mu)\,,\\
\Omega_\mu((\pi_\mu)_{*p}(V), (\pi_\mu)_{*p}(W)) &= \omega_p(V, W)\,,\\
\Omega_\mu(\tilde{V}, \tilde{W}) &= \omega_p(V, W)\,.
\end{aligned} \tag{10.15}$$

◇ *The bilinear form Ω_μ does not depend by the choice of the point p in the orbit and of the vectors V and W of $\mathcal{T}_p J^{-1}(\mu)$.*

Indeed, let

$$G \cdot p = \{\Phi(g, p) : g \in G\}$$

be the orbit of p under the action of G. The tangent spaces $\mathcal{T}_p(G \cdot p)$ and $\mathcal{T}_p(G_\mu \cdot p)$ are

$$\mathcal{T}_p(G \cdot p) = \{\xi_{\mathcal{M}}(p) : \xi \in \mathcal{G}\}$$

and

$$\mathcal{T}_p(G_\mu \cdot p) = \{\xi_{\mathcal{M}}(p) : \xi \in \mathcal{G}_\mu\}\,,$$

where $\xi_{\mathcal{M}}(p)$ is defined in Eq. (10.3) and $\mathcal{G}_\mu$ is the Lie algebra of G_μ.

We can now prove that

$$\mathcal{T}_p(G_\mu \cdot p) = \mathcal{T}_p(G \cdot p) \cap \mathcal{T}_p J^{-1}(\mu)\,. \tag{10.16}$$

As a matter of fact if $\xi_{\mathcal{M}}(p)$ belongs to $\mathcal{T}_p(G \cdot p)$, then Eq. (10.16) can be seen to be equivalent to saying that $\xi_{\mathcal{M}}(p) \in \mathcal{T}_p J^{-1}(\mu)$ if and only if $\xi \in \mathcal{G}_\mu$. Therefore, we may write

$$J_{*p}(\xi_{\mathcal{M}}(p)) = \frac{d}{dt} J(\Phi(e^{t\xi}, p))\bigg|_{t=0} = \frac{d}{dt} J(\Phi_{e^{t\xi}}(p))\bigg|_{t=0}$$
$$= \frac{d}{dt} Ad^*_{e^{-t\xi}}(J(p))\bigg|_{t=0} = \frac{d}{dt} Ad^*_{e^{-t\xi}}\mu\bigg|_{t=0},$$

since J satisfies Eq. (10.7). Because $J(p) = \mu$ for every $p \in J^{-1}(\mu)$,

$$\mathcal{T}_p J^{-1}(\mu) \equiv \ker J_{*p}\,.$$

Then, $\xi_{\mathcal{M}}(p) \in \mathcal{T}_p J^{-1}(\mu) = \ker J_{*p}$ if and only if

$$\frac{d}{dt} Ad^*_{e^{-t\xi}}\mu\bigg|_{t=0} = 0\,,$$

that is, if $\xi \in \mathcal{G}_\mu$.

Let V be a vector in $\mathcal{T}_p\mathcal{M}$; for every $\xi \in \mathcal{G}$, from Eq. (10.4), we have

$$\omega_p(\xi_{\mathcal{M}}(p), V) = (i_{\xi_{\mathcal{M}}}\omega)_p(V) = dJ_\xi|_p(V)\,.$$

If $\sigma(t)$ is the integral curve of V, then

$$\omega_p(\xi_{\mathcal{M}}(p), V) = \frac{d}{dt} J_\xi(\sigma(t))\bigg|_{t=0} = \frac{d}{dt}\langle \xi, J(\sigma(t))\rangle\bigg|_{t=0},$$

so that, $V \in \mathcal{T}_p J^{-1}(\mu) \equiv \ker J_{*p}$ if and only if

$$\frac{d}{dt}\langle \xi, J(\sigma(t))\rangle\bigg|_{t=0} = 0\,, \quad \forall \xi \in \mathcal{G}\,,$$

or equivalently, if and only if

$$\omega_p(\xi_{\mathcal{M}}(p), V) = 0\,, \quad \forall \xi \in \mathcal{G}\,. \tag{10.17}$$

Therefore, the two tangent spaces $\mathcal{T}_p(J^{-1}(\mu))$ and $\mathcal{T}_p(G \cdot p)$ are one the orthogonal complement (with respect to ω) of the other.

Now, because

$$\pi_\mu(\Phi_g(p)) = \pi_\mu(p)\,, \quad \forall g \in G_\mu\,,$$

we have

$$\pi_\mu(\Phi(e^{t\xi}, p)) = \pi_\mu(\Phi_{e^{t\xi}}(p)) = \pi_\mu(p)\,, \quad \forall \xi \in \mathcal{G}_\mu$$

and

$$(\pi_\mu)_{*p}(\xi_{\mathcal{M}}(p)) = 0\,, \quad \forall \xi \in \mathcal{G}_\mu\,. \tag{10.18}$$

In this way the vectors V and W, which correspond to the vectors $\tilde{V}$ and $\tilde{W}$ in Eq. (10.15), are defined up to a vector of $\mathcal{T}_p(G_\mu \cdot p)$; but the addition of a vector of this space to V and W does not modify the right-hand side of Eq. (10.15), because the spaces $\mathcal{T}_p J^{-1}(\mu)$ and $\mathcal{T}_p G \cdot p$ are "orthogonal".

Concerning the independence of Eq. (10.15) from the choice of the point p on the orbit, this depends on the fact that the action Φ is symplectic and on the invariance of $J^{-1}(\mu)$. In fact,

$$(\pi_\mu^* \Omega_\mu)_{\Phi_g(p)}(V', W') = \omega_{\Phi_g(p)}(V', W')\,, \quad V', W' \in \mathcal{T}_{\Phi_g(p)} J^{-1}(\mu)\,,$$

so that, by using the relations

$$V' = (\Phi_g)_{*p}(V)\,, \quad W' = (\Phi_g)_{*p}(W)\,, \quad V, W \in \mathcal{T}_p J^{-1}(\mu)\,,$$

we finally obtain

$$\begin{aligned}(\pi_\mu^* \Omega_\mu)_{\Phi_g(p)}(V', W') &= \omega_{\Phi_g(p)}((\Phi_g)_{*p}(V), (\Phi_g)_{*p}(W)) \\ &= \omega_p(V, W) = (\pi_\mu^* \Omega_\mu)_p(V, W)\,.\end{aligned}$$

◇ *The bilinear form Ω_μ is not degenerate.*

Indeed, if

$$\Omega_\mu(\tilde{V}, \tilde{W}) = 0$$

for every $\tilde{W}$, the representative vector V should be orthogonal to all the vectors of $\mathcal{T}_p J^{-1}(\mu)$, so that it should belong to $\mathcal{T}_p(G \cdot p)$ and, by Eq. (10.16), to $\mathcal{T}_p(G_\mu \cdot p)$. Therefore, by Eq. (10.18), we have

$$\tilde{V} = (\pi_\mu)_{*p}(V) = 0\,.$$

◇ *The differential 2-form Ω_μ is closed.*

In fact, because $d\omega = 0$, we have

$$d\pi_\mu^* \Omega_\mu = 0 .$$

Thus, from

$$d\pi_\mu^* \Omega_\mu = \pi_\mu^* d\Omega_\mu ,$$

we also have

$$d\Omega_\mu = 0 ,$$

since $(\pi_\mu)_*$ is a surjective application.

For more details on reduction processes, see Refs. 106, 41, 153, 128 and 107; last reference also containing an example of noncommutative reduction in the context of noncommutive geometry.[12,33]

10.2 Orbits of a Lie Group in the Coadjoint Representation

In the previous section we have seen how, given a symplectic manifold and a symplectic action of a Lie group on this manifold, which admits a momentum map, under appropriate conditions, we can define a symplectic structure on the reduced phase space. In this section we are going to see how, for the cotangent bundle $\mathcal{T}^*G$ of a Lie group G, we can define a symplectic action and a momentum map, such that the reduced phase space coincides with the orbit of the group in the coadjoint representation.[27]

Let G be a Lie group and consider the action of G on itself given by the left translations L_g

$$\Phi : (g, h) \in G \times G \to \Phi(g, h) = gh \in G , \tag{10.19}$$

that is, by setting

$$\Phi_g \equiv L_g , \quad \forall g \in G .$$

By using Eq. (10.19), we can introduce an action ψ, of G on $\mathcal{T}^*G$

$$\psi : (g, \alpha_h) \in G \times \mathcal{T}^*G \to \psi(g, \alpha_h) = L^*_{g^{-1}}(\alpha_h) \in \mathcal{T}^*G , \tag{10.20}$$

where α_h is an arbitrary point of $\mathcal{T}^*G$, that is a differential 1-form on the tangent space to G at the point h, and

$$(L_{g^{-1}})^*_{gh} : \mathcal{T}^*_h G \to \mathcal{T}^*_{gh} G \tag{10.21}$$

is the transposed operator of the derivative of $L_{g^{-1}}$ at the point gh,

$$(L_{g^{-1}})_{*gh} : \mathcal{T}_{gh} G \to \mathcal{T}_h G .$$

Therefore, Eq. (10.20) gives

$$\psi(e, \alpha_h) = L^*_e(\alpha_h) = \alpha_h ,$$

$$\psi(f, \psi(g, \alpha_h)) = \psi(f, L^*_{g^{-1}}(\alpha_h)) = (L^*_{f^{-1}} \circ L^*_{g^{-1}})(\alpha_h)$$
$$= L^*_{g^{-1}f^{-1}}(\alpha_h) = L^*_{(fg)^{-1}}\alpha_h = \psi(fg, \alpha_h) .$$

By Eq. (10.21), we can see that ψ maps the differential 1-form α_h on $\mathcal{T}_h G$ to a differential 1-form on $\mathcal{T}_{gh} G$. The diffeomorphisms (10.21) preserve the canonical differential 1-form θ on the cotangent bundle.

Moreover, because

$$L^*_{g^{-1}}\omega = -L^*_{g^{-1}} d\theta = -dL^*_{g^{-1}}\theta = -d\theta = \omega , \quad \forall\, g \in G ,$$

where ω is the canonical symplectic form, the action (10.20) on the cotangent bundle is a symplectic action. This allows us to define a map J for ψ as in Eq. (10.10).

Let ξ be an element of $\mathcal{G}$ and consider the action (10.19) of G into itself. The map

$$\Phi^\xi : (t, g) \in \Re \times G \to \Phi^\xi(t, g) = \Phi(e^{t\xi}, g) \in G$$

defines an action of $\Re$ on G.

The vector field

$$\xi_G(g) = \frac{d}{dt}\Phi(e^{t\xi}, g)\bigg|_{t=0} \qquad \forall\, g \in G$$

is a right invariant vector field, because

$$\Phi(e^{t\xi}, g) = e^{t\xi} g = R_g(e^{t\xi}) ,$$

and

$$\xi_G(g) = (R_g)_{*e}(\xi) .$$

From Eq. (10.10), we deduce that the momentum J is the map

$$J : \alpha_g \in \mathcal{T}^*G \to J(\alpha_g) \in \mathcal{G}^* , \tag{10.22}$$

defined by (Appendix E)

$$\langle \xi, J(\alpha_g)\rangle = \alpha_g(\xi_G(g)) = \alpha_g((R_g)_{*e}(\xi)) = (R_g^*\alpha_g)(\xi) , \quad \forall \xi \in \mathcal{G} ,$$

that is

$$J(\alpha_g) = R_g^*\alpha_g . \tag{10.23}$$

Every point μ in $\mathcal{G}^*$ is a regular value for the momentum (10.23); that is, for every $\alpha_g \in J^{-1}(\mu)$, the map

$$J_{*\alpha_g} : \mathcal{T}_{\alpha_g}(\mathcal{T}^*G) \to \mathcal{T}_\mu \mathcal{G}^*$$

is surjective. Indeed if $Y \in \mathcal{T}_\mu\mathcal{G}^*$ and $\mu(t)$ is the integral curve of Y with

$$\mu(0) = \mu ,$$

by applying to $\mu(t)$ the operator $R^*_{g^{-1}}$, we obtain a curve in $\mathcal{T}^*G$ through α_g at $t = 0$. This is so because

$$R_g^*\alpha_g = \mu , \quad \forall \alpha_g \in J^{-1}(\mu)$$

and

$$\frac{d}{dt} J(R^*_{g^{-1}}(\mu(t)))\Big|_{t=0} = \frac{d}{dt} R_g^* R^*_{g^{-1}}\mu(t)\Big|_{t=0} = \frac{d}{dt}\mu(t)\Big|_{t=0} = Y ,$$

which says that, for every $Y \in \mathcal{T}_\mu\mathcal{G}^*$, there exists a vector $X \in \mathcal{T}_{\alpha_g}(\mathcal{T}^*G)$ such that

$$J_{*\alpha_g}(X) = Y .$$

If for every $g \in G$, to the element μ in $\mathcal{G}^*$ we apply the right translation $R^*_{g^{-1}}$, we obtain a right invariant differential 1-form on G,

$$\alpha_\mu(g) = R^*_{g^{-1}}\mu . \tag{10.24}$$

By letting g vary in G, Eq. (10.24) defines all and only the points of $J^{-1}(\mu)$, because of Eq. (10.23). From Eq. (10.24), it is evident that the action of L_g^*

on the cotangent bundle maps points of $J^{-1}(\mu)$ to points of $J^{-1}(\mu)$ for all g belonging to the subgroup G_μ defined by

$$G_\mu = \{g \in G : Ad^*_{g^{-1}}\mu = \mu\}. \tag{10.25}$$

From the relation

$$(gh, \alpha_\mu(gh)) \overset{L^*_{g^{-1}}}{\longleftarrow} (h, \alpha_\mu(h)),\ \forall g \in G_\mu, \tag{10.26}$$

it follows that G_μ can be also expressed in the form

$$G_\mu = \{g \in G : L^*_{g^{-1}}\alpha_\mu = \alpha_\mu\}. \tag{10.27}$$

From Eq. (10.26), we can define an action of G_μ on $J^{-1}(\mu)$, which coincides with the action (10.20), when it is restricted to $G_\mu \times J^{-1}(\mu)$. This action is a proper action: in fact, if $\alpha_\mu(h_n)$ is a sequence of points in $J^{-1}(\mu)$ converging to a point of $J^{-1}(\mu)$, we have

$$\lim_{n\to+\infty} \alpha_\mu(h_n) = \alpha_\mu(h).$$

By the continuity of the map (10.24), we have

$$\lim_{n\to+\infty} \alpha_\mu(h_n) = \alpha_\mu\left(\lim_{n\to+\infty} h_n\right),$$

so that

$$\lim_{n\to+\infty} h_n = h$$

with $h \in G_\mu$, since G_μ is closed.

Let us suppose that the sequence $L^*_{g_n^{-1}}\alpha_\mu(h_n)$ converges to a point of $J^{-1}(\mu)$ for $n \to +\infty$. We can thus write

$$\lim_{n\to+\infty} L^*_{g_n^{-1}}\alpha_\mu(h_n) = \lim_{n\to+\infty} \alpha_\mu(g_n h_n) = \alpha_\mu\left(\lim_{n\to+\infty} g_n h_n\right),$$

where obviously, $g_n \in G_\mu$, $\forall n \in \aleph$. So the sequence $(g_n h_n)_{n\in\aleph}$ converges to a point f of G_μ, because G_μ is closed.

Therefore,

$$\lim_{n\to+\infty} h_n = h, \quad \lim_{n\to+\infty} g_n h_n = f,$$

and the sequence $(g_n)_{n\in\aleph}$ converges to a point of G_μ. The orbit of the point $\alpha_\mu(h)$ of $J^{-1}(\mu)$ under the action of the group G_μ is the set

$$G_\mu \cdot \alpha_\mu(h) = \{L^*_{g^{-1}}\alpha_\mu(h) : g \in G_\mu\}. \tag{10.28}$$

Thus, we have shown that

(a) μ is a regular value of J;
(b) G_μ acts properly on $J^{-1}(\mu)$.

From what has been said in the previous section, conditions (a) and (b) are sufficient to affirm that the set $J^{-1}(\mu)/G_\mu$, that is the set of the orbits of the points of $J^{-1}(\mu)$ under the action of G_μ, is a symplectic manifold. This manifold is the reduced phase space and, of course, can be identified with the orbit of μ under the coadjoint action of the group G, that is

$$G \cdot \mu = \{Ad^*_{g^{-1}}\mu : g \in G\}.$$

Indeed, from Eq. (10.26), the action of G_μ on the points of $J^{-1}(\mu)$ reduces to the left translation of the points on the base. Therefore, with every orbit $G_\mu \cdot \alpha_\mu(h)$ of a point in $J^{-1}(\mu)$ under the action of the group G_μ, we can associate the orbit $G_\mu \cdot h$ of the point h in G under the action of G_μ,

$$G_\mu \cdot \alpha_\mu(h) \to G_\mu \cdot h.$$

In this way the reduced phase space is diffeomorphic to G/G_μ, so that

$$\frac{J^{-1}(\mu)}{G_\mu} \approx \frac{G}{G_\mu}. \tag{10.29}$$

To every orbit $G_\mu \cdot h$ of G/G_μ we can associate a point of the orbit of μ in $\mathcal{G}^*$ under the coadjoint representation

$$G_\mu \cdot h \to Ad^*_{h^{-1}}\mu,$$

so that

$$\frac{G}{G_\mu} \approx G \cdot \mu, \tag{10.30}$$

and then

$$\frac{J^{-1}(\mu)}{G_\mu} \approx G \cdot \mu.$$

Thus, by Eqs. (10.29) and (10.30), the reduced phases space $J^{-1}(\mu)/G_\mu$ can be identified with the orbit $G \cdot \mu$ under the coadjoint action. Hence, the orbit of μ under the coadjoint representation is a symplectic manifold.

Now let us find the expression of the symplectic form Ω_μ on the orbit $G \cdot \mu$; for this purpose let us introduce the map

$$\zeta^* \Omega_\mu = \omega \,. \tag{10.31}$$

Since

$$\alpha_\mu : g \in G \to \alpha_\mu(g) = R^*_{g^{-1}} \mu \in J^{-1}(\mu) \,,$$

the tangent space to $J^{-1}(\mu)$ at the point $\alpha_\mu(g)$ is

$$\mathcal{T}_{\alpha_\mu(g)} J^{-1}(\mu) = \{(\alpha_\mu)_{*g}((R_g)_{*e}(\xi)) : \xi \in \mathcal{G}\} \,.$$

Indeed,

$$(\alpha_\mu)_{*g}((R_g)_{*e}(\xi)) = \frac{d}{dt} \alpha_\mu(e^{t\xi} g)\bigg|_{t=0} \,,$$

where $\alpha_\mu(e^{t\xi} g)$ is a curve in $J^{-1}(\mu)$ through $\alpha_\mu(g)$, so that

$$(\alpha_\mu)_{*g}((R_g)_{*e}(\xi)) \in \mathcal{T}_{\alpha_\mu(g)} J^{-1}(\mu) \,, \quad \forall \xi \in \mathcal{G} \,.$$

Vice versa, if $V \in \mathcal{T}_{\alpha_\mu(g)} J^{-1}(\mu)$, its integral curve is

$$\alpha_\mu(\sigma(t))$$

with $\sigma(0) = g$.

Furthermore,

$$\sigma(t) = \sigma(t) g^{-1} g = \tau(t) g = R_g \tau(t) \,,$$

where $\tau(t)$ is a curve in G passing on e at $t = 0$, so that

$$V = \frac{d}{dt} \alpha_\mu(\sigma(t))\bigg|_{t=0} = \frac{d}{dt} \alpha_\mu(R_g \tau(t))\bigg|_{t=0} = (\alpha_\mu)_{*g}((R_g)_{*e}(\xi)) \,,$$

and where

$$\xi = \frac{d}{dt} \tau(t)\bigg|_{t=0} \in \mathcal{G} \,.$$

Let us evaluate ω and $\zeta^*\Omega_\mu$ on a couple of vectors tangent to $J^{-1}(\mu)$

$$\omega_{\alpha_\mu(g)}(\alpha_\mu)_{*g}((R_g)_{*e}(\xi)),(\alpha_\mu)_{*g}((R_g)_{*e}(\eta)) = (\alpha_\mu^*\omega)_g((R_g)_{*e}(\xi),(R_g)_{*e}(\eta))$$
$$= (\alpha_\mu^*\omega)_g((R_g)_{*e}(\xi),(R_g)_{*e}(\eta))$$
$$= -d\alpha_\mu((R_g)_{*e}(\xi),(R_g)_{*e}(\eta)) .$$

The last step is explicitly performed in the Appendix E.

On the other hand, by Eq. (8.11), we have

$$-d\alpha_\mu((R_g)_{*e}(\xi),(R_g)_{*e}(\eta)) = -\mathcal{L}_{(R_g)_{*e}(\xi)}(\alpha_\mu((R_g)_{*e}(\eta)))$$
$$+\mathcal{L}_{(R_g)_{*e}(\eta)}(\alpha_\mu((R_g)_{*e}(\xi)))$$
$$-\alpha_\mu((R_g)_{*e}(\xi),(R_g)_{*e}(\eta)) .$$

Now, since

$$\mathcal{L}_{(R_g)_{*e}(\xi)}(\alpha_\mu((R_g)_{*e}(\eta))) = \mathcal{L}_{(R_g)_{*e}(\xi)}\mu(\eta) = 0 ,$$
$$\mathcal{L}_{(R_g)_{*e}(\eta)}(\alpha_\mu((R_g)_{*e}(\xi))) = \mathcal{L}_{(R_g)_{*e}(\eta)}\mu(\xi) = 0 ,$$
$$\alpha_\mu((R_g)_{*e}(\xi),(R_g)_{*e}(\eta)) = -\mu([\xi,\eta]) ,$$

we have

$$\omega_{\alpha_\mu(g)}((\alpha_\mu)_{*g}((R_g)_{*e}(\xi)),(\alpha_\mu)_{*g}((R_g)_{*e}(\eta))) = \langle[\xi,\eta],\mu\rangle .$$

For what concerns $\zeta^*\Omega_\mu$ we have

$$(\zeta_\mu^*\Omega_\mu)_{\alpha_\mu(g)}((\alpha_\mu)_{*g}((R_g)_{*e}(\xi)),(\alpha_\mu)_{*g}((R_g)_{*e}(\eta)))$$
$$= \Omega_\mu((Ad^*_{g^{-1}}\mu)_{*g}((R_g)_{*e}(\xi)),(Ad^*_{g^{-1}}\mu)_{*g}((R_g)_{*e}(\eta))) .$$

Let us now evaluate $(Ad^*_{g^{-1}}\mu)_{*g}((R_g)_{*e}(\xi))$:

$$(Ad^*_{g^{-1}}\mu)_{*g}((R_g)_{*e}(\xi)) = \frac{d}{dt}Ad^*_{(e^{t\xi}g)^{-1}}\mu\Big|_{t=0} = \frac{d}{dt}Ad^*_{g^{-1}e^{-t\xi}}\mu\Big|_{t=0}$$
$$= \frac{d}{dt}Ad^*_{e^{-t\xi}}Ad^*_{g^{-1}}\mu\Big|_{t=0} = \frac{d}{dt}Ad^*(e^{t\xi},Ad^*(g,\mu))\Big|_{t=0} ,$$

where Ad^* is the coadjoint action; i.e. $Ad^*(g,\mu) = Ad^*_{g^{-1}}\mu$.

Thus,

$$\begin{aligned}
(Ad^*_{g^{-1}}\mu)_{*g}((R_g)_{*e}(\xi)) &= \frac{d}{dt}Ad^*(e^{t\xi}, Ad^*(g,\mu))\Big|_{t=0} \\
&= \frac{d}{dt}Ad^*(e^{t\xi}g,\mu)\Big|_{t=0} = \frac{d}{dt}Ad^*_{g^{-1}}Ad^*_g Ad^*(e^{t\xi}g,\mu)\Big|_{t=0} \\
&= \frac{d}{dt}Ad^*_{g^{-1}}Ad^*(g^{-1}e^{t\xi}g,\mu)\Big|_{t=0} \\
&= Ad^*_{g^{-1}}\left(\frac{d}{dt}Ad^*(e^{tAd_{g^{-1}}\xi},\mu)\Big|_{t=0}\right) \\
&= Ad^*_{g^{-1}}((Ad_{g^{-1}}\xi)_{\mathcal{G}^*}(\mu)) \\
&= Ad^*_{g^{-1}}Ad^*_g(\xi_{\mathcal{G}^*}(\mu)) = \xi_{\mathcal{G}^*}(\mu)\,,
\end{aligned}$$

where

$$\xi_{\mathcal{G}^*}(\mu) = \frac{d}{dt}Ad^*(e^{t\xi},\mu)\Big|_{t=0}\,.$$

Finally, we can write

$$(\zeta^*\Omega_\mu)_{\alpha_\mu(g)}((\alpha_\mu)_{*g}((R_g)_{*e}(\xi)), (\alpha_\mu)_{*g}((R_g)_{*e}(\eta))) = \Omega_\mu(\xi_{\mathcal{G}^*}(\mu), \eta_{\mathcal{G}^*}(\mu))\,.$$

In this way, we have obtained the formula which defines the symplectic form on the orbit

$$\Omega_\mu(\xi_{\mathcal{G}^*}(\mu), \eta_{\mathcal{G}^*}(\mu)) = \langle \mu, [\xi,\eta]\rangle\,. \tag{10.32}$$

The above relation, of course, also holds for any other point $V = Ad^*_{g^{-1}}\mu$ of the orbit.

Now, since

$$\frac{d}{dt}Ad^*_{e^{-t\xi}}(\mu)\Big|_{t=0} = -ad^*_\xi(\mu)\,,$$

Eq. (10.32) can be written in the following definitive form:

$$\Omega_\mu(ad^*_\xi(\mu), ad^*_\eta(\mu)) = \langle \mu, [\xi,\eta]\rangle\,. \tag{10.33}$$

10.3 The Rigid Body

In this section, we analyze the rigid body motion about a fixed point, in the absence of external forces. The rigid body represents a simple example of Hamiltonian system, whose configurations space is a Lie group. We shall see how, on every orbit of the coadjoint representation, the Euler equation is Hamiltonian, the Hamilton function being given by the kinetic energy.

A rigid body is a system of particles subject to the holonomic constraint defined by the condition that the distance between any two points of the system is constant. The configuration space of a rigid body is the six-dimensional manifold $\Re^3 \times SO(3)$, where $SO(3)$ is the group of the orthogonal matrixes 3×3, if in the considered rigid body there are at least three not-aligned points. Let us consider the problem of determining the motion of a free rigid body. This system is invariant under translations and thus there exist three first integrals which are the three components of the total momentum. Therefore the motion of the centre of mass is a free motion and we can thus choose an inertial system in which the centre of mass is at rest. In this frame a free rigid body rotates about its inertial centre as if it were bound to a fixed point. Thus the problem of the free motion of a rigid body is equivalent to the problem of the rigid body motion about a fixed point with three degrees of freedom. The configurations space is simply $SO(3)$ and the position and the velocity of the body are defined by a point of the tangent bundle $\mathcal{T}SO(3)$. The system is invariant under rotations about the fixed point and, by Noether's theorem, there exist three corresponding first integrals which are the three components J_x, J_y and J_z of the angular momentum. Besides these three integrals there is the total energy of the system, E, which has only the kinetic part. The four first integrals, J_x, J_y, J_z and E, are functions defined on the tangent bundle $\mathcal{T}SO(3)$.

We can define an action of $SO(3)$ on itself with the left translations

$$L_g : h \in SO(3) \to L_g(h) = gh \in SO(3)\,,$$

where gh denotes the matrix product.

The tangent bundle $\mathcal{T}SO(3)$ is isomorphic to $SO(3) \times \mathcal{T}_eSO(3)$, $\mathcal{T}_eSO(3)$ denoting the tangent space to $SO(3)$ at the identity e; i.e. the space of 3×3 antisymmetric matrices.

There are two isomorphisms of $\mathcal{T}SO(3)$ in $SO(3) \times \mathcal{T}_eSO(3)$: the first is defined by the derivative of $L_{g^{-1}}$ as follows:

$$\lambda : \dot{g} \in \mathcal{T}SO(3) \to \lambda(\dot{g}) = (g, (L_{g^{-1}})_{*g}\dot{g}) \in SO(3) \times \mathcal{T}_eSO(3)\,, \tag{10.34}$$

where $\dot{g}$ is a tangent vector to the group at the point g; the second, by the derivative of $R_{g^{-1}}$, the right translation:

$$\rho : \dot{g} \in \mathcal{T}SO(3) \to \rho(\dot{g}) = (g, (R_{g^{-1}})_{*g}\dot{g}) \in SO(3) \times \mathcal{T}_eSO(3)\,. \tag{10.35}$$

The tangent space $\mathcal{T}_eSO(3)$, on its turn, is isomorphic to the Euclidean space $\Re^3$, the isomorphism being given by

$$\mathcal{I} : \begin{pmatrix} 0 & a & b \\ -a & 0 & c \\ -b & -c & 0 \end{pmatrix} \in \mathcal{T}_eSO(3) \to (-c, b, -a) \in \Re^3\,. \tag{10.36}$$

The inner product

$$[\ ,\ ] : (\xi, \eta) \in \mathcal{T}_eSO(3) \times \mathcal{T}_eSO(3) \to [\xi, \eta] = \xi\eta - \eta\xi \in \mathcal{T}_eSO(3)$$

provides the space $\mathcal{T}_eSO(3)$ of a Lie algebra structure; if the internal product in $\Re^3$ is chosen to be the usual vector product, the map (10.36) is a homomorphism of Lie algebras.

10.3.1 *The space and the body angular velocities*

The velocity of the rigid body $\dot{g}$ is a tangent vector to the group at the point g: then the vector

$$\omega_s = (\mathcal{I} \circ \pi_2 \circ \rho)(\dot{g})\,, \tag{10.37}$$

where $\pi_2 : SO(3) \times \mathcal{T}_eSO(3) \to \mathcal{T}_eSO(3)$ is the projection map, is the *angular velocity with respect to the space*, while the vector

$$\omega_c = (\mathcal{I} \circ \pi_2 \circ \lambda)(\dot{g}) \tag{10.38}$$

is the *angular velocity with respect to the body.*

In fact, the element g in $SO(3)$ represents a position of the rigid body obtained by applying the motion g; that is, the left translation L_g, to an arbitrarily chosen initial state (e.g., the unit of the group). The angular velocity vector, ω_s, of the rigid body with respect to a fixed system, is given by

$$\omega_s = \mathcal{I}(\eta), \quad \eta \in \mathcal{T}_e SO(3)$$

and, for every $t \in \Re$, $e^{\eta t}$ is a rotation with angular velocity ω_s. Since, under an infinitesimal rotation $e^{\eta\tau}(\tau \ll 1)$,

$$\dot{g} = \frac{d}{d\tau} e^{\eta\tau} g\bigg|_{\tau=0} = (R_g)_{*e}(\eta), \tag{10.39}$$

we have

$$\eta = (R_{g^{-1}})_{*g}(\dot{g}),$$

from which Eq. (10.37) follows. To the motion $e^{\eta\tau}g$ in the fixed frame it corresponds the infinitesimal rotation $e^{\xi\tau}$ in the body frame obtained by applying $L_{g^{-1}}$ to $e^{\eta\tau}g$,

$$e^{\xi\tau} = g^{-1} e^{\eta\tau} g. \tag{10.40}$$

Of course, $\mathcal{I}(\xi) = \omega_c$ is the angular velocity with respect to the body. Therefore, we can write

$$g e^{\xi\tau} = e^{\eta\tau} g.$$

From Eq. (10.39), we have

$$\dot{g} = \frac{d}{d\tau} e^{\eta\tau} g\bigg|_{\tau=0} = \frac{d}{d\tau} g e^{\xi\tau}\bigg|_{\tau=0} = (L_g)_{*e}(\xi),$$

so that

$$\xi = (L_{g^{-1}})_{*g}(\dot{g})$$

from which Eq. (10.38) follows.

Equation (10.36) allows us to simplify the notation, since we can use, instead of the expressions (10.37) and (10.38) for ω_c and ω_s, the simplified ones

$$\omega_c = (L_{g^{-1}})_{*g}\dot{g} \in \mathcal{G} \tag{10.41}$$

and

$$\omega_s = (R_{g^{-1}})_{*g}\dot{g} \in \mathcal{G}. \tag{10.42}$$

10.3.2 *The space and the body angular momenta*

The Lie algebra $\mathcal{G} = \mathcal{T}_e SO(3)$ of the group $SO(3)$ is the three-dimensional space of the angular velocities of all possible rotations and the Lie bracket of such algebra is given by the usual vector product.

If the tangent bundle $\mathcal{T}SO(3)$ is the space of the rigid body velocities, the cotangent bundle $\mathcal{T}^*SO(3)$ is the space of the angular momenta J. If the vector J lies in the cotangent space to the group at the point g, in analogy with Eqs. (10.41) and (10.42), it can be transported to the cotangent space $\mathcal{G}^*$ to the group at the identity, either with left translations or with right translations. Thus, we obtain two vectors

$$J_c = L_g^* J \in \mathcal{G}^* \tag{10.43}$$

and

$$J_s = R_g^* J \in \mathcal{G}^* \,. \tag{10.44}$$

The vector J_c is the angular momentum with respect to the body and J_s is the angular momentum with respect to the space.

Actually, the algebras $\mathcal{G}$ and $\mathcal{G}^*$ can be identified, since it can be easily proven that

$$\mathcal{I}(\xi) \cdot \mathcal{I}(\eta) = -\frac{1}{2} Tr(\xi\eta) \,, \tag{10.45}$$

where ξ and η are elements of $\mathcal{G}$, the dot $\cdot$ denotes the Euclidean scalar product and Tr the trace operator. The above equation defines an isomorphism between the spaces $\mathcal{G}$ and $\mathcal{G}^*$. We can thus consider the angular velocity and angular momenta vectors as lying in the same space. However, in what follows we shall not make this identification, and we will continue to consider the angular velocity vectors as belonging to $\mathcal{G}$ and the momenta vectors as belonging to $\mathcal{G}^*$.

10.4 Rigid Body Equations

As we saw in the previous section, the total angular momentum J_s of the rigid body is a constant of the motion, so that

$$\frac{dJ_s}{dt} = 0 \,. \tag{10.46}$$

In Sec. 8.2.4, we introduced, for every $\xi \in \mathcal{G}$, the linear operator ad^*_ξ whose action on an element α in $\mathcal{G}^*$ is defined as follows:

$$(ad^*_\xi \alpha)(\eta) = \langle ad^*_\xi \alpha, \eta\rangle = \langle \alpha, ad_\xi \eta\rangle = \langle \alpha, [\xi, \eta]\rangle \tag{10.47}$$

for every $\eta \in \mathcal{G}$. The above equation can be written in the form

$$\langle]\xi, \alpha[, \eta\rangle = \langle \alpha, [\xi, \eta]\rangle\,, \quad \forall \xi, \eta \in \mathcal{G},\ \forall \alpha \in \mathcal{G}^*\,,$$

where the bracket $]\cdot\,,\cdot[$ is defined by

$$]\xi, \alpha[\, \equiv ad^*_\xi \alpha\,. \tag{10.48}$$

Once $\mathcal{G}$ is identified with $\mathcal{G}^*$, the bracket $]\cdot\,,\cdot[$ reduces, up a sign, to the Lie bracket of the algebra $\mathcal{G}$.

If $g(t)$ is a curve in $SO(3)$, the relation $\alpha(t) = Ad^*_{g(t)^{-1}}\alpha$ defines, for every $\alpha \in \mathcal{G}^*$, a curve in $\mathcal{G}^*$. The following relation

$$\frac{d}{dt}\alpha(t) = -](L_{g(t)^{-1}})_{*g(t)}\dot{g}, \alpha(t)[\tag{10.49}$$

is proven in Appendix F.

From Eqs. (10.43) and (10.44), we have

$$J_c(t) = Ad^*_{g(t)} J_s(t)\,, \tag{10.50}$$

so that, by means of Eq. (10.41), Eq. (10.49) gives

$$\frac{dJ_c}{dt} =]\omega_c, J_c[\,. \tag{10.51}$$

Equation (10.51) is called the *Euler equation.* An important consequence of Eq. (10.50) is that the flow defined by the Euler equation maps points of a given orbit of the coadjoint representation to points belonging to the same orbit. It follows that the orbits of the coadjoint representation in the dual space of the algebra are invariant manifolds for the flow defined by the Euler Eq. (10.51).

It is well-known that the vectors ω_c and J_c are related by the following equation:

$$J_c = \mathcal{I}\omega_c\,,$$

where

$$\Im : \mathcal{G} \to \mathcal{G}^*$$

is the inertial operator. Since $\Im$ is linear and symmetric operator, we can define a Riemannian metrics on $SO(3)$. Indeed, by setting

$$(\xi, \eta) = \langle \eta, \Im\xi \rangle, \quad \forall\, \xi, \eta \in \mathcal{G},$$

we define, on every tangent space $\mathcal{T}_g SO(3)$, the metric tensor

$$(X, Y)_g = \langle X, \Im_g Y \rangle, \quad X, Y \in \mathcal{T}_g SO(3),$$

where

$$\Im_g = L^*_{g^{-1}} \Im L_{g^{-1}*}, \quad \forall\, g \in SO(3),$$

which defines a metric tensor field on $SO(3)$.

The kinetic energy T is

$$\begin{aligned} T &= \frac{1}{2}(\dot{g}, \dot{g})_g = \frac{1}{2}\langle \dot{g}, \Im_g \dot{g} \rangle = \frac{1}{2}\langle (L_g)_{*e}\omega_c, \Im_g (L_g)_{*e}\omega_c \rangle \\ &= \frac{1}{2}\langle (L_{g^{-1}})_{*g}(L_g)_{*e}\omega_c, \Im (L_{g^{-1}})_{*g}(L_g)_{*e}\omega_c \rangle = \frac{1}{2}\langle \omega_c, \Im\omega_c \rangle . \end{aligned}$$

Since $\omega_c = \Im^{-1} J_c$, the kinetic energy can be expressed as a function of the angular momentum

$$T = \frac{1}{2}\langle \omega_c, \Im\omega_c \rangle = \frac{1}{2}(\omega_c, \omega_c) = \frac{1}{2}\langle \Im^{-1} J_c, J_c \rangle . \tag{10.52}$$

Thus, the kinetic energy can be considered to be a function defined on the dual space $\mathcal{G}^*$ of the algebra.

Every tangent vector to the orbit V at the point J can be written as follows:

$$]\xi, J[,$$

with $\xi \in \mathcal{G}$. On the other hand, since

$$dT = \omega_c ,$$

the right hand side of the Euler equation can be written in the form

$$]dT, J[,$$

where the differential 1-form dT, being the exterior derivative of a function defined on $\mathcal{G}^*$, belongs to the dual space of $\mathcal{G}^*$, that is $\mathcal{G}$.

The symplectic structure on V is given by Eq. (10.33). In terms of the bracket $]\cdot,\cdot[$ here introduced, it can be written as

$$\Omega_\mu(ad^*_\xi(J), ad^*_{dT}(J)) = \langle [\xi, dT], J\rangle = \langle]\xi, J[, dT\rangle = dH(ad^*_\xi(J)),$$

where H denotes the restriction of T to an orbit of the coadjoint representation.

From the above equation, it follows

$$L_{ad^*_{dT}}\Omega_\mu = 0,$$

which says that the Euler equation is a Hamiltonian equation, H being the Hamilton function.

Chapter 11

Classical Electrodynamics

Particles and fields are the fundamental concepts of classical physics. Particles are identified as point particles and fields are tensor-valued functions on space-time. Given sources and initial conditions, fields are ruled by the Maxwell* equations.

11.1 Maxwell's Equations

The phenomenological equations of the electrodynamics are given by the following Maxwell equations:

$$
\boxed{
\begin{array}{ll}
\displaystyle\int_{\partial U} \vec{B}\cdot\vec{n}d\sigma = 0 & \textit{magnetic poles do not exist} \\
\displaystyle\frac{1}{c}\frac{d}{dt}\int_S \vec{B}\cdot\vec{n}d\sigma = -\int_{\partial S}\vec{E}\cdot d\vec{l} & \textit{Faraday's law} \\
\displaystyle\int_{\partial U} \vec{D}\cdot\vec{n}d\sigma = Q & \textit{Gauss' law} \\
\displaystyle\frac{1}{c}\frac{d}{dt}\int_S \vec{D}\cdot\vec{n}d\sigma + \frac{4\pi}{c}\int_S \vec{J}\cdot\vec{n}d\sigma = \int_{\partial S}\vec{H}\cdot d\vec{l} & \textit{Ampère's law},
\end{array}
}
$$

*James Clerk Maxwell was born in Edinburgh in 1831 and died in Cambridge in 1979. He has been a professor of physics at Cambridge University from the year 1871. Famous physicist and mathematician, he gave a mathematical expression to Faraday's intuitive and experimental views in the classic *Treatise on Electricity and Magnetism* (London, 1873) in which, as a consequence of his electromagnetic theory of the light, he foresaw electromagnetic waves later detected by Hertz and applied by Marconi.

where

- c is the light velocity in vacuum,
- S is a regular surface in $\Re^3$, which may change in time, with a given orientation defining the exterior normal $\vec{n}$,
- ∂S is the boundary of S with the orientation induced by the one of S,
- U is a regular submanifold (a volume) of $\Re^3$ with a given orientation and ∂U is the boundary (a surface) of U with the orientation induced by the one of U.

$\vec{E}$ and $\vec{B}$ are the electric vector field and the magnetic induction vector field that can also be defined by the Lorentz force $\vec{F}$, which acts on a particle with electric charge e and velocity $\vec{v}$:

$$\vec{F} = e\left(\vec{E} + \frac{1}{c}\vec{v} \wedge \vec{B}\right) . \tag{11.1}$$

Maxwell's equations represent the synthesis of discoveries by Faraday, Gauss and Ampère.†

†Michael Faraday was born at Newington, United Kingdom in 1791. In 1813 he was engaged as laboratory assistant by H. Davy. Against the current idea on the *action of forces*, he introduced the concept of *force lines* to explain the propagation of electric or magnetic effects. Today they are known as *integral curves of electric* or *magnetic fields*. In the year 1831, he discovered the *electromagnetic induction phenomenon* and constructed the first electric generator. He also discovered the effects of the magnetic field on the light plane polarization and, in chemistry, two fundamental laws on the propagation of the electric current in chemical solutions. In the first, he established the direct proportionality between the amount of transformed matter and the amount of electric charge passing trough the electrolyte; in the second, he established the proportionality between the amounts of different substances and their equivalent weights. In order to describe the experiments and to explain the results, Faraday invented the words *ion, cathode, anode, electrolyte.* He died at Hampton Court in 1867.

Karl Frederick Gauss was born at Braunschweig in 1777 and died in Göttingen in 1855. From the year 1807 he has been professor at Göttingen University and Director of the Göttingen Astronomic Observatory. Founder of the differential geometry of surfaces, mathematician, physicist and astronomer, he was called *princeps mathematicorum.* In his works, he adopted the motto *pauca sed matura* (little and deep). Indeed, his works are celebrated also for the excellence of the form. However, the *pauca* fill up eleven big volumes.

André Marie Ampère was born at Lyons in 1775, and died at Marseilles on June 10, 1836. Mathematician, chemist, physicist, man fascinated by the mystery, Ampère attempted to find in nature an answer to his need of universality. From the year 1809 he has been professor of mathematics at the *École Polytechnique* in Paris. His papers, concerning the connection between electricity and magnetism, were written in 1820.

When a charge density ρ can be introduced, such that

$$Q = \int_U \rho dv\,,$$

the Maxwell equations take the form:

$$\begin{aligned}
&\int_{\partial U} \vec{B}\cdot\vec{n}d\sigma = 0\\
&\frac{1}{c}\frac{d}{dt}\int_S \vec{B}\cdot\vec{n}d\sigma = -\int_{\partial S}\vec{E}\cdot\vec{dl}\\
&\int_{\partial U} \vec{D}\cdot\vec{n}d\sigma = \int_U \rho dv\\
&\frac{1}{c}\frac{d}{dt}\int_S \vec{D}\cdot\vec{n}d\sigma + \frac{4\pi}{c}\int_S \vec{J}\cdot\vec{n}d\sigma = \int_{\partial S}\vec{H}\cdot\vec{dl}\,,
\end{aligned} \tag{11.2}$$

where $\vec{J} = \rho\vec{v}$ is the current density.

When S is a stationary surface and fields are sufficiently regular, we can use Stokes' theorem to get the Maxwell equations in the following differential form:

$$\boxed{\begin{aligned}
&div\,\vec{B} = 0\\
&rot\,\vec{E} = -\frac{1}{c}\frac{\partial\vec{B}}{\partial t}\\
&div\,\vec{D} = 4\pi\rho\\
&rot\,\vec{H} = \frac{4\pi}{c}\vec{J} + \frac{4\pi}{c}\frac{\partial\vec{D}}{\partial t}\,.
\end{aligned}} \tag{11.3}$$

11.2 Geometrical Identification of Fields on $\Re^3$

Phenomenological Eqs. (11.2) are important to understand the geometrical meaning of fields $(\vec{E},\vec{B},\vec{H},\vec{D})$. Indeed, Eqs. (11.2) show that $\vec{B}$, $\vec{D}$ and $\vec{J}$ have to define differential 2-forms on $\Re^3$, since they are integrated on a 2-dimensional manifold ∂U, while $\vec{E}$ and $\vec{H}$ have to define differential 1-forms on $\Re^3$, since they are integrated on a 1-dimensional manifold ∂S. Finally, ρ has to define a differentiable 3-form, since it is integrated on a volume U.

Actually, because of transformation properties of electromagnetic fields, determined on physical grounds, $\vec{B}$ and $\vec{H}$ have to define *twisted*[56,48] rather than *even* differential forms.

Indeed, the transformation laws of the fields $\vec{E}$ and $\vec{B}$, under space transformations

$$t \to t' = t\,,$$

$$x^i \to x'^i = f^i(x)\,,$$

or time inversion

$$t \to t' = -t\,,$$

$$x^i \to x'^i = x^i\,,$$

may be obtained from the expression (11.1) of the Lorentz force.

It is clear that, owing to the presence of a vector product in the expression of the Lorentz force, $\vec{B}$ must change sign under time inversion. Therefore, $\vec{B}$ will be effected, by such a transformation, even if it does not depend on time.

Clearly, if we consider only coordinate transformations with positive Jacobian determinant, the identification of electromagnetic fields with even differential forms would be possible.

We can resume the above discussion saying that:

Fields entering Ampère's law must be identified with twisted differential forms while fields entering Faraday's law will be identified with even differential forms.

From Eqs. (11.2), we can try the following identification:

$E = E_x dx + E_y dy + E_z dz$	*electric field differential* 1*-form*
$B = B_x dy \wedge dz + B_y dz \wedge dx + B_z dx \wedge dy$	*magnetic induction differential* 2*-form*
$H = H_x dx + H_y dy + H_z dz$	*magnetic field differential* 1*-form*
$D = D_x dy \wedge dz + D_y dz \wedge dx + D_z dx \wedge dy$	*electric induction differential* 2*-form*
$J = J_x dy \wedge dz + J_y dz \wedge dx + J_z dx \wedge dy$	it current differential 2-form
$R = \varrho dx \wedge dy \wedge dz$	*charge differential* 3*-form*

which allows us to rewrite Maxwell's equations in the following form:

$$\int_{\partial U} B = 0$$

$$\frac{1}{c}\frac{d}{dt}\int_S B = -\int_{\partial S} E$$

$$\int_{\partial U} D = 4\pi \int_U R$$

$$\frac{1}{c}\frac{d}{dt}\int_S D + \frac{4\pi}{c}\int_S J = \int_{\partial S} H\,.$$

If S is stationary and fields are regular, we can apply Stokes' theorem to get Maxwell's equations in the following differential form:

$$dB = 0$$

$$dE = \frac{1}{c}\dot{B}$$

$$dD - 4\pi R = 0$$

$$dH - \frac{1}{c}(\dot{D} + 4\pi J) = 0\,.$$

As a consequence of transformation properties, in particular of the behavior of $\vec{B}$ under time inversion, it seems more appropriate to consider electromagnetic fields directly on space-time.

11.3 Geometrical Identification of Electromagnetic Field in Space-Time

By introducing the following differential forms:

$$F = dx^0 \wedge E - B \quad \textit{Faraday's 2-form}$$

$$G = dx^0 \wedge H + D \quad \textit{Ampère's 2-form}$$

$$I = dx^0 \wedge J - R \quad \textit{charge-current 3-form}$$

with $x^0 = ct$, Maxwell's equations can be written simply as follows:

$$\begin{aligned} dF &= 0\,, \\ dG &= I\,. \end{aligned} \tag{11.4}$$

From the last equation, since $d^2 = 0$, we have the *continuity equation*

$$dI = 0\,.$$

11.3.1 *The vector potential and the gauge transformation*

The first equation in Eqs. (11.4) says that F is a closed differential 2-form and, then, that there exists, locally, a differential 1-form A such that

$$F = dA\,.$$

With the differential 1-form A we can associate, by using the metrics, a vector field, which is called the *vector potential.*

The differential 1-form A is not uniquely defined, since we can add to A any exact differential 1-form df:

$$F = dA = d(A + df)\,.$$

The map

$$A \to A' = A + df$$

is called a *gauge transformation.*

Under space inversion

$$P : (x^0, x^1, x^2, x^3) \to (x^0, -x^1, -x^2, -x^3)\,,$$

we have

$$P : (A_0, A_1, A_2, A_3) \to (A_0, -A_1, -A_2, -A_3)\,.$$

Under time inversion

$$P : (x^0, x^1, x^2, x^3) \to (-x^0, x^1, x^2, x^3)\,,$$

we have

$$P : (A_0, A_1, A_2, A_3) \to (A_0, -A_1, -A_2, -A_3)\,.$$

Thus, the potential A behaves like an ordinary differential 1-form under parity transformation and like a twisted differential 1-form under time reversal.

These properties are in agreement with *CPT theorem* according to which *photons are charge-odd particles*; that is, the potential A is a twisted differential 1-form under *charge conjugation*:

$$C: A_0 dx^0 + A_i dx^i \to A'_0 dx'^0 + A'_i dx'^i = -(A_0 dx^0 + A_i dx^i)\,.$$

11.3.2 *Constitutive equations*

Fields $(\vec{E}, \vec{B}, \vec{H}, \vec{D})$ are not independent entities but are connected through phenomenological relations

$$D = D[\vec{E}, \vec{B}]\,,$$
$$H = H[\vec{E}, \vec{B}]\,,$$

which depend on the specific media under consideration.[‡] When conducting media are also considered, there is a generalized Ohm's law

$$J = J[\vec{E}, \vec{B}]\,.$$

If we restrict ourselves to linear response media, the constitutive equations have the following form:

$$\begin{aligned} \vec{D} &= \varepsilon(\vec{E})\,, \\ \vec{B} &= \mu(\vec{H})\,, \end{aligned} \tag{11.5}$$

where ε and μ are invertible linear maps called *dielectric map* and *magnetic permeability map*, respectively.

In the case of specific isotropic media, as the vacuum, Eqs. (11.5) takes the very simple form:

$$\begin{aligned} \vec{D} &= \varepsilon_0 \mathcal{I} \vec{E}\,, \\ \vec{B} &= \mu_0 \mathcal{I} \vec{H}\,, \end{aligned} \tag{11.6}$$

where $\mathcal{I}$ is the identity map.

[‡]The square bracket has been used to remind that the relations $D = D[\vec{E}, \vec{B}]$ and $H = H[\vec{E}, \vec{B}]$ may not be local (hysteresis) and may not be linear.

In order to put Eqs. (11.6) in a provisional geometrical form, we need a linear map between differential 1-forms and differential 2-forms, from E to D and from H to B.

If we think to E, B, H, D as differential forms on the manifold $\mathcal{M} \equiv \Re^3$, the linear map, we are speaking about, can be given by the Hodge dual $* : \Lambda(\Re^3) \rightarrow \Lambda^2(\Re^3)$. Thus, the experimental relations (11.6) determine the Euclidean metric in $\Re^3$ and can be written in the following geometrical form:

$$\vec{D} = \varepsilon_0 * \vec{E}\,,$$

$$\vec{B} = \mu_0 * \vec{H}\,,$$

where the Hodge dual $*$ is constructed out from the volume form $\Omega = dx \wedge dy \wedge dz$ and the Euclidean metric $g_{ij} = \delta_{ij}$.

However, special relativity forces to think F and G as differential 2-forms on the space-time manifold $\mathcal{M} \equiv \Re^4$. Thus, the linear map is still given by a Hodge dual but, this time, constructed out from the volume form $\Omega = cdt \wedge dx \wedge dy \wedge dz$ and the Minkowski metric $g_{ij} = \eta_{ij}$. Therefore, we obtain the constitutive equations in the following form:

$$G = \sqrt{\frac{\varepsilon_0}{\mu_0}} * F\,.$$

In four dimensions, the Hodge dual

$$* : \Lambda^2(\Re^4) \rightarrow \Lambda^2(\Re^4)$$

determines the metrics up to a scalar function, so that the constitutive properties of the vacuum fix up the conformal Lorentzian structure of space-time.

Remark 22 *It is advisable to observe that Maxwell's equations, written in the form*

$$dF = 0\,,$$

$$dG = I\,,$$

are invariant for any transformation in space-time. If we require that such transformations preserve the constitutive equations

$$G = \sqrt{\frac{\varepsilon_0}{\mu_0}} * F\,,$$

then the symmetry transformation group reduces to the conformal group.

Finally, by using the codifferential operator δ, which, in $\mathcal{M} = \Re^4$, can be expressed as follows:

$$\delta = - * d* \,,$$

we have

$$*I = *dG = \sqrt{\frac{\varepsilon_0}{\mu_0}} * d * F = -\sqrt{\frac{\varepsilon_0}{\mu_0}}\delta F \,,$$

so that Maxwell's equations in the vacuum can be written in the following form:

$$dF = 0 \,,$$
$$\delta F = -\sqrt{\frac{\mu_0}{\varepsilon_0}} * I \,,$$

or

$$dF = 0 \,,$$
$$d * F = \sqrt{\frac{\mu_0}{\varepsilon_0}} I \,. \tag{11.7}$$

11.3.3 *The wave equation*

By replacing $F = dA$ in the second Eq. (11.7), we have

$$d * dA = \sqrt{\frac{\mu_0}{\varepsilon_0}} I \,,$$

or

$$\delta dA = -\sqrt{\frac{\mu_0}{\varepsilon_0}} * I \,.$$

We can also write

$$\Box A = -\sqrt{\frac{\mu_0}{\varepsilon_0}} * I + d\delta A \,,$$

where $\Box$ is the Laplace–Beltrami operator which in the Minkowski space-time is given by

$$\Box = \frac{\partial^2}{\partial x_0^2} - \frac{\partial^2}{\partial x_1^2} - \frac{\partial^2}{\partial x_2^2} - \frac{\partial^2}{\partial x_3^2} \,.$$

We, thus, find that the solutions of Maxwell's equations are closely related with the study of the wave equation.

11.3.4 *Plane waves*

Let us look for a solution F of Maxwell's equations in the empty-space

$$dF = 0\,,$$
$$d * F = 0\,,$$

of travelling wave type; that is, such that the fields $\vec{E}$ and $\vec{B}$ are functions of $\xi = x - ct$:

$$\vec{E} = \vec{E}(x - ct)\,, \quad \vec{B} = \vec{B}(x - ct)\,.$$

We have

$$dF = dx^0 \wedge dE - dB\,,$$

where

$$E = E_x dx + E_y dy + E_z dz\,, \quad B = B_x dy \wedge dz + B_y dz \wedge dx + B_z dx \wedge dy\,.$$

Thus,

$$\begin{aligned} dE &= \frac{\partial E_x}{\partial \xi} d\xi \wedge dx + \frac{\partial E_y}{\partial \xi} d\xi \wedge dy + \frac{\partial E_z}{\partial \xi} d\xi \wedge dz \\ &= -c\frac{\partial E_x}{\partial \xi} dt \wedge dx + \frac{\partial E_y}{\partial \xi} d(x - ct) \wedge dy + \frac{\partial E_z}{\partial \xi} d(x - ct) \wedge dz\,, \end{aligned}$$

and

$$\begin{aligned} dB &= \frac{\partial B_x}{\partial \xi} d\xi \wedge dy \wedge dz + \frac{\partial B_y}{\partial \xi} d\xi \wedge dz \wedge dx + \frac{\partial B_z}{\partial \xi} d\xi \wedge dx \wedge dy \\ &= \frac{\partial B_x}{\partial \xi} d(x - ct) \wedge dy \wedge dz - c\frac{\partial B_y}{\partial \xi} dt \wedge dz \wedge dx - c\frac{\partial B_z}{\partial \xi} dt \wedge dx \wedge dy\,. \end{aligned}$$

Therefore,

$$dx^0 \wedge dE = c\frac{\partial E_y}{\partial \xi} dt \wedge dx \wedge dy + c\frac{\partial E_z}{\partial \xi} dt \wedge dx \wedge dz\,,$$

and

$$\begin{aligned}
dF &= dx^0 \wedge dE - dB \\
&= c\frac{\partial E_y}{\partial \xi} dt \wedge dx \wedge dy + c\frac{\partial E_z}{\partial \xi} dt \wedge dx \wedge dz \\
&\quad - \frac{\partial B_x}{\partial \xi} d(x - ct) \wedge dy \wedge dz - c\frac{\partial B_y}{\partial \xi} dt \wedge dz \wedge dx - c\frac{\partial B_z}{\partial \xi} dt \wedge dx \wedge dy \\
&= c\left(\frac{\partial E_y}{\partial \xi} - \frac{\partial B_z}{\partial \xi}\right) dt \wedge dx \wedge dy + c\left(\frac{\partial E_z}{\partial \xi} + \frac{\partial B_y}{\partial \xi}\right) dt \wedge dx \wedge dz \\
&\quad - \frac{\partial B_x}{\partial \xi} dx \wedge dy \wedge dz + c\frac{\partial B_x}{\partial \xi} dt \wedge dy \wedge dz \,.
\end{aligned}$$

In this way,

$$dF = 0 \Rightarrow B_x = 0\,, \quad B_z = E_y\,, \quad B_y = -E_z\,.$$

Similarly,

$$d * F = 0 \Rightarrow E_x = 0\,, \quad B_z = E_y\,, \quad B_y = -E_z\,.$$

It follows that a plane electromagnetic wave has transverse electric and magnetic fields that are determined by two independent functions, corresponding to the two independent polarization states.

Further Readings

- R. Aldovrandi and J. G. Pereira, *An Introduction to Geometrical Physics* (World Scientific, 1995).
- N. V. Balasubramanian, J. W. Linn and D. P. Sen Gupta, *Differential forms on Electromagnetic Networks* (Butterworths, London, 1970).
- W. L. Burke, *Applied Differential Geometry* (Cambridge University Press, 1985).
- H. Flanders, *Differential Forms* (Academic Press, New York, 1963).
- C. Godbillon, *Gèomètrie Differentielle et Mècanique Analytique* (Hermann, Paris, 1969).
- V. Guillemin and S. Sternberg, *Symplectic Techniques in Physics* (Cambridge University Press, 1984).
- B. Kostant, "Orbits, symplectic structures and representation theory," *Proc. U.S.–Japan Seminar in Differential Geometry, Kyoto, 1965* (Nippon Hyoronisha Tokyo, Japan, 1965), p. 71.

- R. S. Ingarden and A. Jamiolkowski, *Classical Electrodynamics* (Elsevier, Amsterdam, 1985).
- P. Libermann and C.-M. Marle, *Symplectic Geometry and Analytical Mechanics* (Dordrecht, Reidel, 1987).
- A. Lichnerovich, "New geometrical dynamics," in *Lecture Notes in Mathematics* (Springer, 1975), p. 570.
- G. M. Marle, *Symplectic Manifolds, Dynamical Groups and Hamiltonian Mechanics* (Reidel, Boston, 1976).
- J. Moser, "Various Aspects of Integrable Hamiltonian Systems," in *Progr. Math.*, Vol. 8 (Birkhäuser, Boston, 1980).
- G. Pichon, *Groupes de Lie: répresentation linéaires et applications* (Hermann Paris-Collection Méthodes).
- W. Thirring, *A course in Mathematical Physics* (Springer Verlag, 1978).
- C. Von Westenholtz, *Differential forms in Mathematical Physics* (North-Holland, 1981).

Part IV

Integrable Field Theories

The last few decades have shown the exciting prospects of tackling nonlinear field theories (in two dimensions) nonperturbatively by exploiting their complete integrability properties.[92,93,89,116]

Let us recall that the concept of completely integrable Hamiltonian systems with finitely many degrees of freedom goes back to the last century.[135,164] Some qualitative features of these systems remain true in some special classes of infinite-dimensional Hamiltonian systems expressed by nonlinear evolution equations; a short list of them being:

$u_t + uu_x + u_{xxx} = 0$	*Korteweg–deVries equation*
$u_t + u^2 u_x + u_{xxx} = 0$	*modified Korteweg–de Vries equation*
$iu_t + \frac{1}{2}u_{xx} + \lvert u\rvert^2 u = 0$	*nonlinear Schrödinger equation*
$u_{xt} + e^u = 0$	*Liouville equation*
$\vec{u}_t - \vec{u} \wedge (\vec{u}_{xx} + \vec{J}u) = 0$	*Landau–Liftchitz equation*
$u_{tt} - (12uu_x + a_{xxx})_x = 0$	*Boussinesque equation*
$u_{xt} + \sin u = 0$	*sine–Gordon equation*
$u_t - 2uu_x - Hu_{xx} = 0$	*Benjamin–Oto equation*
$(u_t + uu_x + u_{xxx})_x + 3\alpha^2 u_y y = 0$	*Kodamtsev–Petviashivili equation*
$\phi = \frac{a}{b}e^{br} + ar \quad (ab > 0)$	*Toda potential*
$iu_t + u_{xx} + i(\lvert\alpha\rvert^2 u)_x = 0$	*derivative-nonlinear Schrödinger equation*
$u_t - (u^{-1/2})_{xxx} = 0$	*Harry–Dym equation*

A further remarkable example of integrable evolution equation is given by the Burgers equation

$$u_t = 2uu_x + u_{xx}\,, \tag{11.8}$$

which describes the heat diffusion and it is integrable by the Hopf–Cole transformation.

It is worth observing that the evolution equation (11.8) does not correspond to a Hamiltonian dynamics but to a dissipative one. Nevertheless, also its integrability may be explained in terms of an invariant mixed tensor field.[79,80]

A relevant progress in the study of these systems with an infinite-dimensional phase manifold $\mathcal{M}$, was the introduction of the *Lax representation*,[131] which played an important role in formulating the *inverse scattering method*, universally recognized as one of the most remarkable result of theoretical physics in the last decades, and of the "*AKNS scheme*."[59] This method allows the integration of nonlinear dynamics, both with a finitely or infinitely many degrees of freedom, for which a Lax representation can be given,[101,8] this being both of physical and mathematical relevance.[156]

On the other hand, the natural arena, for the analysis of the integrability of dynamical systems, is represented by the phase space that is endowed with a natural symplectic structure. In terms of this structure, the scattering data are interpretable as action-angle type variables.

We shall see how the integrability of nonlinear field theories can be naturally explained in terms of mixed tensor fields, rather than symplectic structures, and how such tensors are linked to the Lax operators. The approach leads to a theorem of integrability that does not assume a finite number of degrees of freedom and, for a dynamical system with finitely many degrees of freedom, is equivalent to the classical Liouville theorem.

As a working example we will use the Korteweg–de Vries equation (KdV), which is the most-known completely integrable nonlinear field theory.

Chapter 12

KdV Equation

The equation

$$u_t + uu_x + u_{xxx} = 0\,, \tag{12.1}$$

where $u : (x, t) \in \Re^2 \to \Re$ is a numerical function depending on the variables x, t, and the indices denote partial derivatives, was derived by D. J. Korteweg and G. de Vries[120] in 1985 in order to describe shallow water waves moving in a channel without any change of shape,[170] and it is known as *KdV equation*, or simply, *KdV*.

12.1 An Existence and Uniqueness Theorem

A travelling wave type solution of KdV, of the form

$$u(x,t) = s(x - ct)\,,$$

can be easily found in the hypothesis that s vanishes at the infinity together with its space derivatives.

Indeed, setting $\xi = x - ct$, KdV equation becomes

$$-cs' + ss' + s''' = 0\,,$$

where the apex denotes the derivative with respect to ξ. Thus, integrating and using the boundary condition, we obtain

$$-cs + \frac{1}{2}s^2 + s'' = 0\,.$$

A first integral of the above equation can be found multiplying by s', so that

$$-css' + \frac{1}{2}s^2 s' + s'' s' = 0\,,$$

and integrating once again

$$-cs^2 + \frac{1}{3}s^3 + s'^2 = 0\,,$$

whose solution, given by

$$(x - ct) = \frac{3c}{\cosh^2\left[\frac{\sqrt{c}}{2}(x - ct)\right]}\,,$$

is called a *solitary wave.*

Moreover, a uniqueness theorem can be easily proven[131] in the same hypothesis; that is, by assuming that u goes to zero at the infinity together with its space derivatives.

Indeed, if u and v are two such solutions,

$$u_t + uu_x + u_{xxx} = 0\,,$$

$$v_t + vv_x + v_{xxx} = 0\,,$$

their difference $w = u - v$ will satisfy the equation

$$w_t + uw_x + v_x w + w_{xxx} = 0\,.$$

Thus, multiplying by w, we obtain

$$ww_t + wuw_x + w^2 v_x + ww_{xxx} = 0\,,$$

so that

$$\int_{-\infty}^{+\infty} (ww_t + wuw_x + w^2 v_x + ww_{xxx})dx = 0\,,$$

or

$$\int_{-\infty}^{+\infty} \frac{1}{2}\frac{dw^2}{dt}dx + \int_{-\infty}^{+\infty} \frac{u}{2}\frac{dw^2}{dx}dx + \int_{-\infty}^{+\infty} w^2 v_x dx + \int_{-\infty}^{+\infty} w w_{xxx} dx = 0\,. \tag{12.2}$$

Integrating by parts, it is easily seen that the last integral vanishes

$$\int_{-\infty}^{+\infty} w w_{xxx} dx = 0\,,$$

while

$$\int_{-\infty}^{+\infty} \frac{u}{2}\frac{dw^2}{dx}dx = -\int_{-\infty}^{+\infty} \frac{u_x}{2} w^2 dx\,.$$

Therefore, Eq. (12.2) becomes

$$\frac{d}{dt}\int_{-\infty}^{+\infty} w^2 dx + \int_{-\infty}^{+\infty} (2v_x - u) w^2 dx = 0\,. \tag{12.3}$$

Thus, setting

$$E(t) = \int_{-\infty}^{+\infty} w^2 dx\,, \quad M = \max_{x\in\Re} |2v_x - u|\,,$$

Eq. (12.3) implies

$$\frac{d}{dt}E(t) \le ME(t)\,.$$

Therefore,

$$E(t) \le E(0)e^{Mt}\,,$$

with

$$E(0) = \int_{-\infty}^{+\infty} w^2(x,0)dx = \int_{-\infty}^{+\infty} [u(x,0) - v(x,0)]^2 dx\,.$$

It follows that

$$u(x,0) = v(x,0) \Rightarrow E(0) = 0 \Rightarrow E(t) = 0 \Rightarrow u(x,t) = v(x,t)\,.$$

We can resume saying that

For any given initial condition, $u_0(x)$, in the class of C^∞ functions defined on the real line and vanishing at infinity together with all space derivatives, *there exists one and only one solution $u(x,t)$ of KdV equation satisfying the initial condition, $u(x,0) = u_0(x)$.*

12.2 Symmetries

12.2.1 *Space-time symmetries*

It is easy to see that KdV equation is invariant under Galilei transformation

$$\begin{cases} x' = x + \lambda t\,, \\ t' = t\,. \end{cases}$$

Indeed, we have

$$\frac{\partial u}{\partial x} = \frac{\partial u'}{\partial x'}\frac{\partial x'}{\partial x} + \frac{\partial u'}{\partial t'}\frac{\partial t'}{\partial x} = \frac{\partial u'}{\partial x'}\,,$$
$$\frac{\partial u}{\partial t} = \frac{\partial u'}{\partial x'}\frac{\partial x'}{\partial t} + \frac{\partial u'}{\partial t'}\frac{\partial t'}{\partial t} = \lambda\frac{\partial u'}{\partial x'} + \frac{\partial u'}{\partial t'}\,,$$

where u' denote the composite function

$$u'(x',t') = u(x' - \lambda t', t')\,.$$

Therefore, KdV becomes

$$\lambda\frac{\partial u'}{\partial x'} + \frac{\partial u'}{\partial t'} + u'\frac{\partial u'}{\partial x'} + \frac{\partial^3 u'}{\partial x'^3} = 0\,,$$

or written in terms of the function $\bar{u} \equiv u' + \lambda$,

$$\bar{u}_{t'} + \bar{u}\bar{u}_{x'} + \bar{u}_{x'x'x'} = 0\,.$$

In conclusion, KdV equation is invariant under the transformation

$$\begin{cases} x \to x + \lambda t\,, \\ t \to t\,, \\ u \to u - \lambda\,, \end{cases} \tag{12.4}$$

12.2.2 *Bäcklund transformation*

Internal symmetries, as usual, also play a relevant role in the analysis of dynamical systems, as the following example well shows.

Example 36 *Let us consider the Burger equation, given by*

$$u_t = 2uu_x + u_{xx}\,.$$

It is easy to verify that, under the map

$$u = \frac{v_x}{v}\,, \tag{12.5}$$

Burgers' equation becomes the heat equation

$$v_t = v_{xx}\,. \tag{12.6}$$

The map given by Eq. (12.5) is called the Hopf–Cole transformation, and can be used as follows.

First, let us observe that from Eq. (12.5) we can obtain

$$v_x = uv\,,$$

and then

$$v_{xx} = u_x v + u v_x = u_x v + u^2 v = (u_x + u^2)v\,. \tag{12.7}$$

Second, if v is a solution of Eq. (12.6), then also v_x is a solution of the same equation. The same is, of course, true for all higher derivatives.

Therefore, starting with a solution v of heat equation, we can construct, at least, two solutions, namely u and $\bar{u}$, of Burgers' equation:

$$u = \frac{v_x}{v}\,, \qquad \bar{u} = \frac{v_{xx}}{v_x}\,,$$

so that

$$\bar{u}u = \frac{v_{xx}}{v}\,. \tag{12.8}$$

By comparing Eqs. (12.7) and (12.8), we may write

$$\bar{u} = \frac{u_x + u^2}{u}\,,$$

which allows us to obtain a new solution from a given one, and constitutes an example of so-called Bäcklund transformations. It expresses the invariance of Burgers' equation under translations along the x axis.

Similarly, Miura observed that, by performing the transformation

$$u = v_x - \frac{1}{6}v^2 ,$$

KdV transforms in the so-called modified KdV equation (mKdV):

$$v_t - \frac{1}{6}v^2 v_x + v_{xxx} = 0 ,$$

which is manifestly invariant under the interchange

$$v \to -v ,$$

so that, if v is a solution of mKdV, then $-v$ is again a solution.

As a consequence, given a solution v of mKdV, we obtain *two* solutions u and $\bar{u}$ of KdV:

$$u = v_x - \frac{1}{6}v^2 , \quad \bar{u} = v_x + \frac{1}{6}v^2 .$$

From previous relations, we may write

$$\bar{u} - u = \frac{1}{3}v^2 , \quad \bar{u} + u = 2v_x ,$$

so that

$$(\bar{u} - u)_x = \frac{2}{3}vv_x = \sqrt{\frac{1}{3}}\sqrt{\bar{u} - u}(\bar{u} + u) .$$

The above equation allows us to obtain a new solution from a given one, and constitutes another example of Bäcklund transformation.

By performing a Hopf–Cole transformation

$$v = -6\frac{\psi_x}{\psi} ,$$

on the modified KdV, we have

$$\psi_t + \psi_{xxx} - 3\frac{\psi_{xx}}{\psi}\psi_x = 0 . \tag{12.9}$$

Then, by composing Miura's and Hopf–Cole's transformation, we obtain the map

$$u = -6\frac{\psi_{xx}}{\psi},$$

for which the KdV equation becomes the more unpleasant Eq. (12.9).

However, the look of Eq. (12.9) can be deceiving. Indeed, we stated that KdV can be written in the following form:

$$\begin{cases} u = -6\dfrac{\psi_{xx}}{\psi}, \\ \psi_t + \psi_{xxx} - 3\dfrac{\psi_{xx}}{\psi}\psi_x = 0, \end{cases}$$

which can be written as follows:

$$\begin{cases} \psi_{xx} + \dfrac{1}{6}u\psi = 0, \\ \psi_t + \left(4\partial_{xxx} + u\partial_x + \dfrac{1}{2}u_x\right)\psi = 0, \end{cases}$$

where the shorthand notation

$$\partial_x \equiv \frac{\partial}{\partial x}$$

has been introduced.

Moreover, a change for the better can be done by using the Galilei invariance of KdV expressed by Eq. (12.4):

$$\begin{aligned} &x \to x + 6\lambda t, \\ &t \to t, \\ &u \to u - 6\lambda, \end{aligned}$$

In fact, it is easy to see that this change allows us to write the above system in the following form:

$$\begin{cases} \psi_{xx} + \dfrac{1}{6}u\psi = \lambda\psi, \\ \psi_t + \left(4\partial_{xxx} + u\partial_x + \dfrac{1}{2}u_x\right)\psi = 0, \end{cases}$$

so that, by introducing the operators

$$L = \partial_{xx} + \frac{1}{6}u\,,$$

$$B = -4\partial_{xxx} - u\partial_x - \frac{1}{2}u_x\,,$$

KdV equation assumes the following remarkable form:

$$\begin{cases} L\psi = \lambda\psi\,, \\ \dot{\psi} = B\psi\,. \end{cases} \tag{12.10}$$

12.3 Conservation Laws

From KdV equation we have

$$\int_{-\infty}^{+\infty} u_t dx = -\int_{-\infty}^{+\infty} (uu_x + u_{xxx})dx = -\int_{-\infty}^{+\infty} \frac{\partial}{\partial x}\left(\frac{u^2}{2} + u_{xx}\right) dx = 0\,,$$

so that

$$\frac{d}{dt}\int_{-\infty}^{+\infty} udx = 0\,.$$

Thus, the functional

$$K_1[u] \equiv \int_{-\infty}^{+\infty} udx\,,$$

is a first integral of KdV.

Another first integral is easily obtained by multiplying KdV by u and by applying the same procedure. After an integration by parts, we obtain

$$\begin{aligned} \int_{-\infty}^{+\infty} uu_t dx &= -\int_{-\infty}^{+\infty} (u^2u_x + uu_{xxx})dx \\ &= -\int_{-\infty}^{+\infty} (u^2u_x - u_xu_{xx})dx \\ &= -\int_{-\infty}^{+\infty} \frac{\partial}{\partial x}\left(\frac{u^3}{3} - \frac{u_x^2}{2}\right) dx = 0\,. \end{aligned}$$

Thus, we may write a second conservation law

$$K_2[u] \equiv \frac{1}{2}\int_{-\infty}^{+\infty} u^2 dx\,.$$

A third conservation law is given by

$$K_3[u] \equiv \frac{1}{2}\int_{-\infty}^{+\infty} \left(\frac{u^3}{3} - u_x^2\right) dx\,.$$

Let us observe that the gradients $G_i(u) = \delta K_i/\delta u$ of previous functionals are given by

$$G_1(u) = 1\,,$$

$$G_2(u) = u\,,$$

$$G_3(u) = \left(\frac{u^2}{2} + u_{xx}\right)\,,$$

and that the following Lenard's recursive formula holds[131]

$$\frac{\partial}{\partial x} G_{n+1} = E_k G_n\,, \quad n = 1, 2, 3\,, \tag{12.11}$$

where the operator E_k, expressed by

$$E_k = \partial_{xxx} + \frac{2}{3} u \partial_x + \frac{1}{3} u_x\,, \tag{12.12}$$

is antisymmetric with respect to the L_2 scalar product.

Moreover, Eq. (12.10) suggests that the eigenvalues of the Schrödinger operator

$$L = \frac{d^2}{dx^2} + \frac{1}{6} u(x, t)\,,$$

corresponding to a "potential" given by a solution $u(x, t)$ of KdV, do not depend on time, so that these eigenvalues, considered as functionals $\lambda[u]$ of the potential u, are first integrals of KdV.

The direct proof was given by Gardner *et al.*,[101,155] algebraically solving the eigenvalue equation

$$\psi_{xx} + \frac{1}{6} u(x, t)\psi = \lambda(t)\psi$$

with respect to u:

$$u = 6\lambda(t) - 6\frac{\psi_{xx}}{\psi},$$

and then by replacing the above expression in KdV equation.

Thus,

$$\lambda_t\psi^2 - (\psi Q_x - \psi_x Q)_x = 0, \tag{12.13}$$

where

$$Q = \psi_t + \psi_{xxx} + 3\left(\frac{1}{6}u + \lambda\right)\psi_x.$$

If ψ vanishes when $|x| \to \infty$, Eq. (12.13), integrated with respect to x, gives

$$\lambda_t \int_{-\infty}^{+\infty} \psi^2 dx = 0,$$

and then

$$\lambda = \text{constant}.$$

12.3.1 *Lax representation*

By taking the time derivative of the first equation of the system (12.10), we can write

$$\dot{L}\psi + L\dot{\psi} = \lambda\dot{\psi},$$

and by using the second equation, namely $\dot{\psi} = B\psi$, we have

$$\dot{L}\psi + LB\psi = \lambda B\psi = BL\psi,$$

so that we obtain

$$\dot{L}\psi = [B, L]\psi,$$

where the bracket $[\cdot,\cdot]$ denotes the usual commutator between operators.

The reader can easily check that KdV is just represented by the equation

$$\dot{L} = [B, L]. \tag{12.14}$$

The above equation is called *Lax representation* of KdV and can be introduced for many other systems with an infinite-dimensional phase manifold $\mathcal{M}$.

The introduction of Lax representation[131] has been a relevant progress in the study of integrable systems and has played an important role in formulating the inverse scattering method, universally recognized as one of the standard integration techniques.[101,8]

Shortly, it consists in the following.

Let $\mathcal{M}$ be some space of functions, chosen so that, for each t, the solution $u(t)$ of a generic evolution equation

$$\dot{u}(t) = \Delta(u)\,, \tag{12.15}$$

lies in $\mathcal{M}$.

Suppose that, with each function u in $\mathcal{M}$, we can associate a self-adjoint operator $L(u \to L)$ over some Hilbert space, in such a way that, if u changes with t according to Eq. (12.15), the operators $L(t)$, which also change with t, remain unitary equivalent to themselves:

$$L(t) = U(t)L(0)U(t)^{-1}\,, \tag{12.16}$$

with $U(t)$ denoting a 1-parameter family of unitary operators.

By taking the "time derivative" of the above equation, we have

$$\dot{L} = [B, L]\,, \tag{12.17}$$

where the skew-symmetric operator

$$B = \dot{U}U^{-1}$$

is the generator of the family $U(t)$. The above equation, is called Lax representation of the dynamics given by Eq. (12.15), and the pair (B, L) is called a Lax pair.

A consequence of Eq. (12.16) is that the eigenvalues of the operators $L(t)$ do not depend on t.

Indeed, let us consider the eigenvalue equation for the Lax operator at time $t = 0$:

$$L(0)\psi(0) = \lambda(0)\psi(0)\,.$$

By using Eq. (12.15) and $\dot{U} = BU$, we obtain

$$L(t)(U(t)\psi(0)) = \lambda(0)(U(t)\psi(0))\,, \tag{12.18}$$

or equivalently

$$L(t)\psi(t) = \lambda(t)\psi(t)\,,$$

where

$$\lambda(t) = \lambda(0)\,, \quad \psi(t) = U(t)\psi(0)\,.$$

Therefore,

$$\dot{\lambda} = 0\,, \quad \dot{\psi}(t) = B\psi(t)\,. \tag{12.19}$$

For this reason, Eq. (12.16) is called an *isospectral flow.*

12.3.2 *The inverse scattering method*

Let us show how, by taking advantage of the Gel'fand–Levitan–Marchenko formula[102,143,118,14] Lax representation allows us to solve the given evolution equation (12.15).

Let $u_0(x)$ be the initial condition; i.e. $u_0 = u(x,0)$, and L_0 be the associated Lax operator. Suppose that we are able to solve the corresponding eigenvalue problem

$$L_0\psi^0 = k^2\psi^0\,,$$

that is to find

- the *free states* (continuous spectrum); i.e. the states, corresponding to $k^2 > 0$, represented by waves $\psi^0(x,k)$, whose asymptotic behavior is given by

$$\psi^0(x,k) \underset{x\to-\infty}{\sim} C^0(k)\exp[-ikx]\,,$$

$$\psi^0(x,k) \underset{x\to+\infty}{\sim} C^0_-(k)\exp[-ikx] + C^0_+(k)\exp[ikx]\,,$$

  where

$$T^0(k) = \frac{C^0(k)}{C^0_-(k)} \text{ and } R^0(k) = \frac{C^0_+(k)}{C^0_-(k)}\,, \quad \forall\, k \in \Re\,,$$

  are called the *transmission coefficient* and the *reflection coefficient*, respectively;

- the *bound states* (point spectrum); i.e. the states, corresponding to $k^2 < 0$, represented by eigenfunctions $\psi_n^0(x, k_n)$, with $k_n^2 = -\chi_n^2$ (or better, $k_n = i\chi_n$, $\chi_n > 0$), whose asymptotic behavior is given by

$$\psi_n^0(x, i\chi_n) \underset{x \to -\infty}{\sim} \exp[\chi_n x] \,,$$

$$\psi_n^0(x, i\chi_n) \underset{x \to \infty}{\sim} c_n^0(\chi_n) \exp[-\chi_n x] \,,$$

where the coefficients $c_n^0(\chi_n)$ are called the *normalization constants.*

The set

$$S^0 \equiv \{\chi_n, \, c_n^0(\chi_n), \, R^0(k) \; \forall \, k \in \Re\}$$

is called the *set of scattering data.*

Of course, before solving the evolution equation, only the form of the operator B generating the isospectral flow is known, but not its explicit dependence on (x, t). However, as it will be explicitly shown in the case of KdV, the simple knowledge of the asymptotic behavior of the operator B is enough to determine the scattering data, namely

$$S \equiv \{\chi_n, \, c_n(\chi_n, t), \, R(k, t) \; \forall \, k \in \Re\}$$

of the Lax operator L, associated with the solution $u(x, t)$ of Eq. (12.15), corresponding to the given initial value u_0.

The knowledge of scattering data of the operator L allows us, by means of the Gel'fand–Levitan–Marchenko formula (GLM), to explicitly write L and then $u(x, t)$.

We can synthesize the described procedure with the following picture:

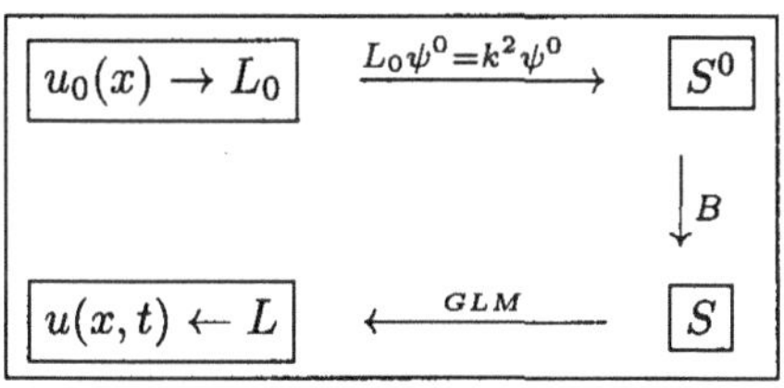

The KdV case

For KdV, the Lax pair is given by

$$L = \partial_{xx} + \frac{1}{6} u(x, t) \,,$$

$$B = -4\partial_{xxx} - u(x,t)\partial_x - \frac{1}{2}u_x(x,t)\,.$$

Let us consider the eigenvalue problem

$$L\psi = k^2\psi\,,$$

which will have as solutions

- *free states*, corresponding to $k^2 > 0$, represented by waves $\psi^0(x,k,t)$, whose asymptotic behavior is given by

$$\psi(x,k,t) \underset{x\to-\infty}{\sim} C(k,t)\exp[-ikx]\,,$$

$$\psi(x,k,t) \underset{x\to+\infty}{\sim} C_-(k,t)\exp[-ikx] + C_+(k,t)\exp[ikx]\,,$$

  where

$$T(k,t) = \frac{C(k,t)}{C_-(k,t)} \text{ and } R(k,t) = \frac{C_+(k,t)}{C_-(k,t)}\,, \quad \forall\, k \in \Re$$

  are the unknown *transmission coefficient* and the *reflection coefficient*, respectively;
- *bound states*, corresponding to $k^2 < 0$, represented by eigenfunctions $\psi_n^0(x,k_n)$, with $k_n^2 = -\chi_n^2$ (or better, $k_n = i\chi_n$, $\chi_n > 0$), whose asymptotic behavior is given by

$$\psi_n(x,i\chi_n,t) \underset{x\to-\infty}{\sim} \exp[\chi_n x]\,,$$

$$\psi_n^0(x,i\chi_n,t) \underset{x\to\infty}{\sim} c_n(\chi_n,t)\exp[-\chi_n x]\,,$$

  where the coefficients $c_n^0(\chi_n,t)$ are the unknown *normalization constants.*

The set

$$S \equiv \{\chi_n, c_n(\chi_n,t), R(k,t)\,, \quad \forall\, k \in \Re\}$$

is the *set of scattering data* which can be determined by the asymptotic behavior

$$B \underset{x\to\infty}{\sim} B_\infty \equiv -4\partial_{xxx}\,,$$

of the operator B.

Indeed, the equation

$$\dot{\psi} = B\psi\,,$$

given in Eq. (12.19), will be true also asymptotically

$$\dot{\psi}_\infty = B_\infty \psi_\infty\,,$$

so that we can write

- for *bound states*

  $$\dot{c}_n(\chi_n, t)\exp[-\chi_n x] = 4\chi_n^3 c_n(\chi_n, t)\exp[-\chi_n x]\,,$$

  which is trivially integrated to

  $$c_n(\chi_n, t) = c_n^0(\chi_n)\exp[4\chi_n^3 t]\,;$$

- for *free states*

  $$\dot{C}(k,t)e^{-ikx} = -4ik^3 C(k,t)e^{-ikx}\,,$$

  $$\dot{C}_-(k,t)e^{-ikx} + \dot{C}_+(k,t)e^{ikx} = -4ik^3[C_-(k,t)e^{-ikx} - C_+(k,t)e^{+ikx}]\,,$$

  which gives

  $$\dot{C}(k,t) = -4ik^3 C(k,t)\,,$$

  $$\dot{C}_-(k,t) = -4ik^3 C_-(k,t)\,,$$

  $$\dot{C}_+(k,t) = 4ik^3 C_+(k,t)\,,$$

  and then,

  $$T(k,t) = T^0(k)\,,$$

  $$R(k,t) = R^0(k)\exp[8ik^3 t]\,.$$

At this point, we know the time evolution of scattering data when the "potential" changes according to the KdV equation. We also know that the number of bound states does not change in time and is determined by the initial condition $u_0(x)$.

In our time-dependent case, the GLM formula (see Appendix G) must be written in the form

$$A(x,y,t) + F(x+y,t) + \int_x^\infty A(x,z,t)F(z+y,t)dz = 0\,,$$

$$F(x,t) = \frac{1}{2\pi}\int_{-\infty}^{+\infty} R(k,t)e^{ikx}dk + \sum_j c_j^2(\chi_j,t)e^{-\chi_j x}\,,$$

and the solution of KdV will be given by

$$u(x,t) = -2\frac{d}{dx}A(x,x,t)\,.$$

The one-soliton solution. Let us consider a *transparent potential*; i.e. an initial condition $u_0(x)$, such that the direct scattering problem gives $R^0(k) = 0$. Suppose, moreover, that there exists only one bound state, with eigenvalue $-\chi^2$ and normalization constant c_0.

The kernel F of GLM will be given by

$$F(x,t) = c_0^2 e^{4\chi^3 t - \chi x}\,,$$

so that GLM gives

$$A(x,y,t) + c_0^2 e^{4\chi^3 t - \chi(x+y)} + \int_x^\infty A(x,z,t)c_0^2 e^{4\chi^3 t - \chi(z+y)}dz = 0\,,$$

or

$$A(x,y,t) + c_0^2 e^{4\chi^3 t - \chi(x+y)} + c_0^2 e^{4\chi^3 t - \chi y}\int_x^\infty A(x,z,t)e^{-\chi z}dz = 0\,.$$

Since this kernel is separable, we can try to find a solution of the following form:

$$A(x,y,t) = h(x,t)\exp[-\chi y]\,,$$

so that the equation simplifies to

$$h(x,t) + c_0^2 e^{4\chi^3 t - \chi x} + c_0^2 e^{4\chi^3 t}h(x,t)\frac{1}{2\chi}e^{-2\chi x} = 0\,,$$

and algebraically, we obtain

$$h(x,t) = -\frac{c_0^2 e^{4\chi^3 t - \chi x}}{1 + \dfrac{c_0^2}{2\chi}e^{4\chi^3 t - 2\chi x}}\,.$$

Thus, we can write

$$A(x,y,t) = -\frac{c_0^2 e^{4\chi^3 t-\chi(x+y)}}{1+\dfrac{c_0^2}{2\chi}e^{4\chi^3 t-2\chi x}},$$

and since $A(x,y,t)$ is continuous and differentiable at $y=x$,

$$A(x,x,t) = -\frac{c_0^2 e^{4\chi^3 t-2\chi x}}{1+\dfrac{c_0^2}{2\chi}e^{4\chi^3 t-2\chi x}}.$$

Finally, after simple calculations, we can write

$$u(x,t) = -2\frac{d}{dx}A(x,x,t) = -\frac{2\chi^2}{\cosh^2[\chi(x-4\chi^2 t)-\delta]}.$$

The N-solitons solution. Let us again consider a *transparent potential*, but this time, with N eigenvalues $-\chi_j^2$ and normalization constant c_j^0.

Then, the kernel of GLM equation is given by

$$F(x,t) = \sum_j c_j^2(\chi_j,t)e^{-\chi_j x},$$

and we can try to find a solution of the following form:

$$A(x,y,t) = \sum_j h_j(x,t)e^{-\chi_j y}.$$

By replacing the above formula in the GLM equation, we obtain

$$\sum_j e^{-\chi_j y}\left[h_j(x,t)+e^{-\chi_j x}+\sum_i h_i(x,t)c_i^2\int_x^\infty e^{-(\chi_j+\chi_i)z}dz\right]=0.$$

The vanishing of the above sum implies the vanishing of the factor in the square brackets, so that performing the integral where contained, we obtain

$$h_j(x,t)+e^{-\chi_j x}+\sum_i h_i(x,t)c_i^2\frac{e^{-(\chi_j+\chi_i)x}}{\chi_j+\chi_i}=0,\ \forall j=1,\ldots,N.$$

The above equation can be rewritten as follows:

$$\sum_i\left(\delta_{ij}+c_i^2\frac{e^{-(\chi_j+\chi_i)x}}{\chi_j+\chi_i}\right)h_i(x,t) = -e^{-\chi_j x},\ \forall j=1,\ldots,N,$$

or by introducing the matrices

$$M = (m_{ij}), \quad K = (e^{-\chi_j x}), \quad H = (h_i(x,t)),$$

with

$$m_{ij} \equiv \delta_{ij} + c_i^2 \frac{e^{-(\chi_j+\chi_i)x}}{\chi_j + \chi_i},$$

in the following compact form:

$$MH = -K.$$

Since $\det M \neq 0$, we have

$$H = M^{-1}K.$$

The reader can perform remaining calculations to have the following simple final solution:

$$u(x,t) = -2\partial_{xx} \ln|\det M|.$$

12.4 KdV as a Hamiltonian Dynamics

It was observed by Gardner that KdV is a Hamiltonian dynamics with infinitely many degrees of freedom. Indeed, Eq. (12.1) can be written in the following form:

$$u_t = -\partial_x \frac{\delta K_3}{\delta u},$$

where the functional $K_3[u]$ is given by

$$K_3[u] \equiv \frac{1}{2}\int_{-\infty}^{+\infty}\left(\frac{u^3}{3} - u_x^2\right)dx.$$

Thus, the "time derivative" of any Frechet differentiable functional $F[u]$ is given by

$$\frac{d}{dt}F[u] = \int_{-\infty}^{+\infty}\frac{\delta F}{\delta u}u_t dx = -\int_{-\infty}^{+\infty}\frac{\delta F}{\delta u}\partial_x\frac{\delta K_3}{\delta u}dx = (G_F, \partial_x G_3),$$

where $G_F \equiv \delta F/\delta u$ and $G_3 \equiv \delta K_3/\delta u$ are the gradients of F and K_3, respectively, and $(\cdot,\cdot)$ is the L_2 scalar product.

It is easy to check that, for any two Frechet differentiable functional F and F', the bracket

$$\{F, F'\} \equiv (G_F, \partial_x G_{F'}),$$

on the chosen class of functions, is antisymmetric. Moreover, it satisfies the Jacobi identity and is a derivation on the associative product of functionals. Thus, we can conclude that KdV is a Poisson dynamics whose Poisson bivector field Λ has components, in the formal basis $e(x) \equiv \delta/\delta u(x)$, given by the operator ∂_x.

The operator ∂_x has a kernel given by constants $c \in \Re$, so that it does not have an inverse, and KdV is not strictly a Hamiltonian dynamics.

However, on the quotient manifold, namely $\mathcal{M}$, KdV is a Hamiltonian dynamics; indeed, the manifold $\mathcal{M}$ can be endowed the symplectic structure

$$\omega_{[u]} = \int_{-\infty}^{+\infty} dx \int_{-\infty}^{x} dy[\delta u(x) \wedge \delta u(y)],$$

which having constant coefficients, is trivially closed.

Moreover, it is easy to show that functionals K_n, whose gradients are constructed by means of Lenard's sequence (12.11),

$$\frac{\partial}{\partial x} G_{n+1} = E_k G_n, \quad n = 1, 2, 3,$$

are pairwise in involution. Indeed,

$$\begin{aligned}\{K_n, K_m\} &= (G_n, \partial_x G_m) = (G_n, E_k G_{m-1}) = -(E_k G_n, G_{m-1}) \\ &= -(\partial_x G_{n+1}, G_{m-1}) = (G_{n+1}, \partial_x G_{m-1}) = \{K_{n+1}, K_{m-1}\}.\end{aligned}$$

Thus, if $n < m$, an index k can be found such that

$$\{K_n, K_m\} = (G_k, AG_k),$$

with A, one of the two antisymmetric operators ∂_x or E_k, so that

$$\{K_n, K_m\} = 0, \quad \forall n, m \geq 1.$$

A further important step on the study of KdV, is represented by the following result by Faddeev and Zakharov.[92]

The symplectic form ω and the Hamiltonian K_3, once expressed in terms of the scattering data $S = \{\Re(k), \chi_n^2, c_n(k_n)\}$ via the GLM transformation,

$$u \to S,$$

become

$$\omega_S = \int_{-\infty}^{+\infty} \delta J(k) \wedge \delta\Phi(k) dk + \sum_i \delta J_i \wedge \delta\varphi_i$$

and

$$K_3[S] = 8 \int_{-\infty}^{+\infty} k^3 J(k) dk - \frac{32}{5} \sum_i J_i^{\frac{5}{2}},$$

respectively, where

$$J(k) = -\frac{k}{\pi} \ln(1 - |R(k)|^2)\Phi(k) = \arg C_+(k,t),$$

$$J_i = \chi_i^2, \quad \varphi_i = 2 \ln b_i,$$

with

$$b_j = ic_j \frac{da(k)}{dk}\bigg|_{k=i\chi_j} = ic_j \frac{d\frac{C}{C_-}(k)}{dk}\Bigg|_{k=i\chi_j}.$$

Thus, $(J(k), \Phi(k), J_i, \varphi_i)$ are Darboux' coordinates for the symplectic form ω. The expression of the Hamiltonian K_3 suggests that they play the same role of usual action-angle coordinates for Hamiltonian systems with finitely many degrees of freedom.

12.5 KdV as a Completely Integrable Hamiltonian Dynamics

A crucial step in the study of KdV was finally performed by F. Magri[137,138] who observed that KdV is a Poisson dynamics also with respect to the Poisson bivector field Λ_k whose components are given, in the formal basis $e(x) \equiv \delta/\delta u(x)$, by Lenard's operator E_k. Indeed, KdV can be written in the form

$$u_t + E_k \frac{\delta K_2}{\delta u} = 0,$$

and the bracket

$$\{F, F'\} \equiv (G_F, E_k G_{F'})$$

satisfies the Jacobi identity, so that the above bracket, being E_k an antisymmetric operator, is a Poisson bracket.

The Jacobi identity can be verified directly or, indirectly, by inverting the operator E_k on the phase manifold quotiented by its null space.

Thus, on the quotient manifold $\mathcal{N}$, we can define a nondegenerate functional 2-form by

$$\omega'_{[u]}(X[u], Y[u]) = (X(u), E_k^{-1} Y(u)), \tag{12.20}$$

where $X(u)$ and $Y(u)$, shortly denoted by X and Y, are C^∞ numerical functions defined on $\mathcal{N}$ and representing the components of vector fields $X[u]$, $Y[u] \in \mathcal{T}_u N$ in the basis $e(x) \equiv \delta/\delta u(x)$; i.e.

$$X[u] = \int_{-\infty}^{+\infty} X(u) \frac{\delta}{\delta u(x)} dx,$$

$$Y[u] = \int_{-\infty}^{+\infty} Y(u) \frac{\delta}{\delta u(x)} dx.$$

The exterior derivative $\delta\omega'$ of the functional 2-form ω' is given by

$$\delta\omega'_{[u]}(X_1[u], X_2[u], X_3[u]) = \sum_{i,j,k=1}^{3} \varepsilon_{ijk}(X_i, (E_k^{-1})_u(X_j, X_k)),$$

where

$$(E_k^{-1})_u(X_i, X_j) = \frac{d}{d\lambda}(E_k^{-1}(u + \lambda X_i))X_j|_{\lambda=0}. \tag{12.21}$$

Since $E_k E_k^{-1} X = X$, we have

$$(E_k)_u(X_i, E_k^{-1} X_j) + E_k(E_k^{-1})_u(X_i, X_j) = 0,$$

so that

$$(E_k^{-1})_u(X_i, X_j) = -E_k^{-1}(E_k)_u(X_i, E_k^{-1} X_j).$$

By using the expression of E_k, given by Eq. (12.12), we obtain

$$(E_k)_u(X_i, E_k^{-1} X_j) = \frac{2}{3} X_i(\partial_x E_k^{-1} X_j) + \frac{1}{3}(E_k^{-1} X_j)(\partial_x X_i),$$

and then

$$\delta\omega'_{[u]}(X_1[u], X_2[u], X_3[u])$$
$$= \frac{1}{3}\sum_{i,j,k=1}^{3} \varepsilon_{ijk}[2(E_k^{-1}X_i, X_j(\partial_x E_k^{-1}X_k)) + (E_k^{-1}X_i, (\partial_x X_j)(E_k^{-1}X_k))],$$

so that

$$\delta\omega'_{[u]} = 0\,.$$

Thus, the $\omega'_{[u]}$ is a symplectic structure on $\mathcal{N}$ and the interior product $i_\Delta\omega'$ is given by

$$(i_\Delta\omega')_u = \int_{-\infty}^{+\infty} dx \frac{\delta k_2}{\delta u}\delta u = (\delta K_2)_u\,, \quad \forall\, u \in \mathcal{M}\,,$$

that is

$$i_\Delta\omega' = \delta K_2\,,$$

so that KdV is, on $\mathcal{N}$, a Hamiltonian dynamics also with respect to ω'.

We can define an operator $\check{T}$ such that

$$E_k = \partial_x\check{T}\,,$$

which allows us to write the Lenard recursion of gradients of conserved functionals $K_i[u]$ in the following form:

$$\check{T}G_n = G_{n+1}\,, \quad (G_1 = 3)\,, \quad \forall n \geq 1\,. \tag{12.22}$$

On the quotient manifold $\mathcal{M}$, the operator ∂_x can be inverted to give

$$D^{-1}\varphi(x) \equiv \frac{1}{2}\left(\int_{-\infty}^{x}\varphi(x)dx - \int_{x}^{+\infty}\varphi(x)dx\right)\,, \quad \forall \varphi \in \mathcal{M}\,,$$

and we have

$$\check{T}\varphi(x) = D^{-1}E_k\varphi(x) = \partial_{xx}\varphi(x) + \frac{2}{3}u\varphi(x) - \frac{1}{3}D^{-1}(u_x\varphi(x))\,, \quad \forall \varphi \in \mathcal{M}\,. \tag{12.23}$$

Between the operator $\check{T}$ and the Lax operator $L = \partial_{xx} + (1/6)u$, there exists the following remarkable relation:

$$L\psi = \lambda\psi \Longrightarrow \check{T}\psi^2 = 4\lambda\psi^2\,, \tag{12.24}$$

that is, if ψ is an eigenstate of the Schrödinger operator belonging to the eigenvalue λ, then ψ^2 is an eigenstate of $\check{T}$ belonging to the eigenvalue 4λ.

Indeed, if $L\psi = \lambda\psi$, we have

$$\psi_{xx} + \frac{1}{6}u\psi = \lambda\psi$$

and

$$\begin{aligned}
\check{T}\psi^2 &= \partial_{xx}\psi^2 + \frac{2}{3}u\psi^2 - \frac{1}{3}\int_{-\infty}^{x} u_y\psi^2 dy \\
&= \partial_x(2\psi\psi_x) + \frac{2}{3}u\psi^2 - \frac{1}{3}u\psi^2 + \frac{2}{3}\int_{-\infty}^{x} u\psi\psi_y dy \\
&= 2\psi_x^2 + 2\psi\psi_{xx} + \frac{1}{3}u\psi^2 + 4\int_{-\infty}^{x}\left(\frac{1}{6}u\psi\right)\psi_y dy \\
&= 2\psi_x^2 + 2\lambda\psi^2 - \frac{1}{3}u\psi^2 + \frac{1}{3}u\psi^2 + 4\lambda\int_{-\infty}^{x}\psi\psi_y dy - 4\int_{-\infty}^{x}\psi_y\psi_{yy}dy \\
&= 2\psi_x^2 + 4\lambda\psi^2 - 2\psi_x^2 = 4\lambda\psi^2\,.
\end{aligned}$$

Let us observe that the relation, expressed by Eq. (12.24), does not depend on the fact that u satisfies the KdV equation. Moreover, by supposing that the ψ's are normalized, we have

$$\lambda = \lambda(\psi,\psi) = (\psi,\lambda\psi) = (\psi, L\psi)\,,$$

so that

$$\delta\lambda = (\psi, \delta L\psi) = \left(\psi, \frac{1}{6}\delta u\psi\right) = \left(\frac{1}{6}\psi^2, \delta u\right),$$

and this implies that

$$\frac{\delta\lambda}{\delta u} = \frac{1}{6}\psi^2\,.$$

We can conclude that the eigenvalues of L are eigenvalues of $\check{T}$ and that the associated eigenstates of $\check{T}$ are the gradients of the eigenvalues

$$\check{T}\frac{\delta\lambda}{\delta u} = 4\lambda\frac{\delta\lambda}{\delta u}.$$

Let us now consider the Poisson bracket of two functionals K and K',

$$\{k, k'\} = (G, \partial_x G'), \tag{12.25}$$

and suppose that the functionals are in involution; that is, their Poisson bracket is vanishing:

$$(G, \partial_x G') = 0.$$

If G_{*u} and G'_{*u} denote the derivatives at the point u of G and G', the derivative, with respect to u, of the above equation gives

$$(G_{*u}(\delta u), \partial_x G') + (G, \partial_x G'_{*u}(\delta u)) = 0, \quad \delta u \in \mathcal{T}_u\mathcal{M}.$$

The operator $G_* : \mathcal{TM} \to \mathcal{TM}$ is symmetric with respect to the L_2 scalar product, so that

$$(\delta u, G_*\partial_x G') - (G'_{*u}\partial_x G, \delta u) = (\delta u, G_{*u}\partial_x G' - G'_{*u}\partial_x G) = 0,$$

or equivalently

$$G_{*u}\partial_x G' = G'_{*u}\partial_x G. \tag{12.26}$$

If G' is the gradient of K_3, the left-hand side of Eq. (12.26) is $-\dot{G}$, where the dot denotes the derivatives of G along integral curves given by the solutions of KdV, while the right hand side is given by FG, where $F = \partial_{xxx} + u\partial_x$.

Thus, we may write

$$\dot{G} = -A^\dagger G, \tag{12.27}$$

where $A = -\partial_{xxx} - u\partial_x - u_x$, the adjoint of the operator F, is the derivative of dynamics.

The "time derivative" of the sequence expressed by Eq. (12.22) gives

$$\dot{G}_{n+1} = \dot{\check{T}}G_n + \check{T}\dot{G}_n,$$

so that, by using Eq. (12.27), we have

$$\dot{\check{T}}G_n = [\check{T}, A^\dagger]G_n,$$

where the bracket $[\cdot, \cdot]$ denotes the usual commutator. The reader can easily check that

$$\dot{\check{T}} = [-A^\dagger, \check{T}]\,. \tag{12.28}$$

The analogy between L and $\check{T}$ can be resumed as follows:

$$\begin{aligned} L\psi &= \lambda\psi\,, & \check{T}\frac{\delta\lambda}{\delta u} &= 4\lambda\frac{\delta\lambda}{\delta u}\,, \\ \dot{L} &= [B, L]\,, & \dot{\check{T}} &= [-A^\dagger, \check{T}]\,, \\ \dot{\psi} &= B\psi\,, & \dot{G} &= -A^\dagger G\,. \end{aligned}$$

Equation (12.28) can also be written in the following form:

$$\dot{\hat{T}} = [A, \hat{T}]\,, \tag{12.29}$$

where the operator $\hat{T}$ is the adjoint, with respect to the scalar product L_2, of $\check{T} = D^{-1}E_k$:

$$\hat{T}\cdot = \partial_{xx} \cdot + \frac{2}{3}u \cdot + \frac{1}{3}u_x D^{-1} \cdot\,. \tag{12.30}$$

Chapter 13

General Structures

In spite of its success as an integration algorithm, a compact *a priori* criterion of integrability in terms of Lax pairs is, to date, lacking.

On the other hand, the inverse scattering method being a transformation from generic coordinates (potentials) to action-angle variables,[92] makes it only natural for us to state an integrability criterion, for soliton equations, by looking at them as dynamical system on an infinite-dimensional phase manifold.[137,138,103,179,168,81,82,144,78,80,147,73,100,166] This point of view is also suggested by the occurrence in these remarkable systems of a peculiar operator[174,175,137,132,138,103,179,104,81,82,144,78,80,147,73,100,20,162,98,99,117,67] relevant for the effectiveness of the methods, which naturally fits in this geometrical setting as a mixed tensor field on the phase manifold $\mathcal{M}$.

13.1 Notation and Generalities

Many geometrical concepts, introduced in Part II, can be extended to infinite dimensional manifolds, whose local model is an infinite-dimensional topological vector space, if the necessary care, connected to the passage from finite-dimensional case to the infinite-dimensional one, is taken.

Many properties of the finite-dimensional case, still hold, in the infinite dimension, only if the considered manifold is a Banach manifold; that is, a manifold locally homeomorphic to a Banach space. The reason is that the *implicit functions theorem* does not hold in an arbitrary topological vector space.

Given a nonlinear operator Δ; i.e. a function of u and its space derivatives, the generic evolution equation

$$\frac{\partial u}{\partial t} = \Delta(u) \tag{13.1}$$

will be considered as a dynamical equation on the functional space $\mathcal{M}$ of field functions $u(x,t)$ regarded as functions of the space coordinates only, defined on the whole real axis, and satisfying suitable boundary conditions.

The rate of change, along the solutions of Eq. (13.1), of any functional $F[u]$ will be given by

$$\frac{d}{dt}F[u] = \int_{-\infty}^{+\infty} \frac{\delta F}{\delta u} \cdot \frac{\partial u}{\partial t} dx = \int_{-\infty}^{+\infty} dx \Delta(u) \cdot \frac{\delta F}{\delta u(x)}\,, \tag{13.2}$$

and then, by the action of the operator

$$\Delta[u] = \int_{-\infty}^{+\infty} dx \Delta(u) \cdot \frac{\delta}{\delta u(x)}\,,$$

which will be called the *dynamical vector field.*

In the functional space $\mathcal{M}$, first order differential operators $\delta/\delta u(x)$ constitute a *basis* for vector fields. The *dual basis* is given by the variations $\delta u(x)$ and, as it is usual

$$\frac{\delta u(y)}{\delta u(x)} = \delta(y-x)\,,$$

where δ is the *Dirac delta function.*

Then, any vector field $X[u]$ can be written in the following form:

$$X[u] = \int_{-\infty}^{+\infty} dx X(u) \frac{\delta}{\delta u(x)}\,,$$

and *dual vectors* or *covectors* $\alpha[u]$ as follows:

$$\alpha[u] = \int_{-\infty}^{\infty} dy \alpha(u) \delta u(y)\,.$$

The contraction $\langle \alpha, X\rangle$ between vectors and covectors will be given by

$$\langle \alpha, X\rangle = \int_{-\infty}^{\infty} X(u)\alpha(u) dx\,.$$

The rate of change, along the solutions of Eq. (13.1), of a vector field $X[u]$ is given by

$$\frac{d}{dt}X[u] = \int_{-\infty}^{+\infty} [X_u \cdot \Delta(u) - \Delta_u \cdot X(u)] \frac{\delta}{\delta u(x)}, \tag{13.3}$$

where the operators X_u and Δ_u, defined by

$$X_u\varphi := \frac{d}{d\varepsilon}X(u+\varepsilon\varphi)|_{\varepsilon=0}, \quad \Delta_u\varphi := \frac{d}{d\varepsilon}\Delta(u+\varepsilon\varphi)|_{\varepsilon=0},$$

are the *week derivatives*, or *Gateaux derivatives* of $X(u)$ and $\Delta(u)$, respectively.

Let us observe that Eqs. (13.2) and (13.3) correspond to the usual *Lie derivatives*, with respect to $\Delta[u]$, of $F[u]$ and $X[u]$, respectively.

So such *time derivatives* will be denoted* by $\mathcal{L}_\Delta F$ and $\mathcal{L}_\Delta X$; $\mathcal{L}_\Delta$ just being the Lie derivative operator with respect to Δ.

We notice that Eq. (13.3) can be written in the form

$$\mathcal{L}_\Delta X := \frac{d}{dt}X[u] = [\Delta, X],$$

where the bracket $[\cdot,\cdot]$ denotes the usual commutator between differential operators.

The tangent space and the cotangent space of $\mathcal{M}$ in u, will be denoted by $\mathcal{T}_u\mathcal{M}$ and $\mathcal{T}_u^*\mathcal{M}$, respectively.

In the continuous (formal) frame $\delta/\delta u(x)$ and coframe $\delta u(x)$, the evolution equation can be regarded as an ordinary differential equation,

$$\frac{du}{dt} = \Delta[u].$$

In order to simplify notations and formulae, in the following a vector field $X[u]$ will be identified with its components $X(u)$ and a mixed tensor field T with its associated endomorphisms $\hat{T}$, or $\check{T}$ defined by

$$T(\alpha, X) = \langle \alpha, \hat{T}X \rangle = \langle \check{T}\alpha, X \rangle.$$

These endomorphisms will be, in general, represented as operators acting on vector fields or their dual.

*Henceforth, to avoid confusion with the Lax operator L, the Lie derivative with respect to a vector field X will be denoted by $\mathcal{L}_X$.

Thus, with abuse of saying, the Lie derivative $\mathcal{L}_\Delta X$ of a vector field X with respect to Δ will be identified with $X_u \cdot \Delta(u) - \Delta_u \cdot X(u)$, and the symmetries X of a given dynamics Δ[†] will be given by the solutions of the following linear differential equation:

$$X_u \cdot \Delta(u) - \Delta_u \cdot X(u) = 0\,.$$

The Lie derivative, with respect to Δ, of an operator $\hat{T}$; that is, of an endomorphism on vector fields associated with a mixed tensor field, will be given by the operator or endomorphism $\widehat{\mathcal{L}_\Delta T}$ given by

$$\widehat{\mathcal{L}_\Delta T}\varphi = \hat{T}_u(\Delta, \varphi) - [\Delta_u, \hat{T}]\varphi\,,$$

where

$$\hat{T}_u(\Delta, \varphi) := \frac{d}{d\varepsilon}\hat{T}(u + \varepsilon\Delta)\varphi|_{\varepsilon=0}\,. \tag{13.4}$$

Therefore, the *invariance* with respect to the dynamics of such a tensor field will be expressed as[‡]

$$\hat{T}_u(\Delta, \varphi) = [\Delta_u, \hat{T}]\varphi\,.$$

13.1.1 *Backward to KdV*

In the case of the KdV equation, $\mathcal{M}$ is the manifold of C^∞ field functions u, considered as functions only of x, and vanishing at the infinity together with its space derivatives.

The *dynamics* is given by the vector field

$$\Delta[u] = -\int_{-\infty}^{\infty} (uu_x + u_{xxx}) \frac{\delta}{\delta u(x)} dx\,,$$

so that the solutions of KdV correspond to the *integral curves* of Δ.

Let us observe that Eq. (12.29) is simply the expression, in local coordinates, of the invariance, under the *KdV flow*, of the mixed tensor field T, defined by

---

[†]A very general and fundamental approach to the analysis of symmetries of nonlinear partial differential equations, is described in Refs. 182 and 30.

[‡]Equation (13.4), in spite of its form, does not correspond, generally, to the Lax representation. A possible tensorial version of this has been given by several authors, some of them in the context of alternative Lagrangians[147,73] or in reading it has the vanishing, along the dynamics, of the covariant derivative of a section of an $\mathcal{M}$-based bundle.[81]

$$T(\alpha, X) = \langle \hat{T}X, \alpha \rangle = \langle X, \hat{T}\alpha \rangle \,, \quad \alpha \in T^*\mathcal{M} \,, \quad X \in T\mathcal{M} \,. \tag{13.5}$$

Indeed, Eq. (12.29) can be written, in geometrical terms, as follows:

$$\mathcal{L}_\Delta T = 0 \,, \tag{13.6}$$

where $\mathcal{L}_\Delta$ is the Lie derivative with respect to Δ.

The tensor T, which in local coordinates can be written in the form

$$T[u] = \int_{-\infty}^{+\infty} dx \hat{T}(u) \frac{\delta}{\delta u} \otimes \delta u \,, \tag{13.7}$$

satisfies the condition

$$\hat{T}_u(\hat{T}X, Y) - \hat{T}_u(\hat{T}X, Y) = \hat{T}[\hat{T}_u(X, Y) - \hat{T}_u(X, Y)] \,, \tag{13.8}$$

which is the analogue, in this infinite-dimensional setting, of Eq. (6.48). The above condition can also be written as follows:

$$(\mathcal{L}_{\hat{T}X} T)^\wedge Y = \hat{T}(\mathcal{L}_X T)^\wedge Y \,, \quad X, Y \in T_u\mathcal{M} \,. \tag{13.9}$$

We recall that Eq. (13.9), or Eq. (13.8), is called the *Nijenhuis condition* or the *Nijenhuis bracket*, and that the tensor field

$$\mathcal{N}_T[u](\alpha, X, Y) = \langle \alpha, (\mathcal{L}_{\hat{T}X} T)^\wedge Y - \hat{T}(\mathcal{L}_X T)^\wedge Y \rangle \,, \tag{13.10}$$

with $\alpha \in T_u^*\mathcal{M}$, $X, Y \in T_u\mathcal{M}$, is called the *Nijenhuis torsion* of T. Thus, Nijenhuis' condition (13.9) is expressed by

$$\mathcal{N}_T = 0 \,. \tag{13.11}$$

A consequence of Eq. (13.11) is that the vector fields of the sequence

$$\Delta_{n+1} = \hat{T}\Delta_n \,, \quad (\Delta_1 = -u_x) \,, \quad \forall n \geq 1$$

close on an Abelian Lie algebra of symmetries for KdV, and KdV being a Hamiltonian dynamics, the sequence

$$\frac{\delta K_{n+1}}{\delta u} = \check{T} \frac{\delta K_n}{\delta u} \,, \quad \left(\frac{\delta k_1}{\delta u} = 3 \right) \,, \quad \forall n \geq 1$$

is a sequence of gradients of conserved functionals. In other words, Eq. (13.11) ensures that the endomorphism $\check{T}$ generates a sequence of closed 1-forms, in the sense that

$$(\delta\alpha = 0 \,, \quad \delta\check{T}\alpha = 0) \Longrightarrow \delta(\check{T}^n \alpha) = 0 \,, \quad \alpha \in T_u^*\mathcal{M} \,, \quad \forall n \geq 1 \,.$$

In our case, the functional 1-forms are exacts; that is, they are exterior derivatives of functionals which, since T is Δ-invariant, are first integrals of KdV.

13.2 Strongly and Weakly Symplectic Forms

At this point, it is advisable to spend some words about the definition of symplectic form on an infinite-dimensional manifold, since in this case, a distinction between strongly symplectic forms and weakly symplectic forms must be introduced.

We say that a differential 2-form ω, on an infinite-dimensional manifold $\mathcal{M}$, is a *strongly symplectic structure*, if

(a) ω is closed, that is $d\omega = 0$;

(b) $\forall p \in \mathcal{M}, \omega_p : \mathcal{T}_p\mathcal{M} \times \mathcal{T}_p\mathcal{M} \to \Re$ is a nondegenerate bilinear form; i.e. the map

$$\mathcal{I} : \mathcal{T}_p\mathcal{M} \to \mathcal{T}_p^*\mathcal{M}, \tag{13.12}$$

which with every vector $X \in \mathcal{T}_p\mathcal{M}$ associates the differential 1-form $\mathcal{I}(X)$ on $\mathcal{T}_p\mathcal{M}$, defined as below

$$(I(X))(Y) = \omega_p(X,Y), \quad \forall Y \in \mathcal{T}_p\mathcal{M},$$

is *injective* and *surjective*. In other words, $\mathcal{I}$ is an isomorphism between the spaces $\mathcal{T}_p\mathcal{M}$ and $\mathcal{T}_p^*\mathcal{M}$.

If the map (13.12) is only injective, then the differential 2-form ω is said to be a *weakly symplectic structure*.

Such a distinction has not been done in finite dimensions, since an injective map between two finite dimensional vector space, with the same dimension, is also surjective.

In infinite dimensions, the distinction is instead important. Indeed, let us consider a locally Hamiltonian vector field X and a strongly symplectic form ω; then

$$\mathcal{L}_X\omega = \delta i_X\omega = 0.$$

If $i_X\omega$ is also an exact differential form; i.e.

$$i_X\omega = -\delta H, \tag{13.13}$$

the vector field X is a globally Hamiltonian vector field and H is the Hamilton function.

Vice versa, if H is a differentiable function on $\mathcal{M}$

$$H : \mathcal{M} \to \Re\,,$$

there exists a vector field X on $\mathcal{M}$ such that Eq. (13.13) holds, since the map (13.12) is an isomorphism; but, if ω is only weakly symplectic, the vector field X cannot exist.

13.3 Invariant Endomorphism

All the evolution equations, introduced at the beginning (p. 267), apart from the Burgers' equation, are Hamiltonian systems with respect to a symplectic structure. Actually, many of them are Hamiltonian dynamics with respect to two symplectic structures,[137,138,103] namely ω_1 and ω_2.

For instance,

- in the case of KdV equation, we have

  $$\omega_1(X,Y) = (X(u), D^{-1}Y(u))\,, \quad \omega_2(X,Y) = (X(u), E_k^{-1}Y(u))$$

  with

  $$D^{-1} = \frac{1}{2}\left(\int_{-\infty}^{x} dx - \int_{x}^{\infty} dx\right), \quad E_k = \partial_{xxx} + \frac{2}{3}u\partial_x + \frac{1}{3}u_x\,,$$

  where the bracket $(\cdot,\cdot)$ denotes the L_2 scalar product. The Lenard sequence of gradients of conserved functionals is established, in terms of the operators $D = \partial/\partial x$ and E_k, as follows:

  $$DG_{n+1} = E_k G_n\,;$$

- in the case of the *sine–Gordon* equation[§]

  $$v_{xt} + \sin v = 0\,,$$

  we have

  $$\omega_1(X,Y) = (X(v), DY(v))\,, \quad \omega_2(X,Y) = (X(v), E_s^{-1}Y(v))\,,$$

[§]Here x, t denote light-cone coordinates.

where

$$E_s = D^{-1} + D + v_x D^{-1} v_x \,.$$

Indeed,

$$E_s \sin v = D^{-1} \sin v \,,$$

so that the sine–Gordon equation can also be written in the following form:[138]

$$v_t + E_s \sin v = 0 \,.$$

Thus, a Lenard's type recursion of gradients of conserved functionals can be written as follows:

$$\tilde{G}_{n+1} = (D^2 + v_x^2 + v_{xx} D^{-1} v_x)\tilde{G}_n \,, \quad \tilde{G}_1 = \frac{3}{2} v_x^2 \,.$$

Many of the previous systems, including the Burgers' equation, admit, in conclusion, an operator; i.e. an endomorphism on the module of vector fields, namely $\hat{T}$, which is invariant under the dynamics and responsible for the construction of (infinitely many) Abelian symmetries (vector fields) or, for the Hamiltonian ones, of infinitely many conservation laws.

Thus, the endomorphism $\hat{T}$, or its associated tensor field

$$T(\alpha, X) = \langle \alpha, \hat{T} X \rangle$$

appears to be the most interesting object in the analysis of integrability of field theories. In fact, as it has been shown in Part III, it is possible to characterize the complete integrability of systems with finitely many degrees of freedom (Liouville integrability) in terms of mixed tensor field T satisfying suitable conditions.

Example 37 *The sine–Gordon equation*

$$v_{xt} + \sin v = 0 \,,$$

admits the invariant endomorphism

$$\check{T}_s = D^2 + v_x^2 + v_{xx} D^{-1} v_x \,,$$

which is related to the one $\check{T}_k$ of KdV by the similarity transformation

$$\check{T}_s = V \check{T}_k V^{-1} \,,$$

with

$$V \equiv 3(\sqrt{-1}D^2 - v_{xx} - 3v_x D)$$

and where the tilde indicates that the transformation

$$u = \frac{3}{2}(v_x^2 + \sqrt{-4}v_{xx}) \tag{13.14}$$

has been performed.

Then, T_s and T_k are the same tensor field referred to two different coordinate systems and KdV equation corresponds, in the same reading, to the Hamiltonian dynamics generated by the second conserved functional of sine–Gordon equation.[179] *It follows that the conserved functionals of the sine–Gordon equation can be obtained from the ones of KdV equation simply by using the transformation* (13.14). *For instance,*

$$\int u dx \to \frac{3}{2} \int v_x^2 dx \,,$$

$$\frac{1}{2} \int u^2 dx \to \frac{9}{8} \int (v_x^4 - 4v_{xx}^2) dx \,,$$

and so on.

Example 38 *The Liouville equation*

$$\sigma_{xt} + \exp \sigma = 0$$

admits the invariant endomorphism

$$\check{T}_L = D^2 - D\sigma_x D{-}1\sigma_x + a^2 \,, \quad a \equiv \lim_{x \to +\infty} \sigma_x \,,$$

which is related to the one $\check{T}_k$ of KdV equation by the similarity transformation

$$\check{T}_L = J\tilde{\check{T}}_k J^{-1}$$

with

$$J \equiv 3(-D^2 + \sigma_{xx} + \sigma_x D) \,,$$

and where the tilde indicates that the following transformation

$$u = -\frac{3}{2}(\sigma_x^2 + 2\sigma_{xx} - a^2)$$

has been performed.

Then T_L and T_k are the same tensor field referred to two different coordinate systems and KdV equation corresponds, in the same reading, to the Hamiltonian dynamics generated by the second conserved functional of Liouville's equation.[180]

Example 39 *The Burgers' equation*

$$u_t = 2uu_x + u_{xx}$$

admits the invariant endomorphism

$$\hat{T}_B = D + DuD^{-1},$$

which generates an Abelian sequence of symmetries of the dynamics.

The next sections will be devoted to analyze the properties of our phenomenological tensor fields.

13.3.1 *Dynamical invariance*

Because of the Lenard's sequence and of the bi-Hamiltonian structure of (some) evolution equations, the first relevant property of the tensor field T is given by

$$\mathcal{L}_\Delta T = 0.$$

This characterization of the dynamics is very suggestive because of the similitude

Dynamics		**Invariant structure**
Symplectic	ω	*a not degenerate, skewsymmetric, closed* $\binom{0}{2}$ *tensor field*
Geodesical	Γ	*a connection 2-form*
Killing	g	*a symmetric, not degenerate* $\binom{2}{0}$ *tensor field*
Hamiltonian	Λ	*a skewsymmetric* $\binom{2}{0}$ *tensor field, fulfilling Jacobi identity*
Liouville	Ω	*a volume form*
Lax	T	*a* $\binom{1}{1}$ *tensor field with vanishing torsion*

13.3.2 *Nijenhuis torsion*

The second relevant property, coming by the Lenard sequence, is $\delta(\check{T}^n\alpha) \equiv 0$ if α is δ-closed and δ_T-closed; that is, if $\delta\alpha = 0$ and $\delta(\check{T}\alpha) = 0$.

We know that such a property is ensured by

$$\mathcal{N}_T(\alpha, X, Y) = 0\,,$$

where[159,110,96,97,160]

$$\mathcal{N}_T(\alpha, X, Y) \equiv \langle \alpha, \mathcal{H}_T(X, Y)\rangle$$

and

$$\mathcal{H}_T(X, Y) \equiv [(\mathcal{L}_{\hat{T}X}T)^\wedge - \hat{T}(\mathcal{L}_X T)^\wedge]Y\,.$$

13.3.3 *Bidimensionality of eigenspaces of T (KdV and sG)*

Since T is a $(1,1)$-tensor field, we can put a corresponding eigenvalue problem for the associated endomorphism $\check{T}$ on $\Lambda(\mathcal{M})$:

$$\check{T}G_\lambda = \lambda G_\lambda\,.$$

It is not difficult to see that for each λ there exist two (generalized) eigenvectors, namely G_λ^1, G_λ^2 such that

$$\check{T}G_\lambda^1 = \lambda G_\lambda^1\,, \quad \check{T}G_\lambda^2 = \lambda G_\lambda^2 + G_\lambda^1\,;$$

this corresponds to Jordan's normal form for a finite matrix.

Explicitly, we have

$$G_\lambda^1 = e^{2\ell_j}[f_2(ik_j, x)]^2\,, \quad G_\lambda^2 = e^{2\ell_j}\frac{\partial}{\partial k_j}[f_2(ik_j, x)]^2\,,$$

where $f(k, x)$ are the Jost solutions of the Lax operator L

$$L^2 f = -k^2 f\,, \quad k^2 = -\lambda\,.$$

13.4 Invariant Endomorphisms and Liouville's Integrability

It has been shown that the properties

- $\mathcal{L}_\Delta T = 0$
- $\mathcal{N}_T = 0$, with $\mathcal{N}_T(\alpha, X, Y,) \equiv \langle \alpha, [(\mathcal{L}_{\hat{T}X}T)^\wedge - \hat{T}(\mathcal{L}_X T)^\wedge]Y\rangle$

- $d = \dim$ (eigenspaces of T) $= 2$

seem to be verified by all evolution equations integrable by the Inverse Scattering Method.

We recall that in Part III, by using the first two properties but assuming diagonalizability instead of the third property,[78,80] a geometrical integrability scheme was constructed according to which it was stated that:

A dynamical vector field Δ which admits an invariant ($\mathcal{L}_\Delta T = 0$) mixed, diagonalizable tensor field T, with vanishing Nijenhuis tensor field ($\mathcal{N}_T = 0$) and doubly degenerate eigenvalues λ without stationary points ($\delta\lambda \neq 0$), is separable, integrable and Hamiltonian; i.e. a separable completely integrable Hamiltonian system.[80]

The proof was performed by showing that

- $\mathcal{N}_T = 0$ implies the Frobenius integrability of the eigenspaces of T.
- $\mathcal{L}_\Delta T = 0$ implies the separability of Δ along the eigenmanifolds in dynamics with 1 degree of freedom, each of them with a first integral.

The construction of a symplectic form (actually infinitely many), with respect to which Δ is a Hamiltonian vector field, was then easily accomplished.

In spite of the relevance of the diagonalizable case, the third property is a characteristic feature of soliton theories. We want to state here an *a priori* separability criterion, based on this new spectral hypothesis and worth using for soliton equations. As far as solitonic dynamics is concerned, integrability is proven without further hypotheses, while for background-radiation dynamics, a compact *a priori* integrability criterion is, to date, lacking.

The present results should naturally lead to the corresponding ones in terms of Lax pairs (these are considered in the context of bundles just based on phase manifold),[81] once the relationship between them and the above operator, now only analytically understood, will be translated into clear geometrical terms.

We can prove the following integrability criterion[83]:

A dynamical vector field Δ which admits an invariant mixed tensor field T, with vanishing Nijenhuis tensor $\mathcal{N}_T$ and bidimensional eigenspaces, completely separates in 1 degree of freedom dynamics. The ones associated to those degrees of freedom, whose corresponding eigenvalues λ are not stationary, are integrable and Hamiltonian.

Indeed, denote by λ^i the generic discrete eigenvalue of T and assume that the continuous spectrum of T consists of the real semiaxis $\Re^+$. Then the vanishing of the Nijenhuis torsion $\mathcal{N}_T$, associated to T, means that for all $\alpha \in \Lambda^{(1)}(\mathcal{M})$ and $X, Y \in \mathcal{TM}$,

$$\mathcal{N}_T(\alpha, Y, X) \equiv \langle \alpha, [(\mathcal{L}_{TX}T) - T(\mathcal{L}_X T)]Y \rangle = 0 . \tag{13.15}$$

According to our assumptions, a basis

$$(e_i, \varepsilon_i, f_{1,(k)}, f_{2,(k)}) , \quad i = 1, 2, \ldots, n , \quad k \in \Re^+$$

of $\mathcal{TM}$ exists, such that

$$\begin{aligned} Te_i &= \lambda^i e_i , \\ T\varepsilon_i &= \lambda^i \varepsilon_i + e_i , \qquad i = l, 2, \ldots, n , \\ Tf_{l,(k)} &= kf_{l,(k)} , \qquad l = 1, 2, \ k \in \Re^+ . \end{aligned}$$

Now introduce the corresponding dual basis

$$\{\vartheta^i, \tau^i, \gamma^{1,(k)}, \gamma^{2,(k)}, \quad i = 1, 2, \ldots, n , \quad k \in \Re^+\} .$$

of $\Lambda^1(\mathcal{M})$ that is a basis, for which

$$\begin{aligned} \langle \vartheta^i, e_j \rangle &= \langle \tau^i, \varepsilon_j \rangle = \delta^i_j , \\ \langle \gamma^{l,(k)}, f_{p,(h)} \rangle &= \delta^{l,(k)}_{p,(h)} , \\ \langle \vartheta^i, \varepsilon_j \rangle &= \langle \tau^i, e_j \rangle = \langle \vartheta^i, f_{l,(k)} \rangle = 0 , \\ \langle \tau^i, f_{l,(k)} \rangle &= \langle \gamma^{l,(k)}, e_i \rangle = \langle \gamma^{l,(k)}, \varepsilon_i \rangle = 0 , \end{aligned} \tag{13.16}$$

where $i, j = 1, \ldots, n$, $\delta^{l,(k)}_{p,(h)} \equiv \delta^l_p \delta(k - h)$.

The relations corresponding to Eqs. (13.16) in terms of differential 1-forms read

$$\begin{aligned} \check{T}\vartheta_i &= \lambda^i \vartheta_i + \tau^i , \\ \check{T}\tau_i &= \lambda^i \tau_i , \quad i = l, 2, \ldots, n , \\ \check{T}\gamma^{l,(k)} &= k\gamma^{l,(k)} , \quad l = 1, 2 , \quad k \in \Re^+ , \end{aligned}$$

where $\check{T}$ denotes the transposed of $\hat{T}$. As it will be shown, no more ingredients are needed to prove the separability in 1 degree of freedom dynamics, and (except for nowhere stationarity of the λ^i's) integrability of the discrete part of it. The analysis starts by observing that an explicit transcription of condition

(13.15) is the following:

$$
\begin{aligned}
&L_{e_i}\lambda^j = 0\,, \quad L_{f_{l,(k)}}\lambda^i = 0\,, \quad (\lambda^i - \lambda^j)L_{\varepsilon_i}\lambda^j = 0\,,\\
&(T-\lambda^i)(T-\lambda^i)[e_i, e_j] = 0\,,\\
&(T-\lambda^i)(T-\lambda^j)^2[e_i, \varepsilon_j] = 0\,,\\
&(T-\lambda^i)^2(T-\lambda^j)^2[\varepsilon_i, \varepsilon_j] = 0\,,\\
&(T-k)(T-h)[f_{l,(k)}, f_{p,(h)}] = 0\,,\\
&(T-\lambda^i)(T-k)[e_i, f_{l,(k)}] = 0\,,\\
&(T-\lambda^i)(T-k)[\varepsilon_i, f_{l,(k)}] = 0\,.
\end{aligned}
\tag{13.17}
$$

As a matter of fact, it is easily seen that Eq. (13.17) are equivalent to¶

$$\tau^i \wedge \delta\tau^i = \tau^i \wedge \vartheta^i \wedge \delta\vartheta^i = \gamma^{1,(k)} \wedge \gamma^{2,(k)} \wedge \delta\gamma^{l,(k)} = 0\,, \tag{13.18}$$

this implying, by the Frobenius theorem, that without loss of generality, the τ's, ϑ's and γ's can be considered to be closed differential forms, or equivalently, the basis

$$\{e_i, \varepsilon_i, f_{1,(k)}, f_{2,(k)}\,, \quad i = 1, 2, \ldots, n\,, \quad k \in \Re^+\}$$

can be chosen to be a holonomic frame.

On the other hand, the first line of Eq. (13.17) is equivalent to $\delta\lambda^i = (L_{\varepsilon_i}\lambda^i)\tau^i$, this implying that

$$\check{T}\delta\lambda^i = \lambda^i\delta\lambda^i\,. \tag{13.19}$$

It particularly means that the τ's can be chosen as equal to the $\delta\lambda^i$'s if, as will be assumed, the λ^i's are nowhere stationary. Furthermore, holonomicity implies that the set of functions $\lambda^1, \lambda^2, \ldots, \lambda^n$ can be completed to form a coordinate system‖

$$(\lambda^1, \lambda^2, \ldots, \lambda^n, \varphi^1, \varphi^2, \ldots, \varphi^n, \psi^{1,k}, \psi^{2,k}\,, \quad k \in \Re^+)$$

in such a way that

$$e_i = \frac{\delta}{\delta\varphi^i}\,, \quad \varepsilon_i = \frac{\delta}{\delta\lambda^i}\,, \quad f_{l,(k)} = \frac{\delta}{\delta\psi^{1,(k)}}\,.$$

¶Here and in the following, δ denotes the exterior derivative and $\wedge$ the usual wedge product.

‖Some of them may not be global but only periodic ones.

Just to fix our ideas, the tensor operator T acquires the following *canonical* form:

$$T = \sum_i \lambda_i \left(\frac{\delta}{\delta\lambda^i} \otimes \delta\lambda^i + \frac{\delta}{\delta\varphi^i} \otimes \delta\varphi^i \right) + \frac{\delta}{\delta\varphi^i} \otimes \delta\lambda^i + \sum_{\ell=1}^{2} \int_0^\infty dkk \frac{\delta}{\delta\psi_k^\ell} \otimes \delta\psi^\ell(k) .$$

It is now easily proved that for such a T, the Δ invariance, namely $L_\Delta T = 0$, gives

$$\begin{aligned}
&\text{(a)} \quad \langle \delta\lambda^i, \Delta \rangle = 0 , \\
&\text{(b)} \quad \frac{\delta}{\delta\varphi^j} \langle \delta\varphi^i, \Delta \rangle = 0 , \\
&\text{(c)} \quad \frac{\delta}{\delta\psi^{l,(k)}} \langle \delta\varphi^i, \Delta \rangle = 0 , \\
&\text{(d)} \quad (\lambda^i - \lambda^j) \frac{\delta}{\delta\lambda^j} \langle \delta\varphi^i, \Delta \rangle = 0 , \\
&\text{(e)} \quad \frac{\delta}{\delta\varphi^i} \langle \delta\psi^{l,(k)}, \Delta \rangle = 0 , \\
&\text{(f)} \quad \frac{\delta}{\delta\lambda^i} \langle \delta\psi^{l,(k)}, \Delta \rangle = 0 , \\
&\text{(g)} \quad (k - h) \frac{\delta}{\delta\psi^{l,(h)}} \langle \delta\psi^{p,(k)}, \Delta \rangle = 0 ,
\end{aligned} \tag{13.20}$$

from which separability and integrability follow. More specifically, Eq. (13.20(a)) means the vanishing of "λ components" of Δ; Eq. (13.20(b)) the independence of the φ components on the φ's; Eq. (13.20(c)) the independence of the continuous coordinates; and Eq. (13.20(d)) just means that each φ component can only depend on the corresponding λ. On the other hand, Eq. (13.20(e)) shows that the continuous components cannot depend on discrete variables; and Eq. (13.20(f)) that each continuous component can only be a function of the continuous variables with the same continuous index. The most general form of Δ is then

$$\Delta = \sum_{i=1}^{n} \Delta^i(\lambda^i) \frac{\delta}{\delta\varphi^i} + \sum_{\ell=1}^{2} \int_0^\infty dk \Delta^\ell(k)(\psi^1(k), \psi^2(k)) \frac{\delta}{\delta\psi^l(k)} .$$

The dynamical equation then decouples in the following second-order systems for the continuous degrees of freedom (*background radiation dynamics*):

$$\dot{\psi}^1(k) = \Delta^{1,k}(\psi^{1,(k)}, \psi^{2,(k)}),$$
$$\dot{\psi}^{2,(k)} = \Delta^{2,k}(\psi^{1,(k)}, \psi^{2,(k)}),$$

and the following trivially integrable ones:

$$\dot{\phi}^i = \Delta^i(\lambda^i),$$
$$\dot{\lambda}^i = 0,$$

for the discrete part (soliton dynamics). Incidentally, the discrete part of the dynamics is Hamiltonian with respect to all symplectic forms

$$\omega_0 = \sum_i f_i(\lambda^i)\delta\lambda^i \wedge \delta\varphi^i$$

for the discrete part of the spectrum, f's being arbitrary nonvanishing functions.

Remark 23 *The vector field Δ is not supposed to define a Hamiltonian dynamics. Its Hamiltonian structure arises from the supposed bidimensionality of eigenspaces of T and the requirement $\delta\lambda \neq 0$.*

13.5 Recursion Operators in Dissipative Dynamics

We have seen that a nonlinear evolution equation $u_t = \Delta[u]$; i.e. the equation defining integral curves of the vector field Δ, is integrable once that a mixed tensor field T on $\mathcal{M}$ exists satisfying the following conditions:

- T is Δ invariant; i.e. $\mathcal{L}_\Delta T = 0$,
- T satisfies Nijenhuis condition; i.e. $[\mathcal{L}_{TX}T - T\mathcal{L}_X T]Y = 0$, for any two vector fields X and Y,
- T is diagonalizable with doubly degenerate eigenvalues λ without stationary points.

These assumptions on T not only ensure generic integrability, but also the existence of symplectic forms with respect to which dynamics is Hamiltonian and integrability is the usual one in terms of action-angle variables. On the other hand, there are many physically relevant cases in which the dynamics is not Hamiltonian, and nevertheless a suitable generalization of the above geometrical scheme could still be useful. The aim of the present example is to explore the possibility of using invariant mixed tensor fields to analyze dissipative dynamics. In order to do that, it is natural to begin by removing only the last condition on T, as it is the one leading to the existence of constants of motion. An instance of a dynamics which admits an invariant mixed tensor field T which satisfies Nijenhuis condition, but which is not diagonalizable without complexification and whose eigenvalues are trivially constant, is given by Burgers' equation.

This equation is just the simplest one combining both nonlinear propagation and diffusive effects, and it can be used as the working example for our analysis.

13.5.1 *The Burgers' hierarchy*

It is well-known[114,72] that the Burgers' equation can be linearized through the transformation

$$u = \frac{v_x}{v}, \tag{13.21}$$

where v satisfies the heat equation

$$v_t = v_{xx}.$$

It can easily be shown[70] that the Burgers' equation is a member of a whole hierarchy of nonlinear evolution equations which linearize, by using the same transformation (13.21), to equations of the type

$$v_t = D^n v, \quad n = 1, 2, \ldots, \tag{13.22}$$

with D denoting x derivative. The even elements of Eq. (13.22) obviously define dissipative dynamics, while the odd ones are integrable Hamiltonian evolution equations with respect to the following symplectic form:

$$\omega = \int_{-\infty}^{+\infty} \delta_1 v(x)(D^{-1}\delta_2 v)(x)dx, \tag{13.23}$$

where

$$(D^{-1}f)(x) = \int_{-\infty}^{+\infty} f(y)dy,$$

with Hamiltonian functionals given by

$$H_p = \frac{1}{2}\int_{-\infty}^{+\infty}(D^p v)^2 dx\,. \tag{13.24}$$

In order that Eqs. (13.23) and (13.24) make sense, some assumptions on the functional space $\mathcal{M}$ must be made, for example to assume that $\mathcal{M}$ consists of fast decreasing infinitely differentiable functions. Then clearly

$$T[v] = D$$

is a Nijenhuis Δ-invariant tensor operator for the heat equation hierarchy. In the present geometrical approach, Eq. (13.21) plays the role of a coordinate transformation, and thus a Nijenhuis Δ-invariant tensor operator for the Burgers' hierarchy is readily obtained from $\hat{T}[v]$ by[94]

$$T[u] = \left(\frac{\delta v}{\delta u}\right)^{-1} T[v]\left(\frac{\delta v}{\delta u}\right),$$

which easily yields

$$T[u] = D + DuD^{-1}\,. \tag{13.25}$$

The Burgers' hierarchy is then obtained by repeated applications, on the translation group generator $\Delta_0 = u_x$, of the tensor operator expressed by Eq. (13.25),

$$\Delta_k = T^k\Delta_0\,. \tag{13.26}$$

The first fields of the hierarchy are

$$\begin{aligned}
\Delta_0 &= u_x\,,\\
\Delta_1 &= 2uu_x + u_{xx}\,,\\
\Delta_2 &= (3u^3 + 3uu_x + u_{xx})_x\,.
\end{aligned}$$

This hierarchy is just the transcription in the new coordinate frame of the linear one and, apart from some technical points on the phase manifold $\mathcal{M}$, one can translate what has been said for Eq. (13.22) to the Burgers' hierarchy. More precisely, Eq. (13.26) splits into the following two sub hierarchies:

- *Dissipative hierarchy*

$$T\Delta_0, T^3\Delta_0, \ldots, T^{2n+1}\Delta_0, \ldots,$$

- *Hamiltonian hierarchy*

$$\Delta_0, T^2\Delta_0, \dots, T^{2n}\Delta_0, \dots ,$$

which are, respectively, a sequence of dissipative and Hamiltonian vector fields. The foregoing statement can be understood by examining the spectral properties of T, whose *block diagonal form*[79] is

$$T = \int dk (e_{(k)} \otimes \vartheta'^k - e'_k \otimes \vartheta^k) k \,,$$

where the vector fields

$$e_{(k)}[u] = \int_{-\infty}^{\infty} dx(-u \cos kx - k \sin kx) \exp\left[-\int_{-\infty}^{x} u dy\right] \frac{\delta}{\delta u(x)} \,,$$

$$e'_{(k)}[u] = \int_{-\infty}^{\infty} dx(-u \sin kx + k \cos kx) \exp\left[-\int_{-\infty}^{x} u dy\right] \frac{\delta}{\delta u(x)} \,,$$

are a basis of a generic invariant subspace

$$\begin{aligned} Te_{(k)} &= -ke'_{(k)} \,, \\ Te'_{(k)} &= ke_{(k)} \,, \\ \langle \vartheta'^{(k)}, e'_{(k)} \rangle &= \langle \vartheta^{(k)}, e_{(k)} \rangle = \delta(h-k) \,, \\ \langle \vartheta'^{(k)}, e_{(h)} \rangle &= \langle \vartheta^{(k)}, e'_{(h)} \rangle = 0 \,. \end{aligned}$$

The conditions

$$[e_{(k)}, e_{(h)}] = [e'_{(k)}, e'_{(h)}] = [e'_{(k)}, e_{(h)}] = 0$$

imply the holonomicity of the frame; i.e. the existence of coordinates

$$(q^{(k)}, p^{(k)}) \,,$$

such that

$$e_{(k)} = \frac{\delta}{\delta q^{(k)}} \,, \quad e'_{(k)} = \frac{\delta}{\delta p^{(k)}} \,.$$

In the bidimensional integral manifold of $\{e_{(k)}, e'_{(k)}\}$, the operator T can be projected to

$$\delta\varphi^{(k)} \otimes \frac{\delta}{\delta J^{(k)}} - \delta J^{(k)} \frac{\delta}{\delta \varphi^{(k)}} \,, \quad \textit{no sum over } k \,,$$

where

$$J^{(k)} = \frac{1}{2}(q^{(k)^2} + p^{(k)^2}), \quad \varphi^{(k)} = \arctan\frac{q^{(k)}}{p^{(k)}}$$

are action-angle type variables. Then, $\hat{T}$ transforms a dissipative integrable field of the type

$$X_D^{(k)} = \Delta(J^{(k)})\frac{\partial}{\partial J^{(k)}}$$

into a Hamiltonian one

$$X_H^{(k)} = \Delta(J^{(k)})\frac{\partial}{\partial \varphi^{(k)}}$$

and *vice versa.*

This alternating character of T is responsible for the splitting of hierarchy (13.26) into two subhierarchies. Furthermore, we observe that

- $\hat{T}$ has a bidimensional invariant spaces, but is not diagonalizable without complexification.
- $\hat{T}^2$, which characterizes the Hamiltonian subhierarchy, is diagonalizable with doubly degenerate constant eigenvalues.

Thus, for none of the subhierarchies one can use the integrability criterion to establish their integrability.

However, we observe that the projections of dissipative dynamics on the bidimensional invariant spaces simply are 1 degree of freedom dynamics, while for the Hamiltonian ones, the existence of a functional $J^{(k)}[u]$, which is not trivially conserved on each bidimensional space, ensures its integrability. It is worthwhile remarking that this same functional $J^{(k)}[u]$ obviously plays the role of a Ljapunov** functional for the projection of the dissipative dynamics on the bidimensional invariant submanifold, thus ensuring the asymptotic stability of the solution $J^{(k)}[u] = 0$.

The Hamiltonian subhierarchy

We discuss in more details the Hamiltonian character of subhierarchy (13.22). In order to do that, some care is needed for the appropriate choice of the

**Alexander Ljapunov was born in Jaroslav (central Russia) in 1857 and died in S. Petersburg in 1918. He has been professor of mathematics at Kharkov University and after, member of the S. Petersburg Academy of Science.

functional space $\mathcal{M}$ on which dynamics is defined. The most natural one would be to take $\mathcal{M}$ as the functional space whose elements u go to a constant as $x \to \pm\infty$, as it is the space on which there lies the typical solitary wave of Burgers' hierarchy. However, with such a choice it would not be possible to introduce a Hamiltonian structure on $\mathcal{M}$.

This can be understood easily by going back to the linear hierarchy for which $\mathcal{M}$ becomes, via the transformation (13.21), the space of functions which as $x \to \pm\infty$ behave like $\exp[kx]$, and the Hamiltonian becomes meaningless. One is then tempted to restrict $\mathcal{M}$ in such a way, that both symplectic structures and the Hamiltonian one be well-defined. This can be accomplished by considering only function $v(x)$ tending to some nonvanishing fixed constants as $x \to \pm\infty$ or, equivalently, functions $u(x)$ vanishing as $x \to \pm\infty$, whose integral has fixed value. More precisely, as for what refers to tangent spaces, the derivative of the Hopf–Cole map is a bijection $\delta v \to \delta u$ between $\mathcal{S}(\Re)$; i.e. the space of all *fast decreasing* test functions, and the space of functions which are derivatives of elements of $\mathcal{S}(\Re)$, this ensuring the existence of a symplectic structure with respect to which the subhierarchy is Hamiltonian.

The previous analysis shows the role played by the spectral hypothesis on the invariant mixed tensor field T in characterizing dynamical systems. The violation of the diagonalizability hypothesis allowed the inclusion of dissipative dynamics into the geometrical scheme. Moreover, the example shows that even if the eigenvalues of T^2 are trivially constant, sequences of constants of motion can be constructed by it.

Chapter 14

Meaning and Existence of Recursion Operators

Some confusion exists in the literature about recursion operators. This chapter will be addressed to clarify the meaning and the existence of recursion operators for completely integrable Hamiltonian systems.

In previous chapters it has been shown that completely integrable Hamiltonian dynamical systems may admit more than one Hamiltonian description. It has been also shown that, usually, with these alternative descriptions, one associates a (1,1)-tensor field which can be used under suitable conditions, as a recursion operator, namely as an operator which generates enough constants of the motion in involution. It seems to be an open question whether it is possible to find a recursion operator for any completely integrable system.

In the hypothesis of nonresonance, it has been shown that a recursion operator can always be constructed, even for some infinite dimensional systems.[80] Some authors claimed however that this is not always the case.

So it seems to us that it is of some interest to further comment on possible meanings of recursion operators and to show that, in condition of nonresonance, any integrable system can be reduced to a linear normal form via a nonlinear noncanonical transformation. For these normal forms, it is straightforward to construct recursion operators.[130]

14.1 Integrable Systems

Let $\mathcal{M}$ be a smooth $2n$-dimensional manifold. Let us suppose we can find n vector fields $X_1, \ldots, X_n \in \mathcal{X}(\mathcal{M})$ and n functions $F_1, \ldots, F_n \in \mathcal{F}(\mathcal{M})$ with the following properties:

$$[X_i, X_j] = 0\,, \tag{14.1}$$

$$\mathcal{L}_{X_i} F^j = 0\,, \quad i, j \in \{1, \ldots, n\}\,. \tag{14.2}$$

Let us suppose also that, on an open dense submanifold of $\mathcal{M}$, we have

$$X_1 \wedge \cdots \wedge X_n \neq 0\,, \tag{14.3}$$

$$dF^1 \wedge \cdots \wedge dF^n \neq 0\,. \tag{14.4}$$

We shall show that any dynamical system Δ on $\mathcal{M}$, which is of the form

$$\Delta = \sum_{i=1}^{n} \nu^i X_i\,, \quad \nu^i = \nu^i(F^1, \ldots, F^n)\,, \tag{14.5}$$

is explicitly integrable on the submanifold on which Eqs. (14.3) and (14.4) are satisfied.

We assume finally, that the level sets of the submersion

$$\mathbf{F} : \mathcal{M} \to \Re^n\,, \quad \mathbf{F} = (F^1, \ldots, F^n) \tag{14.6}$$

are compact. Then the vector fields X_i are complete on each leaf $\mathbf{F}^{-1}(\mathbf{a})$, $\mathbf{a} \in \Re^n$, and they integrate to a locally free action of the Abelian group $\Re^n$. Moreover, each leaf is parallelizable and we can find closed differential 1-forms $\alpha^1, \ldots, \alpha^n$, $d\alpha^i = 0$, such that

$$\alpha^i(X_j) = \delta^i_j\,, \quad i, j \in \{1, \ldots, n\}\,. \tag{14.7}$$

With all previous construction, the vector field Δ in Eq. (14.5) can be explicitly integrated in a neighborhood of each leaf $\mathbf{F}^{-1}(\mathbf{a})$, where we take as coordinates the functions $\{F^i, \varphi^j\}$ with $d\varphi^j = \alpha^j$. The equations of motion of Δ are given by

$$\begin{aligned} \dot{\phi}^i &= \nu^i(F^1, \ldots, F^n)\,, \\ \dot{F}^i &= 0\,. \end{aligned} \tag{14.8}$$

Therefore, the corresponding solutions are

$$\varphi_i(t) = t\nu^i(\mathbf{F}(p_0)) + \varphi^i(p_0)\,,$$
$$F_i(t) = F_i(p_0)\,, \tag{14.9}$$

with $p_0 \in \mathcal{M}$ the initial point. We see that the functions ν^i play the role of frequencies.

We stress the fact that up to now we have not used any Hamiltonian structure. For an algebraic characterization of complete integrability, see Refs. 77 and 126.

14.1.1 *Alternative Hamiltonian descriptions for integrable systems*

We shall now investigate under which conditions a dynamical system, which is integrable in the sense stated before, admits infinitely many alternative Hamiltonian descriptions.

With the n-functions $F^1, \ldots, F^n$ obeying the condition expressed by Eq. (14.4), we can define a closed differential 2-form by

$$\omega_f = \sum_i df_i(F^j) \wedge \alpha^i\,, \tag{14.10}$$

which is nondegenerate as long as $df_1 \wedge \cdots \wedge df_n \neq 0$. Any one of these symplectic forms makes the action of $\Re^n$ a Hamiltonian one. Indeed, by construction of ω_f,

$$i_{X_j}\omega_f = -df_j\,, \quad j \in \{1, \ldots, n\}\,. \tag{14.11}$$

As for the vector field Δ in Eq. (14.5), we shall have that

$$i_\Delta \omega_f = -\sum_i \nu^i df_i\,. \tag{14.12}$$

A necessary condition for $i_\Delta \omega_F$ to be exact is that it is closed, namely that

$$\sum_i d\nu^i \wedge df_i = 0\,. \tag{14.13}$$

All sets of solutions of this equation for $f^1, \ldots, f^n$ satisfying $df_1 \wedge \cdots \wedge df_n \neq 0$ will give alternative Hamiltonian descriptions for the dynamical systems Δ

in Eq. (14.4). Moreover, any such Δ will be completely integrable in the Liouville–Arnold sense, the functions $f_1, \ldots, f_n$ being constants of the motion (by assumption of Eq. (14.2)) in involution,

$$\{f_i, f_j\}_A = \omega_f(X_i, X_j) = \mathcal{L}_{X_i} f_j = 0\,. \tag{14.14}$$

There are two limiting case where it is easy to exhibit solutions of Eq. (14.13).

The constant case

All the frequencies ν^i are constant numbers so that $d\nu^i = 0$ and Eq. (14.13) is automatically satisfied.

Any differential 2-form in Eq. (14.10) is an admissible symplectic structure, and the corresponding Hamiltonian function is given by

$$\omega_f = \sum_i \nu^i f_i\,. \tag{14.15}$$

An example of system for which this happens is given by the n-dimensional harmonic oscillator written as

$$\Delta = \sum_i \nu^i \Delta_i\,,$$

$$\Delta_i = \frac{1}{\sqrt{m_i k_i}} p_i \frac{\partial}{\partial q^i} - \sqrt{m_i k_i} q_i \frac{\partial}{\partial p^i}\,, \quad \textit{no sum over } i\,,$$

$$\nu^i = \sqrt{\frac{k_i}{m_i}}\,. \tag{14.16}$$

Here m_i and k_i are the mass and the elastic constant of the ith oscillator. Now the functions F^i are just given by the partial Hamiltonians

$$F^i = \frac{1}{2}\left(\frac{p_i^2}{m_i} + k_i q_i^2\right), \quad i \in \{1, \ldots, n\}\,. \tag{14.17}$$

The nonresonant case

None of the frequencies ν^i is constant and we have that $d\nu^1 \wedge \cdots \wedge d\nu^n \neq 0$. In this case we may think of the ν^i as "coordinates" and of the f^j as functions of the ν^i.

In this second case, very simple solutions of Eq. (14.13) are given by linear functions $f_i = \sum_j A_{ij}\nu^j$, $i \in \{1, \dots n\}$, $A_{ij} \in \Re$. The corresponding Hamiltonian description for Δ can given with quadratic Hamiltonian functions by

$$\omega_A = \sum_{ij} A_{ij} d\nu^i \wedge \alpha^j \,, \tag{14.18}$$

$$H_A = \frac{1}{2}\sum_{ij} A_{ij}\nu^i\nu^j \,. \tag{14.19}$$

Moreover, any other symplectic structure of the form

$$\omega_f = \sum_i df_i(\nu^i) \wedge \alpha^j \,, \tag{14.20}$$

in which any f_i depends only on the corresponding frequency ν^i, will be admissible as long as ω_f is nondegenerate; that is, as long as $df_1 \wedge \cdots \wedge df_n \neq 0$. The associated Hamiltonian functions depend on the explicit form of the functions f_i. For instance, if $f_i = (\partial G_i/\partial \tilde{\nu}^i)(\tilde{\nu}^i)$, the corresponding Hamiltonian can be written as

$$H_G = \sum_i \left(G_i - \nu^i \frac{\partial G_i}{\partial \nu^i} \right) \,. \tag{14.21}$$

A simple example for these case is given again by the n-dimensional harmonic oscillator written as

$$\Delta = \sum_i F^i \Delta_i \,, \tag{14.22}$$

where F^i and Δ_i are given by Eqs. (14.17) and (14.16), respectively. Now the partial Hamiltonians F^i play the role of frequencies.

The intermediate cases are more involved. For further comments on them we refer to Ref. 80.

It is worth stressing that there may be admissible Hamiltonian structures for Δ that cannot be derived by using the previous construction.

14.1.2 *Recursion operators for integrable systems*

We shall now show how to construct recursion operators for the integrable systems that we have considered in the previous sections. As we have seen, given the dynamical system expressed by Eq. (14.5), we can construct infinitely many Hamiltonian structures given for instance by Eq. (14.10) or Eq. (14.20).

The constant case: $d\nu^i = 0, \forall i \in \{1, \ldots, n\}$.
Two possible alternative symplectic structures are obtained from Eq. (14.10) as

$$\omega_1 = \sum_{ij} \delta_{ij} dF^i \wedge \alpha^j = \sum_k \omega_k \,, \tag{14.23}$$

$$\omega_f = \sum_{ij} \delta_{ij} f^i(F^i) dF^i \wedge \alpha^j = \sum_k f^k(F^k) \omega_k \,, \tag{14.24}$$

with the condition $df_1 \wedge \cdots \wedge df_n \neq 0$. Given them, we can construct a (1,1)-tensor field T on $\mathcal{M}$ by

$$T = \omega_f \circ \omega_1^{-1} = \sum_k f^k(F^k)\, \mathbf{I}_k \,, \tag{14.25}$$

where $\mathbf{I}_k$ is the identity operator on the kth bidimensional "plane" of $\mathcal{T}^*\mathcal{M}$ with "coordinates" (dF^k, α^k).

The nonresonant case: $d\nu^1 \wedge \cdots \wedge d\nu^n \neq 0$.
In this case two possible alternative symplectic descriptions are obtained from Eq. (14.20) as

$$\omega_1 = \sum_{ij} \delta_{ij} d\nu^i \wedge \alpha^j = \sum_k \omega_k, \tag{14.26}$$

$$\omega_f = \sum_{ij} \delta_{ij} f^i(\nu^i) d\nu^i \wedge \alpha^j = \sum_k f^k(\nu^k) \omega_k \,, \tag{14.27}$$

with the condition $df_1 \wedge \cdots \wedge df_n \neq 0$. Given these structures we can construct a (1,1)-tensor field T on $\mathcal{M}$ by

$$T = \omega_f \circ \omega_1^{-1} = \sum_k f^k(\nu^k)\, \mathbf{I}_k \,, \tag{14.28}$$

where $\mathbf{I}_k$ is the identity operator on the kth bidimensional "plane" of $\mathcal{T}^*\mathcal{M}$ with "coordinates" $(d\nu^k, \alpha^k)$.

From the way they are constructed, one sees that T in Eqs. (14.25) and (14.28) are invariant under Δ, have double degenerate spectrum with eigenvalues without critical points, and vanishing Nijenhuis torsion $\mathcal{N}_T$. Therefore they are recursion operators for the dynamical system Δ.

14.1.3 *Liouville–Arnold integrable systems*

Assume the dynamical vector field Δ on the symplectic manifold $(\mathcal{M}, \omega_0)$ has n constants of the motion $H^1, \ldots, H^n$, which are in involution (with respect to the Poisson structure associated with ω_0), functionally independent, $dH^1 \wedge \cdots \wedge dH^n \neq 0$, and generate complete vector fields $X_1, \ldots, X_n$. We have then an action of $\Re^n$ on $\mathcal{M}$ that is locally free and fibrating.

In this situation, *angle* differential 1-forms $\alpha^1, \ldots, \alpha^n$ can be found, such that

$$\alpha^i(X_j) = \delta^i_j \,, \quad d\alpha^i = 0 \,.$$

Given any function F of the H^j, (or $dF \wedge dH^1 \wedge \cdots \wedge dH^n = 0$) satisfying the condition

$$\det\left(\frac{\partial^2 F}{\partial H^i \partial H^j}\right) \neq 0 \,,$$

the differential 2-form

$$\omega_F = d\left(\frac{\partial F}{\partial H^i}\alpha^i\right)$$

is an admissible symplectic structure for the $\Re^n$ action. In particular, if

$$F = \frac{1}{2}\sum_i H_i^2 \,,$$

we just get back the $\{H^i\}$ as Hamiltonian functions.

With a set of *action-angles* variables (J_k, φ^k), we have that

$$\Delta = \nu^k \frac{\partial}{\partial \varphi^k} \,,$$

$$\omega_0 = dJ_k \wedge d\varphi^k \,,$$

$$i_\Delta \omega = \nu^k dJ_k = \frac{\partial H}{\partial J_k} dJ_k = -dH \,,$$

where $\nu^k = \partial H/\partial J_k$, $k \in \{1, \ldots, n\}$ are the frequencies. In the *nonresonant case* when $d\nu^1 \wedge \cdots \wedge d\nu^n \neq 0$, or equivalently, $\det(\partial\nu^h/\partial J_k) \neq 0$, * we can use the ν^k as variables and write the admissible symplectic structure

$$\omega_\nu = \sum_k d\nu^k \wedge d\varphi^k ,$$

with Hamiltonian a quadratic function

$$H_\nu = \frac{1}{2}\sum_k (\nu^k)^2 .$$

By using the analysis of the previous section, we obtain that a not resonant complete integrable system has infinitely many admissible symplectic structures, some of them having the form

$$\omega_f = \sum_i df_i(\nu^i) \wedge d\varphi^i ,$$

with the condition $df^1 \wedge \cdots \wedge df^n \neq 0$. However, in general, we may not obtain ω_0 in this way. Moreover, such systems do admit recursion operators given by Eq. (14.28).

Example 40 *Let us consider the following* 2*-degrees of freedom, completely integrable system. Take* $\mathcal{M} = \Re^2 \times T^2 = \{(x, y, \vartheta, \eta)\}$ *with symplectic structure* $\omega_0 = dx \wedge d\vartheta + dy \wedge d\eta$*. The dynamical system is described by the Hamiltonian* $H = x^3 + y^3 + xy$*. The corresponding dynamical vector field is given by*

$$\Delta = \nu_\vartheta \frac{\partial}{\partial\vartheta} + \nu_\eta \frac{\partial}{\partial\eta} ,$$

$$\begin{aligned} \nu_\vartheta &= 3x^2 + y , \\ \nu_\eta &= 3y^2 + x . \end{aligned} \tag{14.29}$$

From what we have said before, this system admits infinitely many alternative Hamiltonian descriptions in the dense open submanifold characterized by $d\nu_\vartheta \wedge d\nu_\eta \neq 0$*, namely by* $36xy - 1 \neq 0$*, which coincides with the submanifold on which* H *is nondegenerate. Two such structures are given by*

$$\omega_1 = d\nu_\vartheta \wedge d\vartheta + d\nu_\eta \wedge d\eta , \tag{14.30}$$

*This is also equivalent to the nondegeneracy of the Hamiltonian function.

$$\omega_2 = f(\nu_\vartheta)d\nu_\vartheta \wedge d\vartheta + g(\nu_\eta)d\nu_\eta \wedge d\eta\,, \tag{14.31}$$

where f and g are any two functions such that $df \wedge dg \neq 0$. The corresponding recursion operators, $T = \omega_2 \circ \omega_1^{-1}$, are given by

$$T = f(\nu_\vartheta)\left(d\nu_\vartheta \otimes \frac{\partial}{\partial \nu_\vartheta} + d\vartheta \otimes \frac{\partial}{\partial \vartheta}\right) + g(\nu_\eta)\left(d\nu_\eta \otimes \frac{\partial}{\partial \nu_\eta} + d\eta \otimes \frac{\partial}{\partial \eta}\right). \tag{14.32}$$

We stress the fact that ω_0 is not among the symplectic structures constructed in Eq. (14.31) and that our recursion operators (14.32) cannot be "factorized" through ω_0.

We shall make some more comment on the meaning of recursion operators and on their use in the analysis of complete integrability.[78,80,185,145]

Let us suppose we have a dynamical vector field $\Delta \in \mathcal{X}(\mathcal{M})$ and a compatible (1,1)-tensor T field, namely $\mathcal{L}_\Delta T = 0$, so that the functions $\mathrm{tr}\, T^k$, $k \geq 1$ are constants of the motion. By applying powers of T, we obtain vector fields $\Delta_k = T^k(\Delta)$, which are symmetries of Δ. The Lie algebra $\{\Delta_k, k \geq 0\}$ is Abelian if $\mathcal{N}_T = 0$.

If $F \in \mathcal{F}(\mathcal{M})$ is a constant of the motion for Δ, we say that T is an *F-weak recursion operator* if $\mathcal{N}_T = 0$ and $d(T(dF)) = 0$. If T is an F-weak recursion operator, one can prove that $d(T^k(dF)) = 0$, $\forall k > 1$. Locally, one finds functions $F_k \in \mathcal{F}(\mathcal{M})$ by $dF_k = T^k(dF)$, which are constants of the motion for Δ.

It is worth stressing that a given operator T may be a recursion operator for the constant of the motion F and not a recursion for another constant of the motion G. Moreover, it may also happen that the tensor T is an F-recursion operator but $T^k(dF) \wedge dF = 0$, $\forall k \geq 1$, so that one cannot use T and F to generate new constants of the motion. This is what happens for instance with the Kepler problem if one starts with the standard Hamiltonian function.[150] However, it is always true that $T[d(1/k)\mathrm{tr}\, T^k] = d(1/(k+1)\mathrm{tr}\, T^{k+1})$.

If ω is an admissible symplectic structure for Δ, namely $L_\Delta \omega = 0$, we say that T is a *ω-weak recursion operator* if† $\mathcal{N}_T = 0$ and $d(T(\omega)) = 0$. If T is a ω-weak recursion operator, one proves that $d(T^k(\omega)) = 0$, $\forall k > 1$. All differential 2-forms $\omega_k = T^k(\omega)$ are then admissible symplectic structures for Δ.

---

†Again, we use the same symbol for the extension of T to differential forms.

It is worth stressing that given any two admissible symplectic structures ω_1 and ω_2 for Δ, it need not be true that they are connected by a recursion operator. Moreover, it may happen that $T^k(\omega) \wedge \omega = 0$, $\forall\, k \geq 1$, so that one does not generate new symplectic structures.

If Δ is Hamiltonian with respect to the couple (ω, H), namely $i_\Delta \omega = -dH$, we say that T is a *strong recursion operator* if it is both a H-recursion operator and an ω-recursion operator. If this is the case, any vector field Δ_k is a Hamiltonian one with respect to ω with Hamiltonian function H_k as well as with respect to ω_k with Hamiltonian function H. Moreover, the constants of the motion H_k are pairwise in involution with respect to the Poisson structure constructed by inverting anyone of the symplectic structures ω_k, $k \geq 0$.

Chapter 15

Miscellanea

15.1 A Tensorial Version of the Lax Representation

In this section it is shown that the Lax representation (LR) can be regarded as the vanishing, along the dynamics, of the covariant derivative of a section of an $\mathcal{M}$-based bundle.[81]

Although the Hamiltonian structure of nonlinear field theories leads to an extremely simple method for the construction of sequences of conserved functionals and to a geometrical interpretation of scattering data, it has not played a fundamental role in the construction of the LR. On the other hand, although a deep and effective interpretation is to consider the LR as a linear problem whose integrability condition coincides with the original nonlinear evolution equation, it is not clear how the existence of an LR, in this sense, qualifies the vector field and the manifold. In spite of its connection with the powerful method of the *inverse spectral transform*,[8] the Lax formulation is then lacking of a clear-cut geometrical interpretation; that is, the Lax dynamics is not defined in terms of a geometrical structure it preserves. Preliminary interesting answers to the problem are given by the *loop groups approach* (see, for instance, Ref. 163).

The present geometrical approach is motivated, first of all by the interest *per se* of the possible geometrical structures underlying the Lax representation for a dynamical vector field on a manifold, on the other hand, by our belief that a geometrical understanding can be of help in the extension to more space dimensions.

Once given a vector field Δ on a manifold $\mathcal{M}$,

$$\Delta : \mathcal{M} \to \mathcal{TM}\,, \quad \tau_{\mathcal{M}} \circ \Delta = \mathbf{1}_{\mathcal{M}}\,,$$

where $\tau_{\mathcal{M}}$ is the natural projection of tangent bundle $\mathcal{TM}$, our aim is to translate in geometrical terms the problem of looking for a Lax pair L, B; that is, for a pair of operator fields on $\mathcal{M}$ such that

$$\dot{L} = [B, L]\,.$$

The structure of the Lax equation naturally suggests two simple and appealing geometrical readings. First of all, one can think of it as the explicit form of the equation

$$\mathcal{L}_\Delta L = 0\,, \tag{15.1}$$

once L has been interpreted as a section of the linear frame bundle. In fact, once fixed a frame, L and Δ can be written as

$$L = L^\mu_\lambda e_\mu \otimes \vartheta^\lambda\,, \quad \Delta = \Delta^\mu e_\mu\,,$$

and an equation of the form $\dot{L} = [B, L]$ is obtained by imposing Eq. (15.1) with

$$B^i_j = i_{e_j} d\Delta^i - \Delta^k i_{[e_k, e_j]} \vartheta^i\,.$$

(To be specific in notation here and in the following, except for an infinite dimensional example, $\mathcal{M}$ is supposed to be a finite-dimensional differential manifold with $\Re^n$ as a local model.) On the other hand the Lax equation can be read as the explicit form of an equation of the type

$$D_\Delta L = 0\,, \tag{15.2}$$

where the covariant derivative is taken with respect to a prescribed connection on a fiber bundle based on $\mathcal{M}$, not necessarily the linear frame bundle. To illustrate this possibility, consider the case in which the mentioned fiber bundle coincides with the principal fiber bundle of the structural group $GL(n, \Re)$; i.e. the linear frame bundle. In such a case the connection form ω can be written as

$$\omega = (\omega^\rho_\lambda)\,, \quad \rho, \lambda = 1, \ldots, n\,,$$

where the ω^{ρ}_{λ}'s are real-valued *1*-forms, and

$$DL = (dL^{\mu}_{\lambda} + \omega^{\mu}_{\rho} L^{\rho}_{\lambda} - \omega^{\rho}_{\lambda} L^{\mu}_{\rho}) e_{\mu} \otimes \vartheta^{\lambda} .$$

By contracting with Δ and imposing Eq. (15.2), we obtain

$$\dot{L}^{\mu}_{\lambda} + i_{\Delta}\omega^{\mu}_{\rho} L^{\rho}_{\lambda} - L^{\mu}_{\rho} i_{\Delta}\omega^{\rho}_{\lambda} = 0 ,$$

where the dot denotes the $i_{\Delta}d$ operator. In more compact notation

$$\dot{L} = [B, L] ,$$

where

$$B = i_{\Delta}\omega , \tag{15.3}$$

i.e. $B^{i}_{j} = -\Delta^{\alpha}\Gamma^{i}_{\alpha,j}$, Γ's being the connection coefficients.

As it has been shown in the previous chapter, equations of the first type $\mathcal{L}_{\Delta}T = 0$ are satisfied by $(1,1)$-tensor fields associated with completely integrable nonlinear field theories and play, in connection with symplectic structure and, under some special assumptions, a relevant role in their integrability properties.

The "phenomenology" of integrable nonlinear field theory shows that two distinct operator fields play two different roles in them. One, let us call it T, which generates a sequence of conserved functionals, by its construction is surely an endomorphism of the module $\mathcal{X}(\mathcal{M})$ of vector fields on $\mathcal{M}$ (or by duality of $\mathcal{X}(\mathcal{M})^{*}$) and satisfies the equation $\mathcal{L}_{\Delta}T = 0$. The other one, let us call it L, is the linear operator that is used in the inverse scattering method; it is not *a priori* an endomorphism of $\mathcal{X}(\mathcal{M})$, and once we assume it to be an object of this type, it does not satisfy the equation $\mathcal{L}_{\Delta}L = 0$. It is then natural to assume that the Lax equation must be read as an equation of the type $D_{\Delta}L = 0$. This assumption is confirmed by specific examples showing that, while the equation $\mathcal{L}_{\Delta}T = 0$ is typically a feature that the dynamical vector field shares with a large class of fields, on the contrary the equation $D_{\Delta}L = 0$, once chosen a suitable connection, is able to fix without ambiguity the direction field associated with Δ.

The following example, though elementary, exhibits all the essential features of the exposed idea.

15.1.1 *The LR of the harmonic oscillator as a parallel transport condition*

In a natural chart the dynamical vector field is

$$\Delta = p\frac{\partial}{\partial q} - q\frac{\partial}{\partial p}.$$

Once given the connection form*

$$\omega = \frac{1}{4H}\begin{pmatrix} 0 & qdp - pdq \\ pdq - qdp & 0 \end{pmatrix}, \tag{15.4}$$

where H is the Hamiltonian $H = (1/2)(p^2 + q^2)$, Eq. (15.3) implies

$$B = \frac{1}{2}\begin{pmatrix} 0 & 1 \\ -1 & 0 \end{pmatrix}.$$

It is then straightforward to see that Eq. (15.2) is satisfied; that is, in a chosen frame

$$\dot{L} = [B, L],$$

where L is the tensor field

$$L = p\left(\frac{\partial}{\partial p} \otimes dp - \frac{\partial}{\partial q} \otimes dq\right) + q\left(\frac{\partial}{\partial p} \otimes dq + \frac{\partial}{\partial q} \otimes dp\right). \tag{15.5}$$

Equation (15.2) can also be read, once given L and ω, as an equation for Δ, and in this sense, not only is a property of Δ, but also defines as already mentioned, without ambiguity, the direction field associated with Δ, this being in contrast to the characterization induced by an equation of the type $\mathcal{L}_\Delta T = 0$. In order to elucidate this point let us consider the following examples.

---

*The column vector $\binom{a}{b}$ represents the vector field $a(\partial/\partial q) + b(\partial/\partial p)$.

15.1.2 The Δ-invariant tensor field for the harmonic oscillator

The general solution in Δ of the equation $\mathcal{L}_\Delta T = 0$, given

$$T = \frac{1}{2H}\left(q^2\frac{\partial}{\partial q}\otimes dq + p^2\frac{\partial}{\partial p}\otimes dp\right),$$

can be written in the following form:

$$\Delta(p,q) = f(p,q)\left(p\frac{\partial}{\partial q} - q\frac{\partial}{\partial p}\right) + g(H)\left(q\frac{\partial}{\partial q} + p\frac{\partial}{\partial p}\right),$$

with f and g being arbitrary functions.

In coordinate notation, the equation $\mathcal{L}_\Delta T = 0$ reads

$$\dot{T} = [A, T], \tag{15.6}$$

where

$$A = \begin{pmatrix} \dfrac{\partial \Delta^q}{\partial q} & \dfrac{\partial \Delta^q}{\partial p} \\ \dfrac{\partial \Delta^p}{\partial q} & \dfrac{\partial \Delta^p}{\partial p} \end{pmatrix} = \begin{pmatrix} 0 & 1 \\ -1 & 0 \end{pmatrix}.$$

On the other hand, and this is a general feature of Lax type equations derived by invariance of tensor fields, connections exist such that Eq. (15.6) becomes the coordinate transcription of equation $D_\Delta T = 0$. As matter of fact, the connection form

$$\omega = \frac{1}{3H}\begin{pmatrix} qdq + pdp & qdppdq \\ pdq - qdp & 0 \end{pmatrix}$$

satisfies the relation $A = -i_\Delta \omega$. The general solution of equation $D_\Delta T = 0$, devised as an equation for Δ, is

$$\Delta'(p,q) = f(p,q)\left(p\frac{\partial}{\partial q} - q\frac{\partial}{\partial p}\right)$$

(f being an arbitrary function), i.e. the harmonic oscillator up to parameterization. To avoid misunderstanding, we remark that the Lie derivative along

Δ of tensor fields satisfying Eq. (15.2) is generally a different zero. This is, for instance, the case for the tensor field given by Eq. (15.5).

15.1.3 *The Δ-invariant tensor field for KdV*

We recall that the evolution equation is

$$\dot{u} + uu_x + u_{xxx} = 0,$$

and that a tensor field satisfying the equation

$$\mathcal{L}_\Delta T = 0 \tag{15.7}$$

is given by the operator field

$$T\cdot = \partial_{xx}\cdot + 3u\cdot + \frac{u_x}{3}\int_{-\infty}^{x} \cdot dy,$$

whose adjoint is used for construction of the sequences of conserved functionals and is related by a Miura-like transformation, of tensorial nature, to the analogous operator for the sine–Gordon equation. Equation (15.7), explicitly written acquires the form $\dot{T} = [A, T]$ with $A = \partial_{xxx} - u\partial_x - u_x$.

Remark 24 *The existence of Δ-invariant T is so peculiar: in Lagrangian dynamics, q-equivalent Lagrangians*[148] *always lead to a Δ-invariant T.*

15.1.4 *The Δ-covariant tensor field for KdV*

In order to consider the usual L for KdV we will again adopt the coordinate notation in terms of "local coordinates" $u(x)$: differentials $\delta u(x)$ and functional derivatives $\delta/(\delta u(x))$, as formal elements of the "continuous natural basis" of cotangent and tangent spaces, respectively. The vector field is then written as

$$\Delta = \int_{-\infty}^{\infty} dx \Delta(u) \frac{\delta}{\delta u(x)}, \quad \Delta(u) = -\partial_x\left(\frac{1}{2}u^2 + u_{xx}\right).$$

It is easy to verify that the Lie derivative of the Lax tensor field,

$$L = \iint_{-\infty}^{\infty} dx dy \left[\delta''(x-y) + \frac{1}{6}u(x)\delta(x-y)\right] \frac{\delta}{\delta u(x)} \otimes \delta u(y),$$

corresponding to the Lax operator $L = \partial_{xx} + (1/6)u$, does not vanish.

If, on the other hand, we consider the connection form

$$\omega[x,y] = \int_{-\infty}^{\infty} \Gamma(x,y)\delta u(z)dz\,,$$

with

$$\omega[x,y] = \delta(x-z)\left[4\delta'''(x-y) + u(x)\delta'(x-y) - \frac{1}{2}u_x\delta(x-y)\right](uu_z + u_{zzz})^{-1}\,,$$

we obtain

$$\ker(B) = -i_\Delta\omega[x,y] = 4\delta'''(x-y) + u(x)\delta'(x-y) - \frac{1}{2}u_x\delta(x-y)\,,$$

and hence

$$B = 4\partial_{xxx} - u\partial_x - \frac{1}{2}u_x\,.$$

Therefore, the Lax representation for KdV equation can be written in the form $D_\Delta L = 0$.

In order to illustrate the utility of the geometrical reading of the LR as a condition of parallel transport, consider the transformations of the Lax pair induced by transformations of field variables. This matter is relevant according to the general feeling[138,179,180,105,119] that several integrable nonlinear field theories are equivalent between them, up to inversion problems of transformations. This point of view has, for instance, led to connect the T operators for sine–Gordon and KdV and the T operators for Liouville' equation and KdV. To give a simple example of the transformation method, consider once again the harmonic oscillator. To transform, for instance, the Lax pair

$$L = \begin{pmatrix} -p & q \\ q & p \end{pmatrix}, \quad B = \frac{1}{2}\begin{pmatrix} 0 & 1 \\ -1 & 0 \end{pmatrix},$$

to action-angle variables (J,φ), L must be transformed as $(1,1)$-tensor field and B as the contraction of the connection form (15.4) with the dynamical vector field. The transformation law for ω from the natural basis in x coordinates to that in x' is

$$\omega' = \left(\frac{\partial x'}{\partial x}\right)\omega\left(\frac{\partial x}{\partial x'}\right) + \left(\frac{\partial x'}{\partial x}\right)d\left(\frac{\partial x}{\partial x'}\right).$$

Then, in the new frame

$$L = \begin{pmatrix} -\sqrt{2J}\cos\varphi & -2\sin\varphi \\ -\sqrt{(2J)^3}\sin\varphi & \sqrt{2J}\cos\varphi \end{pmatrix}, \quad B = \begin{pmatrix} 0 & \dfrac{1}{4J} \\ -J & 0 \end{pmatrix},$$

which are, of course, a Lax pair for the dynamics $\dot{J} = 0, \dot{\varphi} = 1$.

15.2 Liouville Integrability of Schrödinger Equation

Some years ago it was suggested[176] the use of complex canonical coordinates in the formulation of a generalized dynamics including classical and quantum mechanics as special cases. In the same spirit, a somehow dual viewpoint can be proposed[151]: rather than to complexify classical mechanics it may be useful to give a formulation of quantum mechanics in terms of *realified* vector spaces.

By using the Stone–von Neumann theorem, a quantum mechanical system is associated with a vector field on some Hilbert space (*Schrödinger picture*) or a vector field; i.e. a derivation, on the algebra of observables (*Heisenberg picture*).

In classical mechanics the analog infinitesimal generator of canonical transformations is a vector field on a symplectic manifold (the *phase space*).

Therefore, if we want to use similar procedures, we need to real-off $L_2(Q, C)$, the Hilbert space of square integrable complex functions defined on the configuration space Q, as a symplectic manifold or, more specifically, as a cotangent bundle. We shall see that it can be considered as $\mathcal{T}^*(L_2(Q, \Re))$; $L_2(Q, \Re)$ denoting the Hilbert space of square integrable real functions defined on Q.

The approach is different from previous ones[124,66,71,69] also dealing with the integrability of quantum mechanical system in the Heisenberg and Schrödinger picture.

In order to make more transparent the geometrical and the physical content of the subject, difficult technical aspects (which are however important in the context of infinite dimensional manifold,[10] as for instance, the distinction between *weakly* and *strongly* not degenerate bilinear forms, or the inverse of a Schrödinger operator and so on) will not be addressed. We shall limit ourselves to again observe that no serious difficulties arise working on an infinite dimensional manifold whose local model is a Banach space, as in that case, the *implicit function theorem* still holds true.

Although in an infinite dimensional symplectic manifold, a Darboux's chart *a priori* does not exist, for the Schrödinger equation

$$i\hbar\frac{\partial\psi}{\partial t} = -\frac{\hbar^2}{2m}\Delta\psi + U(\mathbf{r})\psi\,,$$

natural canonical coordinates p and q can be introduced.

We introduce the real and the imaginary part of the wave function ψ:

$$\begin{cases} p(\mathbf{r},t) = \operatorname{Im}\psi(\mathbf{r},t)\,, \\ q(\mathbf{r},t) = \operatorname{Re}\psi(\mathbf{r},t)\,, \end{cases}$$

and in this way $L_2(Q,C)$ is considered as the cotangent bundle of $L_2(Q,\Re)$.

In these new coordinates, the Schrödinger equation takes the form

$$\frac{d}{dt}\begin{pmatrix} p \\ q \end{pmatrix} = \frac{1}{\hbar}\begin{pmatrix} 0 & 1 \\ -1 & 0 \end{pmatrix}\begin{pmatrix} \dfrac{\delta H_1}{\delta p} \\ \dfrac{\delta H_1}{\delta q} \end{pmatrix},$$

where H_1 is defined by

$$H_1[q,p] := \frac{1}{2}\int d\mathbf{r}\left\{\frac{\hbar^2}{m}[(\nabla p)^2 + (\nabla q)^2] + U(\mathbf{r})(p^2+q^2)\right\},$$

and $\delta H/\delta p$, $\delta H/\delta q$ denote the components of the gradient of $H[q,p]$ with respect to the real L_2 scalar product.

Our system is then a Hamiltonian dynamical system with respect to the Poisson bracket defined for any two functionals $F[q,p]$ and $G[q,p]$ by

$$\Lambda_1(\delta F,\delta G) := \{F,G\}_1 := \frac{1}{\hbar}\int d\mathbf{r}\left(\frac{\delta F}{\delta q}\cdot\frac{\delta G}{\delta p} - \frac{\delta F}{\delta p}\cdot\frac{\delta G}{\delta q}\right). \tag{15.8}$$

What is less known is that the previous one is not the only possible Hamiltonian structure. Indeed, the Schrödinger equation can also be written as

$$\frac{d}{dt}\begin{pmatrix} p \\ q \end{pmatrix} = \frac{1}{\hbar}\begin{pmatrix} 0 & -\mathcal{H} \\ \mathcal{H} & 0 \end{pmatrix}\begin{pmatrix} \dfrac{\delta H_0}{\delta p} \\ \dfrac{\delta H_0}{\delta q} \end{pmatrix},$$

where H_0 is defined by

$$H_0[q,p] := \frac{1}{2}\int d\mathbf{r}(p^2 + q^2)\,,$$

and $\mathcal{H}$ is the Schrödinger operator

$$\mathcal{H} := -\frac{\hbar^2}{2m}\Delta + U(\mathbf{r})\,.$$

It is then again a Hamiltonian dynamical systems with respect to a new Poisson bracket which, for any two functionals $F[q,p]$ and $G[q,p]$, is defined by

$$\Lambda_0(\delta F, \delta G) := \{F, G\}_0 := \int d\mathbf{r}\left(\frac{\delta F}{\delta q}\cdot\mathcal{H}\frac{\delta G}{\delta p} - \frac{\delta F}{\delta p}\cdot\mathcal{H}\frac{\delta G}{\delta q}\right).$$

Thus, with the same vector field, we have the two following choices:

- A phase manifold with a universal symplectic structure

$$\omega_1 := \hbar\int d\mathbf{r}(\delta p \wedge \delta q)$$

  and a Hamiltonian functional depending on the classical potential.
- A phase manifold with a symplectic structure determined by the classical potential

$$\omega_0 := \hbar\int d\mathbf{r}(\mathcal{H}^{-1}\delta p \wedge \delta q)$$

  and the universal Hamiltonian functional representing the quantum probability.

The two brackets satisfy the Jacobi identity, as the associated differential 2-forms are closed for they do not depend on the point $\psi \equiv (p,q)$ of the *phase space*.

We have then the relation

$$\frac{\delta H_1}{\delta u} = \check{T}\frac{\delta H_0}{\delta u}\,, \tag{15.9}$$

where

$$\check{T} := \Lambda_1^{-1}\circ\Lambda_0 = \begin{pmatrix}\mathcal{H} & 0\\ 0 & \mathcal{H}\end{pmatrix}$$

and

$$\frac{\delta H}{\delta u} = \begin{pmatrix} \dfrac{\delta H}{\delta q} \\ \dfrac{\delta H}{\delta p} \end{pmatrix} .$$

Since the tensor field T does not depend on the point $\psi \equiv (p, q)$ of the *phase space*, its torsion is identically zero, so that the relation (15.9) can be iterated to

$$\frac{\delta H_n}{\delta u} = \check{T}^n \frac{\delta H_0}{\delta u} .$$

It turns out that the Schrödinger equation admits infinitely many conserved functionals defined by

$$H_n[q,p] := \frac{1}{2} \int d\mathbf{r}(p\mathcal{H}^n p + q\mathcal{H}^n q) \equiv \int d\mathbf{r}(\bar{\psi}\mathcal{H}^n \psi) .$$

They are all in involution with respect to the previous Poisson brackets:

$$\{H_n, H_m\}_0 = \{H_n, H_m\}_1 = 0 .$$

It is worth stressing that for smooth potentials $U(x)$ in one space dimension, the eigenvalues of the Schrödinger operator $\mathcal{H}$ are not degenerate, so that the eigenvalues of T are double degenerate.

The eikonal transformation

The map

$$\begin{cases} p(\mathbf{r}, t) = A(\mathbf{r}, t) \sin \dfrac{S(\mathbf{r}, t)}{\hbar} , \\ q(\mathbf{r}, t) = A(\mathbf{r}, t) \cos \dfrac{S(\mathbf{r}, t)}{\hbar} , \end{cases}$$

is a canonical transformation between the (p, q) and $(\pi = S(2\hbar)^{-1}\hat{E}, \chi = A^2)$ coordinates, since

$$\delta p \wedge \delta q = \delta \left(\frac{S}{2\hbar} \right) \wedge \delta A^2 .$$

The Hamiltonian H_1 becomes

$$K_1[\chi,\pi] = \int d\mathbf{r} \left\{ \frac{\hbar^2}{2m} \left(\frac{(\nabla\chi)^2}{4\chi} + 4\chi(\nabla\pi)^2 \right) + U\chi \right\},$$

so that Hamilton's equations

$$\begin{cases} \dfrac{\partial \pi}{\partial t} = -\dfrac{1}{\hbar}\dfrac{\delta K_1}{\delta \chi}, \\ \dfrac{\partial \chi}{\partial t} = \dfrac{1}{\hbar}\dfrac{\delta K_1}{\delta \pi}, \end{cases}$$

give

$$\begin{cases} \dfrac{\partial \pi}{\partial t} = \dfrac{\hbar}{2m}\dfrac{\Delta(\sqrt{\chi})}{\sqrt{\chi}} - \dfrac{\hbar}{m}(\nabla\pi)^2 - U\hbar^{-1}, \\ \dfrac{\partial \chi}{\partial t} = -\dfrac{2\hbar}{2m}\mathrm{div}(\chi\nabla\pi), \end{cases}$$

where $P = \chi$ and $\mathbf{J} = \hbar\chi(\nabla S/m)$ represent the *probability density* and the *current density*, respectively.

This transformation being nonlinear will transform previous bi-Hamiltonian descriptions into a mutually compatible pair of nonlinear type.

Finally, it is worth to stress that the Schrödinger equation, in spite of its linearity, shows that the class of completely integrable field theories in higher dimensional spaces is not empty. Moreover, previous analysis appears to be interesting also in the formulation of variational principles[109] for stochastic mechanics.

15.2.1 *Comparison with the nonlinear Schrödinger equation*

The two-dimensional nonlinear Schrödinger equation

$$i\hbar\frac{\partial \psi}{\partial t} = -\frac{\hbar^2}{2m}\psi_{xx} + b|\psi|^2\psi, \tag{15.10}$$

has been analyzed by several authors.[186,122,188,95]

In the canonical coordinates

$$\begin{cases} p(x,t) = \operatorname{Im}\psi(x,t)\,, \\ q(x,t) = \operatorname{Re}\psi(x,t)\,, \end{cases}$$

it takes the form

$$\frac{d}{dt}\begin{pmatrix} p \\ q \end{pmatrix} = \frac{1}{\hbar}\begin{pmatrix} 0 & -1 \\ 1 & 0 \end{pmatrix}\begin{pmatrix} \dfrac{\delta K_1}{\delta p} \\ \dfrac{\delta K_1}{\delta q} \end{pmatrix},$$

where K_1 is defined by

$$K_1[q,p] := \frac{1}{2}\int dx \left\{ -\frac{\hbar^2}{2m}[(\partial_x p)^2 + b(\partial_x q)^2] + (p^2+q^2)^2 \right\}.$$

It is then a Hamiltonian dynamical system with respect to the canonical Poisson bracket Λ_1 defined by Eq. (15.8):

$$\Lambda_1(\delta F, \delta G) := \{F,G\}_1 := \frac{1}{\hbar}\int dx \left(\frac{\delta F}{\delta q}\cdot\frac{\delta G}{\delta p} - \frac{\delta F}{\delta p}\cdot\frac{\delta G}{\delta q} \right).$$

The previous one is not the only possible Hamiltonian structure. As matter of fact the nonlinear Schrödinger equation can also be written as[122,138]

$$\frac{d}{dt}\begin{pmatrix} q \\ p \end{pmatrix} = \mathcal{H}_N \begin{pmatrix} \dfrac{\delta K_0}{\delta q} \\ \dfrac{\delta K_0}{\delta p} \end{pmatrix},$$

where $\mathcal{H}_N$ is the *Poisson operator*

$$\mathcal{H}_N = \frac{1}{\hbar} i(\partial_x \circ + \psi D^{-1}[\psi(\bar{\circ}) + \bar{\psi}(\circ)])\,,$$

or equivalently,

$$\mathcal{H}_N = \frac{1}{\hbar}\begin{pmatrix} -\dfrac{\hbar}{2m}\partial_x + 2\alpha p D^{-1} p & -2\alpha p D^{-1} q \\ -2\alpha q D^{-1} p & -\dfrac{\hbar}{2m}\partial_x + 2\alpha q D^{-1} q \end{pmatrix},$$

with $\alpha = b\sqrt{2m}/\hbar$, and

$$D^{-1} := \frac{1}{2}\left(\int_{-\infty}^{x} dx - \int_{x}^{\infty} dx\right), \quad K_0[q,p] := \frac{\hbar}{\sqrt{2m}} \int dx(qp_x).$$

It is then again a Hamiltonian dynamical systems with respect to a new Poisson bracket† given, for any two functionals $F[q,p]$ and $G[q,p]$, by

$$\Lambda_2(\delta F, \delta G) := \{F,G\}_2 := \int dx \left(\frac{\delta F}{\delta q} \cdot \mathcal{H}_N \frac{\delta G}{\delta p} - \frac{\delta F}{\delta p} \cdot \mathcal{H}_N \frac{\delta G}{\delta q}\right).$$

Once again, with the same vector field, we have two following choices:

- A phase manifold with the canonical symplectic structure

$$\omega_1 \equiv \Lambda_1^{-1} := \hbar \int dx(\delta p \wedge \delta q)$$

  and a Hamiltonian functional accounting for the interaction.
- A phase manifold with a symplectic structure determined by the interaction

$$\omega_2 \equiv \Lambda_2^{-1} := \hbar \int dx(\mathcal{H}_N^{-1} \delta p \wedge \delta q)$$

  and a *free* Hamiltonian functional given by the mean value of the momentum $\hat{p} = -i\hbar\partial_x$.

We have then the relation

$$\frac{\delta K_1}{\delta u} = \check{T}_N \frac{\delta K_0}{\delta u}, \tag{15.11}$$

where

$$\check{T} := \Lambda_1^{-1} \circ \Lambda_2 = \begin{pmatrix} 2\alpha q D^{-1} p & \frac{\hbar}{2m}\partial_x + 2\alpha q D^{-1} q \\ -\frac{\hbar}{2m}\partial_x + 2\alpha p D^{-1} p & -2\alpha p D^{-1} q \end{pmatrix}.$$

It can be shown that the sum $\Lambda_2 + \Lambda_1$ is again a Poisson bracket. This is equivalent to the vanishing of the torsion of the tensor field T_N, so that the

†For simplicity the proof of Jacobi identity for Λ_2 is omitted.

relation (15.11) can be iterated to

$$\frac{\delta K_n}{\delta u} = (\check{T}_N)^n \frac{\delta K_0}{\delta u} \,. \tag{15.12}$$

Therefore, the nonlinear Schrödinger equation admits infinitely many conserved functionals. The first three functionals are

$$K_{-1}[q,p] = \frac{1}{2}\int dx(p^2+q^2) = \int dx(\bar{\psi}\psi)\,,$$

$$K_0[q,p] = \int dx(qp_x) = 2i\int dx(\bar{\psi}\psi_x)\,,$$

$$K_1[q,p] = \frac{1}{2}\int dx \left\{\frac{\hbar^2}{2m}[(\partial_x p)^2 + (\partial_x q)^2] + (p^2+q^2)^2\right\}\,.$$

They are all in involution with respect to the previous Poisson brackets, i.e.

$$\{K_n, K_m\}_0 = \{K_n, K_m\}_1 = 0\,.$$

By observing that

$$\frac{\delta K_0}{\delta u} = \check{T}_N \frac{\delta K_{-1}}{\delta u}\,,$$

the recursion relation (15.12) can be completed to

$$\frac{\delta K_n}{\delta u} = (\check{T}_N)^{n+1} \frac{\delta K_{-1}}{\delta u}\,.$$

In terms of the operators $\hat{T}$, $\hat{T}_G$, $\hat{T}_K$, $\hat{T}_N$ defined by

$$\hat{T}\circ := -\Delta\circ + U\circ \equiv \mathcal{H}\circ\,,$$

$$\hat{T}_G\circ := \partial_{xx}\circ + \psi_x D^{-1}\psi_x\circ\,,$$

$$\hat{T}_K\circ := \partial_{xx}\circ + \frac{2}{3}\psi\circ + \frac{1}{3}\psi_x D^{-1}\circ\,,$$

$$\hat{T}_N\circ := i\{\partial_x\circ + \psi D^{-1}[\psi(\bar{\circ}) + \bar{\psi}(\circ)]\}\,,$$

we have the general scheme on the next page.

Schrödinger Hierarchy

$\vdots$

$\dot{\psi} = i\mathcal{H}^{\,2}\psi$

$\uparrow$ $\hat{T}$ $|$

$\dot{\psi} = -i\mathcal{H}\,\psi$

$\uparrow$ $\hat{T}$ $|$

$\dot{\psi} = -i\psi$

$|$ $\hat{T}_N$ $\downarrow$

s–Gordon Hierarchy $\leftarrow \hat{T}_G -$ $\dot{\psi} = \psi_x$ $- \hat{T}_K \longrightarrow$ *KdV Hierarchy*

$|$ $\hat{T}_N$ $\downarrow$

$\dot{\psi} = -i\psi_{xxx} + |\psi|^2\psi$

$|$ $\hat{T}_N$ $\downarrow$

$\dot{\psi} = -\psi_{xxx} + 3|\psi|^2\psi_x$

$|$ $\hat{T}_N$ $\downarrow$

$\dot{\psi} = -i\psi_{xxxx} + 4|\psi|^2\psi_{xx} + 3\bar{\psi}(\psi_x)^2 + 2\psi|\psi_x|^2 + \psi^2\bar{\psi}_{xx} + \frac{3}{2}|\psi|^4\psi$

$\vdots$

Nonlinear Schrödinger Hierarchy

Remark 25 *It is interesting to observe that K_{-1} is a conserved functional both for the Schrödinger and the nonlinear Schrödinger equations. The same is not true for K_0. This is due the fact that Schrödinger equation is not invariant under space translations and K_0 corresponds to the mean value $\langle \hat{p} \rangle$ of the linear momentum $\hat{p} = -ih\partial_x$. In other words the vector field associated to K_0 via the canonical Poisson bracket Λ_1 is invariant for translation.*

It is worth finally to compare the recursion operators of the Schrödinger, with vanishing potential $U(x)$, and nonlinear Schrödinger, with $\alpha = 0$, hierarchies. It is easy to see that, in this case, $\hat{T} = \hat{T}_N^2$.

15.3 Integrable Systems on Lie Group Coadjoint Orbits

Integrability is also analyzed, by several authors, using the Eulerian approach of coadjoint orbits of Lie groups.

Let $\{e_i\}$ be a basis of a Lie algebra $\mathcal{G}$ with $[e_i, e_j] = c_{ij}^k e_k$ and $\{\vartheta^i\}$ be the dual basis, in $\mathcal{G}^*$ the dual of $\mathcal{G}$.

Moreover, let x be coordinates in $\mathcal{G}^*$ with respect to (ϑ^i), and $\mathcal{F}$ be the set

$$\mathcal{F} = \{f \in C^\infty\,, \quad f : \mathcal{G}^* \to \Re\}\,.$$

Let us define the bracket

$$\{f, g\}(x) = c_{jk}^i x_i \frac{\partial f}{\partial x_j}\frac{\partial g}{\partial x_k} \equiv \langle x, [\nabla f, \nabla g]\rangle\,,$$

where $\forall\, f \in \mathcal{F}$ and $\forall\, x, y \in \mathcal{G}^*$, and the *gradient* ∇f of f is the element of $\mathcal{G}$ defined by

$$\langle y, \nabla f\rangle = \frac{d}{dt} f(x + ty)|_{t=0}\,.$$

The existence on $\mathcal{G}$ of a not degenerate scalar product $(\cdot\,,\cdot)$, which is invariant for the adjoint representation, allows the identification of $\mathcal{G}$ with $\mathcal{G}^*$ according to

$$\langle \tilde{y}, x\rangle = (y, x)\,.$$

On the other hand, the property $([c, b], a) + (b, [c, a]) = 0$ implies that the previous bracket can be also written as

$$\{f, g\}(x) = (\nabla g, [x, \nabla f])\,.$$

So for a dynamics generated in $\mathcal{G}$ by a function $H \in \mathcal{F}$,

$$\frac{df}{dt} = \{H, f\} = (\nabla f, [x, \nabla H]),$$

we have

$$\frac{dx}{dt} = [x, \nabla H].$$

This corresponds to the Euler approach for the rigid body dynamics, and to the Lax representation of KdV too, in reading the phase manifold of KdV as the coadjoint orbit of the ∞-dimensional Lie group of integral operator Fourier integrable on the circle S^1.[61,60,49]

15.4 Deformation of a Lie Algebra

15.4.1 *Deformation*

Let $\mathcal{G}$ be a Lie algebra and X, Y any two elements in $\mathcal{G}$.

A family of brackets

$$[X, Y]_\lambda = [X, Y] + \lambda\omega(X, Y),$$

satisfying the Jacobi identity $\forall\lambda \in \Re$, is called a *deformation*[87] of the Lie algebra $\mathcal{G}$.

Therefore, ω has to satisfy the following conditions:

$$\begin{cases} [X, \omega(Y, Z)] + \omega(X, [Y, Z]) + \text{cyclic permutation of } X, Y, Z = 0, \\ \omega(X, \omega(Y, Z)) + \text{cyclic permutation of } X, Y, Z = 0. \end{cases}$$

Such a deformation is a 2-*cocycle* ω on $\mathcal{G}$, with coefficients in the adjoint representation, that defines a new Lie algebra structure.

A deformation is called *trivial* if there exists an endomorphism $T : \mathcal{G} \to \mathcal{G}$, such that the operator $1 + \lambda T$ is a morphism from the new Lie bracket $[\cdot\,,\cdot]_\lambda$ to the old Lie bracket $[\cdot\,,\cdot]$.

Thus, for a trivial deformation, we have

$$(1 + \lambda T)[X, Y]_\lambda = [(1 + \lambda T)X, (1 + \lambda T)Y].$$

The above equality implies that, for arbitrary λ, the following condition must be satisfied:

$$(1+\lambda T)([X,Y]+\lambda\omega(X,Y)) = [X,Y]+\lambda T[X,Y]+\lambda[X,TY]+\lambda^2[TX,TY],$$

i.e.

$$\omega(X,Y) = [TX,Y]+[X,TY]-T[X,Y],$$
$$T\omega(X,Y) = [TX,TY].$$

Therefore, ω is a *coboundary of the cochain* T with the property

$$H_T(X,Y)=0 \quad \text{with} \quad H_T(X,Y) = [(\mathcal{L}_{\hat{T}X}T)^\wedge - \hat{T}(\mathcal{L}_X T)^\wedge]Y.$$

Moreover,

$$T[X,Y]_T = [TX,TY],$$

with

$$[X,Y]_T \equiv \omega(X,Y).$$

15.4.2 *Lie–Nijenhuis and exterior-Nijenhuis derivatives*

The *Lie–Nijenhuis derivative* is defined on vector fields by

$$\mathcal{L}_X^T Y = [X,Y]_T = \omega(X,Y) = [TX,Y]+[X,TY]-T[X,Y],$$

and on differential forms, by defining the following *exterior-Nijenhuis* derivative:

$$(d_T f)(X) = df(\widehat{T}X), \quad f\in\Lambda(\mathcal{M}),$$
$$(d_T\alpha)(X,Y) = d\alpha(TX,Y)+d\alpha(X,TY)-(d\check{T}\alpha)(X,Y), \quad \alpha\in\Lambda^1(\mathcal{M}).$$

Indeed, the exterior-Nijenhuis derivative has the property

$$[d_T(d_T,f)](X,Y) = df(H_T(X,Y)),$$

so that

$$d_T^2 = 0 \Leftrightarrow H_T = 0.$$

Moreover, if T is invertible, the *Poincaré lemma* holds; that is, if $d_T\alpha = 0$, then locally a differential form β exists, such that $\alpha = d_T\beta$.

We also notice that

$$d_T d = -d d_T \,.$$

Tedious calculations show that

$$(d_T\alpha)(X,Y) = \langle \mathcal{L}_X^T\alpha, Y\rangle - \langle \mathcal{L}_X^T\alpha, X\rangle + \langle \alpha, \mathcal{L}_X^T Y\rangle \,.$$

The vanishing of the Nijenhuis bracket

$$[TX,TY] - T[X,TY] - T[TX,Y] + T^2[X,Y] = 0 \,,$$

for a tensor field R satisfying the condition $R^2 = 1$, gives the *modified classical Yang–Baxter equation*[171,123]

$$[RX,RY] - R[RX,Y] - R[X,RY] + [X,Y] = 0 \,.$$

In this case, the condition on ω can be rewritten in the following form:

$$\omega(X,Y) = [RX,Y] + [X,RY] - R[X,Y] \,,$$

$$\omega(X,Y) = R[RX,RY] \,.$$

Finally, we observe[63] that also the bracket defined by

$$[X,Y]_R = \omega(X,Y) + R[X,Y] \equiv [RX,Y] + [X,RY] \,,$$

satisfies the Jacobi identity.

It follows that, if R solves the modified classical Yang–Baxter equation, all the brackets

$$\omega(X,Y) = [RX,Y] + [X,RY] - R[X,Y] \,,$$

$$[X,Y]_R = [RX,Y] + [X,RY] \,,$$

$$[X,Y]_\lambda = [X,Y]_R + \lambda\omega(X,Y) \,,$$

satisfy the Jacobi identity.

Remark 26 *Different approaches to complete integrability of systems with infinitely many degrees of freedom exist, but a clear connection between them is, up to now, lacking. Perhaps a deeper understanding may provide new tools to tackle the relevant problems of nonlinear quantum theory.*[113,91]

Chapter 16

Integrability of Fermionic Dynamics

There have been several attempts to analyze integrability of fermionic dynamical systems (see for instance, Refs. 31, 142 and 74) and to extend to such systems,[75] in algorithmic sense at least, results and techniques used for Bosonic dynamics and based on the role of recursion operators. In particular, one would like to define a graded Nijenhuis torsion.

In this chapter, we address this issues. We show that a mixed (1,1)-graded tensor field T can act as a recursion operator if and only if T is an even map.[129]

There are dynamical systems, like supersymmetric Witten's dynamics[184] that allow a bi-Hamiltonian description with an even and odd Hamiltonian function and in term of an even and an odd Poisson structure, respectively, so that the dynamical vector field is always even.[183,172] This allows to construct an odd tensor field which could be a good candidate as a recursion operator. We explicitly show that this is not possible.

16.1 Recursion Operators in the Bosonic Case

Here we briefly recall an alternative characterization in term of an invariant (under the dynamical evolution) (1,1)-tensor field T.

We shall deal only with smooth; i.e. C^∞ objects, and notations will follow as close as possible those of Refs. 1 and 41. In particular if $\mathcal{M}$ is an ordinary manifold (finite-dimensional), we denote by $\mathcal{F}(\mathcal{M})$ the ring of real-valued functions on $\mathcal{M}$, by $\mathcal{X}(\mathcal{M})$ the Lie algebra of vector fields, by $\mathcal{X}(\mathcal{M})^*$ its dual of forms and by $\mathcal{T}_1^1(\mathcal{M})$ the mixed (1,1)-tensor fields.

It has been shown that the main property of the tensor field T, in the analysis of complete integrability of its infinitesimal automorphisms, is the vanishing of its Nijenhuis tensor $\mathcal{N}_T = 0$. It is then plausible that a suitable generalization of such a condition could play an important role in analyzing the integrability of dynamical systems with fermionic degrees of freedom. Moreover, it seems natural to think that such a generalization could come from a graded generalization of some of the following relations which are available in the Bosonic case:

(a) $\mathcal{N}_T = 0 \Longrightarrow \operatorname{Im} T$ is a Lie algebra.
(b) $\mathcal{N}_T = 0$, $d(TdH) = 0 \Longrightarrow d(T^k dH) = 0$.
(c) $\mathcal{N}_T = 0 \Longleftrightarrow d_T \circ d_T = 0$, where d_T is the exterior-Nijenhuis derivative.
(d) $T =: \Lambda_1^{-1} \circ \Lambda_2$, $N_T = 0 \Longleftrightarrow \Lambda_1 + \Lambda_2$ satisfies the Jacobi identity. Here Λ_1 and Λ_2 are two Poisson structures.
(e) $\omega(X,Y) =: [TX,Y] + [X,TY] - T[X,Y]\,,\quad T\omega(X,Y) = [TX,TY]$.

One could expect that some, if not all, of the previous relations do not hold true in the *graded situation.*

Before we proceed with the analysis of the *graded Nijenhuis condition* we shall give a brief review of the *graded differential calculus* on *supermanifolds* that will be followed by the study of some simple examples.

16.2 Graded Differential Calculus

We review some fundamentals of supermanifold theory[15,167] while referring to the literature for a mathematically coherent definition.[169,64] In the following, by *graded* we shall always mean Z_2*-graded.*

The basic algebraic object is a real exterior algebra $B_L = (B_L)_0 \oplus (B_L)_1$ with L generators. An (m,n)-dimensional supermanifold is a topological manifold S modeled over the *vector superspace*

$$B_L^{m,n} = (B_L)_0^m \times (B_L)_1^n \tag{16.1}$$

by means of an atlas whose transition functions fulfill a suitable *supersmoothness* condition. A supersmooth function $f : U \subset B_L^{m,n} \to B_L$ has the usual superfield expansion

$$f(x^1 \dots x^m, \vartheta^1 \dots \vartheta^n) = f_0(x) + \sum_{\alpha=1}^{n} f_\alpha(x)\, \vartheta^\alpha + \dots + f_{1\dots n}(x)\, \vartheta^1 \dots \vartheta^n\,, \quad (16.2)$$

where the x's are the even (Grassmann) coordinates, the ϑ's are the odd ones, and the dependence of the coefficient functions $f_{\dots}(x)$ on the even variables is fixed by their values for real arguments.

We shall denote by $\mathcal{G}(S)$ and $\mathcal{G}(U)$ the graded ring of supersmooth B_L-valued functions on S and $U \subset S$, respectively.

The class of supermanifolds which, up to now, turns out to be relevant for applications in physics is given by the De Witt supermanifolds. They are defined in terms of a coarse topology on $B_L^{m,n}$, called the *De Witt topology*, whose open sets are the counterimages of open sets in $\Re^m$ through the body map $\sigma^{m,n} : B_L^{m,n} \to \Re^m$. An (m,n) supermanifold is De Witt if it has an atlas such that the images of the coordinate maps are open in the De Witt topology. A *De Witt (m,n)-supermanifold* is a locally trivial fiber bundle over an ordinary m-manifold S_0 (called the body of S) with a vector fiber.[167] This is not a surprise the fact that, modulo some technicalities, a De Witt supermanifold can be identified with a Berezin–Konstant supermanifold.[121,5]

The graded tangent space TS is constructed in the following manner. For each $x \in S$, let $\mathcal{G}(x)$ be the germs of functions at x and denote by T_xS the space of graded B_L-linear maps $X : \mathcal{G}(x) \to B_L$ that satisfy Leibnitz rule. Then, T_xS is a free graded B_L-module of dimension (m,n), and the disjoint union $\cup_{x \in S} T_xS$ can be given the structure of a rank (m,n) super vector bundle over S, denoted by TS. The sections $\mathcal{X}(S)$ of TS are a graded $\mathcal{G}(S)$-module and are identified with the graded Lie algebra $Der\mathcal{G}(S)$ of derivations of $\mathcal{G}(S)$. Derivations (or vector fields) are said to be even (or odd) if they are even (or odd) as maps (satisfying in addition a graded Leibnitz rule) from $\mathcal{G}(S) \to \mathcal{G}(S)$. A local basis is given by

$$\frac{\partial}{\partial x^1}, \dots, \frac{\partial}{\partial x^m}, \frac{\partial}{\partial \vartheta^1}, \dots, \frac{\partial}{\partial \vartheta^n}\,. \quad (16.3)$$

Remark 27 *Unless explicitly stated, by using a partial derivative we shall always mean a left derivative, namely a derivative acting from left. In general, if $z^i = (x^j, \vartheta^k)$, when acting on any homogeneous function $f \in \mathcal{G}(S)$, left and* *right derivatives are related by*

$$\frac{\overrightarrow{\partial}}{\partial z^i} f = (-1)^{p(z^i)[p(f)+1]} f \frac{\overleftarrow{\partial}}{\partial z^i}, \quad i \in \{1, \dots, m+n\}. \tag{16.4}$$

In a similar way, one defines the cotangent space and bundle. T_x^*S is the space of *graded B_L-linear maps* from $T_x(S) \to B_L$ and $T^*S = \cup_{x \in S} T_x^*S$. T_x^*S is a *free graded B_L-module* of dimension (m, n), while T^*S is a rank (m, n) *super vector bundle over S*.

The *sections* $\mathcal{X}(S)^*$ of T^*S are a graded $\mathcal{G}(S)$ module and are identified with the graded $\mathcal{G}(S)$-linear maps from $Der\mathcal{G}(S) \to \mathcal{G}(S)$. They are the differential 1-forms on S and are said to be even (respectively odd), if they are even (respectively odd) as maps $\mathcal{X}(\mathcal{M}) \to \mathcal{G}(S)$.

In general, a *p covariant and q contravariant graded tensor* is any graded $\mathcal{G}(S)$-multilinear map* $\alpha : \mathcal{X}(S) \times \cdots \times \mathcal{X}(S) \times \mathcal{X}(S)^* \times \cdots \times \mathcal{X}(S)^* \longrightarrow \mathcal{G}(S)$. The collection of all rank (p, q) tensors is a graded $\mathcal{G}(S)$ module.

A *graded p-form* is a skew-symmetric covariant graded tensors of rank p.

We denote by $\Omega^p(S)$ the collection of all differential p forms. The *exterior differential* on S is defined by letting $X \perp df = X(f)$, $\forall f \in \mathcal{G}(S)$, $X \in \mathcal{X}(S)$ and is extended to maps $\Omega^p(S) \to \Omega^{p+1}(S), p \geq 0$, in the usual way, so that $d^2 = 0$.

If $X_i \in \mathcal{X}(S)$ are homogeneous elements,

$$\begin{aligned} &X_1 \wedge \cdots \wedge X_{p+1} \perp d\varphi \\ &\quad = \sum_{i=1}^{p+1} (-1)^{a(i)} X_i(X_1 \wedge \dots \overset{i}{\checkmark} \dots \wedge X_{p+1} \perp \varphi) \\ &\qquad + \sum_{1 \leq i < j \leq p} (-1)^{b(i,j)} [X_i, X_j] \wedge X_1 \wedge \dots \overset{i}{\checkmark} \dots\dots \overset{j}{\checkmark} \dots \wedge X_{p+1} \perp \varphi, \end{aligned} \tag{16.5}$$

where

$$a(i) = 1 + i + p(X_i) \sum_{h=1}^{i-1} p(X_h),$$

*With p-$\mathcal{X}(S)$ factors and q-$\mathcal{X}(S)^*$ factors.

$$b(i,j) = i + j + p(X_i)\sum_{h=1}^{i-1} p(X_h) + p(X_j)\sum_{\substack{h=1\\h\neq i}}^{j-1} p(X_h)\,. \tag{16.6}$$

From definition, one has that $p(d) = 0$.

The Lie derivative $L_{(\cdot)}$ of forms is defined by

$$\begin{aligned} &L_{(\cdot)} : \mathcal{X}(S) \times \Omega^p(S) \to \Omega^p(S)\,,\\ &L_X = X \perp \circ d + d \circ X \perp\,, \quad \forall X \in \mathcal{X}(S)\,. \end{aligned} \tag{16.7}$$

Clearly, $p(L_X) = p(X)$.

The Lie derivative of any tensor product can be defined in an obvious manner by requiring the Leibnitz rule and can be extended to any tensor by using linearity.

Suppose now that we have a tensor $T \in \mathcal{T}_1^1(S)$, which is homogeneous of degree $p(T)$. Again we can define two graded endomorphisms of $\mathcal{X}(S)$ and $\mathcal{X}(S)^*$ by the formulae (in the following two formulae, X, Y are homogeneous elements in $\mathcal{X}(S)$, while α is any element in $\mathcal{X}(S)^*$)

$$\begin{aligned} &\hat{T} : \mathcal{X}(S) \longrightarrow \mathcal{X}(S)\,, \quad \check{T} : \mathcal{X}(S)^* \longrightarrow \mathcal{X}(S)^*\,,\\ &T(X,\alpha) =: \hat{T}X \perp \alpha =: (-1)^{p(X)p(T)} X \perp \check{T}\alpha\,. \end{aligned} \tag{16.8}$$

We could be tempted to define a graded Nijenhuis torsion of T by a relation analogous to usual one of the Bosonic case

$$\begin{aligned} {}^G\mathcal{N}_T(X,Y;\alpha) &=: {}^G\mathcal{H}_T(X,Y) \perp \alpha\,,\\ {}^G\mathcal{H}_T(X,Y) &=: \hat{T}^2[X,Y] + (-1)^{p(T)p(X)}[\hat{T}X,\hat{T}Y] - \hat{T}[\hat{T}X,Y]\\ &\quad - (-1)^{p(T)p(X)}\hat{T}[X,\hat{T}Y]\,. \end{aligned} \tag{16.9}$$

It is easy to see, just by computation, that

The map ${}^G\mathcal{H}_T : \mathcal{X}(S) \times \mathcal{X}(S) \to \mathcal{X}(S)$ defined in Eq. (16.9) is $\mathcal{G}(S)$-linear and graded antisymmetric if and only if $p(T) = 0$.

Remark 28 *When $p(T) = 1$, the map defined in Eq. (16.9) is not antisymmetric nor linear also over even function, also when it is restrict to even vector fields. Therefore Eqs. (16.8) and (16.9) define a graded tensor (which is in addition graded antisymmetric) if and only if $p(T) = 0$.*

16.3 Poisson Supermanifold

We briefly describe how to introduce super Poisson structures on an (m,n)-dimensional supermanifold S.[5,133] For additional results, see also Ref. 68. As before, we shall denote by $z^i = (x^j, \vartheta^k), i \in \{1,\ldots,m+n\}$ the local coordinates on S. By direct calculations it can be proven[5] that

If (ω^{ij}) is an $(m+n)\times(m+n)$ matrix, depending upon the point $z \in S$, with the following properties:

- *the elements ω^{ij} are homogeneous with parity $p(\omega^{ij}) = p(z^i) + p(z^j) + p(\omega)$ and $p(\omega)$ not depending on the indices i and j;*
- $$\omega^{ji} = -(-1)^{[p(z^i)+p(\omega)][p(z^j)+p(\omega)]}\omega^{ij}\,, \tag{16.10}$$
- $$(-1)^{[p(z^i)+p(\omega)][p(z^l)+p(\omega)]}\omega^{is}\,\frac{\overrightarrow{\partial}}{\partial z^s}\,\omega^{jl} + (-1)^{[p(z^l)+p(\omega)][p(z^j)+p(\omega)]}\omega^{ls}\,\frac{\overrightarrow{\partial}}{\partial z^s}\,\omega^{ij} + (-1)^{[p(z^j)+p(\omega)][p(z^i)+p(\omega)]}\omega^{js}\,\frac{\overrightarrow{\partial}}{\partial z^s}\,\omega^{li} = 0\,; \tag{16.11}$$

then, the following bracket

$$\{F,G\} =: F\,\frac{\overleftarrow{\partial}}{\partial z^i}\,\omega^{ij}\,\frac{\overrightarrow{\partial}}{\partial z^i}\,G \tag{16.12}$$

makes $\mathcal{G}(S)$ a Lie superalgebra (Poisson superstructure).

We have two different kind of structures according to the fact that $p(\omega) = 0$ (even Poisson structure), or $p(\omega) = 1$ (odd Poisson structure). Indeed, one can check that the bracket (16.12) has properties:

- $$\{F,G\} = -(-1)^{[p(F)+p(\omega)][p(G)+p(\omega)]}\{G,F\}\,; \tag{16.13}$$
- $$(-1)^{[p(F)+p(\omega)][p(H)+p(\omega)]}\{\{F,G\},H\} + (-1)^{[p(G)+p(\omega)][p(F)+p(\omega)]}\{\{G,H\},F\} + (-1)^{[p(H)+p(\omega)][p(G)+p(\omega)]}\{\{H,F\},G\} = 0\,. \tag{16.14}$$

We infer from Eqs. (16.13) and (16.14) that, when thought of as elements of the Poisson superalgebra, homogeneous elements of $\mathcal{G}(S)$ preserve their parity if $p(\omega) = 0$, while they change it if $p(\omega) = 1$.

If the matrix (ω^{ij}) is regular, then its inverse $(\omega_{ij}), \omega_{ij}\omega^{jk} = \delta_i^k$, gives the components of a symplectic structure $\omega = \frac{1}{2}dz^i \wedge dz^j \omega_{ji}$, namely, ω is closed and nondegenerate with the properties

$$p(\omega_{ij}) = p(z^i) + p(z^j) + p(\omega)\,,$$
$$\omega_{ji} = -(-1)^{p(z^i)p(z^j)}\omega_{ij}\,, \tag{16.15}$$

and ω is homogeneous with parity just equal to $p(\omega)$.

There is also a *Darboux theorem.*[133]

Theorem 41 *Let (S, ω) be an (m, n)-dimensional symplectic manifold with ω homogeneous. Then,*

Proposition 42

- *If $p(\omega) = 0$, then* $\dim S = (2r, n)$ *and there exist local coordinates such that*

$$\omega = dq^i \wedge dp^i + d\xi^j \wedge d\xi^j\,, \quad \omega_{ij} = \begin{pmatrix} 0 & 1_r & 0 \\ -1_r & 0 & 0 \\ 0 & 0 & 1_n \end{pmatrix}. \tag{16.16}$$

- *If $p(\omega) = 1$, then* $\dim S = (m, m)$ *and there exist local coordinates such that*

$$\omega = du^i \wedge d\xi^i\,, \quad \omega_{ij} = \begin{pmatrix} 0 & 1_m \\ -1_m & 0 \end{pmatrix}.$$

By having a Poisson structure, we can deal with Hamilton equations. From Eq. (16.12), if H is the Hamiltonian, the corresponding equations are

$$\dot{z}^i = \omega^{ij}\frac{\partial}{\partial z^j}H\,. \tag{16.17}$$

Now we would like to maintain the possibility of explicitly constructing the flow of Eq. (16.17). This requires that the dynamical evolution be an even vector field. In turn this implies that the Poisson structure and the Hamiltonian function should have the same parity so that in particular, with an odd Poisson structure, we need an odd Hamiltonian function.

Before we analyze the graded Nijenhuis condition, we will study a few examples.

Example 43 (Bosonic–Fermionic oscillator) *The mixed bosonic–fermionic harmonic oscillator in (2,2) dimensions is described with coordinates* (q, p, η, ξ) *and has the following equations of motion:*

$$\begin{cases} \dot{q} = p\,, \\ \dot{p} = -q\,, \\ \dot{\eta} = \xi\,, \\ \dot{\xi} = -\eta\,. \end{cases} \tag{16.18}$$

Equations (16.18) can be given two Hamiltonian descriptions. The Hamiltonians are the usual even one

$$H = \frac{1}{2}(p^2 + q^2) + i\xi\eta\,, \tag{16.19}$$

and an odd one

$$K = p\xi + q\eta\,, \tag{16.20}$$

while the two Poisson structures are respectively

$$\Lambda_H = \begin{pmatrix} 0 & 1 & 0 & 0 \\ -1 & 0 & 0 & 0 \\ 0 & 0 & i & 0 \\ 0 & 0 & 0 & i \end{pmatrix}, \quad \omega_H = \begin{pmatrix} 0 & -1 & 0 & 0 \\ 1 & 0 & 0 & 0 \\ 0 & 0 & -i & 0 \\ 0 & 0 & 0 & -i \end{pmatrix}, \tag{16.21}$$

and

$$\Lambda_K = \begin{pmatrix} 0 & 0 & 0 & 1 \\ 0 & 0 & -1 & 0 \\ 0 & 1 & 0 & 0 \\ -1 & 0 & 0 & 0 \end{pmatrix}, \quad \omega_K = \begin{pmatrix} 0 & 0 & 0 & -1 \\ 0 & 0 & 1 & 0 \\ 0 & -1 & 0 & 0 \\ 1 & 0 & 0 & 0 \end{pmatrix}. \tag{16.22}$$

We can construct a mixed invariant tensor field T by

$$T =: \omega_H \circ \Lambda_K = \begin{pmatrix} 0 & 0 & 1 & 0 \\ 0 & 0 & 0 & 1 \\ 0 & -i & 0 & 0 \\ i & 0 & 0 & 0 \end{pmatrix}. \tag{16.23}$$

However, this odd tensor field ($p(T) = 1$) *is not a recursion operator. Indeed, it is easy to check that*

$$TdH = -i(dq)\xi + i(dp)\eta - i(d\eta)p + i(d\xi)q\,,$$

so that

$$TdK = dH\,, \quad d(TdH) \neq 0\,. \tag{16.24}$$

If we evaluate the Poisson brackets of the coordinate variables, by using the two symplectic structure given by Eqs. (16.21) and (16.22), we find that

$$\{q,p\}_H = 1\,, \quad \{p,q\}_H = -1\,, \quad \{\eta,\eta\}_H = i\,, \quad \{\xi,\xi\}_H = i\,, \tag{16.25}$$

and

$$\{q,\xi\}_K = 1\,, \quad \{\xi,q\}_K = -1\,, \quad \{p,\eta\}_K = -1\,, \quad \{\eta,p\}_K = 1\,, \tag{16.26}$$

the remaining ones being identically zero. We see that the sum $\{\cdot,\cdot\}_+$ *of the two structures is itself a Poisson structure with the property*

$$\{F,G\}_+ = -(-1)^{p(F)p(G)}\{G,F\}_+\,, \tag{16.27}$$

but it has not definite parity. Moreover the bracket $\{\cdot,\cdot\}_+$ *is degenerate.*

Example 44 (Witten dynamics) *Interesting examples come from supersymmetric dynamics. It has been shown*[183,172] *that the dynamics of Witten's Hamilton systems*[184] *can be given a bi-Hamiltonian description with an even Poisson bracket and Grassmann even Hamiltonian, or with an odd bracket and Grassmann odd Hamiltonians. Instead of considering the general case we shall study a supersymmetric Toda chain with coordinates* (q,p,η,ξ).

The even Hamiltonian is given by

$$H = \frac{1}{2}(p^2 + e^q) + \frac{1}{2}i\xi\eta e^{\frac{q}{2}}\,. \tag{16.28}$$

With the even Poisson structure

$$\Lambda_H = \begin{pmatrix} 0 & 1 & 0 & 0 \\ -1 & 0 & 0 & 0 \\ 0 & 0 & i & 0 \\ 0 & 0 & 0 & i \end{pmatrix}, \quad \omega_H = \begin{pmatrix} 0 & -1 & 0 & 0 \\ 1 & 0 & 0 & 0 \\ 0 & 0 & -i & 0 \\ 0 & 0 & 0 & -i \end{pmatrix}, \tag{16.29}$$

the equations of motion read

$$\begin{cases} \dot{q} = p\,, \\ \dot{p} = -\frac{1}{2}e^{q} - \frac{1}{4}i\xi\eta e^{\frac{q}{2}}\,, \\ \dot{\eta} = \frac{1}{2}\xi e^{\frac{q}{2}}\,, \\ \dot{\xi} = -\frac{1}{2}\eta e^{\frac{q}{2}}\,. \end{cases} \tag{16.30}$$

Then the following functions are constants of the motion

$$\begin{aligned} K &= p\xi + e^{\frac{q}{2}}\eta\,, \\ L &= p\eta - e^{\frac{q}{2}}\xi\,, \\ F &= i\xi\eta\,. \end{aligned} \tag{16.31}$$

We can use K(or L) in Eq. (16.31) as an alternative Hamiltonian function. The corresponding symplectic structure is given by

$$\omega_K = dq \wedge d\xi + dp \wedge dq(e^{-\frac{q}{2}}\eta) + dp \wedge d\eta(-2e^{-\frac{q}{2}}) + df \wedge dH\,, \tag{16.32}$$

where $f(q,p,\eta,\xi)$ is a function explicitly given by

$$f = A\xi + B\eta$$

with

$$\begin{aligned} A(q,p) &= \frac{1}{p^2+e^q}\left(\frac{2}{\sqrt{p^2+e^q}}\ln\left(\frac{e^{\frac{q}{2}}}{p+\sqrt{p^2+e^q}}\right) + \frac{2e^{\frac{q}{2}}}{\sqrt{p^2+e^q}} - 2\right), \\ B(q,p) &= \frac{1}{p^2+e^q}\left(\frac{2e^{\frac{q}{2}}}{\sqrt{p^2+e^q}}\ln\left(\frac{e^{\frac{q}{2}}}{p+\sqrt{p^2+e^q}}\right) - \frac{2p}{\sqrt{p^2+e^q}} - 2pe^{-\frac{q}{2}}\right). \end{aligned} \tag{16.33}$$

The symplectic structure ω_K can also be written in the following form:

$$\omega_K = d\{dq(-\xi) + dp(2e^{-\frac{q}{2}}\eta) + fdH\}\,.$$

If Γ is the dynamical vector field of the Toda system, as given by Eq. (16.30), then, the function f will satisfy the relation $i_\Gamma df = e^{-q/2}\eta$ and this, in turn, *ensures that* $i_\Gamma \omega_K - dK$. *It take some algebra to check that the* $(1,1)$*-tensor field*

$$T = \omega_K \circ \Lambda_H \,, \tag{16.34}$$

is such that

$$TdH = dK \,,$$
$$d(T^2 dH) \neq 0 \,. \tag{16.35}$$

Again, the tensor field T *in Eq. (16.34) is not a recursion operator.*

16.3.1 *Super Nijenhuis torsion*

Let us recall that one of the most relevant consequences deriving from a (not graded) $(1-1)$-tensor field T, with a vanishing Nijenhuis torsion, is the possibility to generate sequences of exact differential 1-forms according to

$$\mathcal{N}_T = 0 \,, \quad d(TdF) = 0 \Longrightarrow d(T^k dF) = 0 \,. \tag{16.36}$$

The above relation is a consequence of the identity

$$X \wedge Y \perp d(T^2\alpha) = \{X \wedge TY + TX \wedge Y\} \perp d(T\alpha) - \{TX \wedge TY\} \perp d\alpha$$
$$- \mathcal{H}_T(X,Y) \perp \alpha \,. \tag{16.37}$$

in which α is any differential 1-form.

Indeed, by assuming that both α and $T\alpha$ are closed, Eq. (16.37) implies that $T^2\alpha$ is closed if and only if $\mathcal{H}_T = 0$, namely if and only if the Nijenhuis torsion of T vanishes.

Let us analyze now the graded situation. Suppose T is a graded $(1,1)$-tensor field that is homogeneous of parity $p(T)$. Then, if α is any differential 1-form, by using the definition Eq. (16.5), after some (graded) algebra, the analogue of Eq. (16.37) reads

$$X \wedge Y \perp d(T^2\alpha) = \{(-1)^{p(T)p(Y)} X \wedge TY + (-1)^{p(T)[p(X)+p(Y)]} TX \wedge Y\}$$
$$\perp d(T\alpha) - (-1)^{p(T)[p(X)+p(T)]} TX \wedge TY \perp d\alpha$$
$$- (-1)^{p(T)\,G}\mathcal{H}_T(X,Y) \perp \alpha$$
$$+ (-1)^{p(T)p(X)}[1 - (-1)^{p(T)}] L_{TX}(TY \perp \alpha) \,, \tag{16.38}$$

where ${}^G\mathcal{H}_T$ is defined in Eq. (16.9).

It is clear then, that for a (1, 1) odd tensor field a (1, 2)-tensor field corresponding to its super Nijenhuis torsion can be defined only when $p(T) = 0$. The same result is attained with the use of the general approach $d_T \circ d_T = 0$.

Summing up, we have shown that there are examples of dynamical systems whose dynamical vector field Γ admits two Hamiltonian descriptions, odd and even, respectively, and that the tensor field T, constructed out of the corresponding Poisson structures is not a recursion operator since it cannot generate new integrals of motion after the first two ones.

We have also shown that this fact is general and that for a generic-graded (1, 1)-tensor field T, a graded Nijenhuis torsion cannot be defined unless T is even.

From the nature of the proof it seems plausible that a similar theorem should hold true also in infinite dimensions.

The *no go* theorem we have proved in our paper does not exhausts, obviously, the analysis of complete integrability for graded Hamiltonian systems. Much more attention must be paid, however, in generalizing to the graded case geometrical structures that play a relevant and natural role in the nongraded situation.

Further Readings

- M. Batchelor, "Graded Manifols and Supermanifolds," in *Mathematical Aspects of Superspace*, Clarke, C. J. S., Rosenblum, A., Seifert, H.-J. eds. (Dordrecht, Reidel, 1984).
- M. Dubois-Violette, "Dérivation et calcul differéntiel non commutatif," *C. R. Acad. Sci. Paris* **307** *Série I* **403** (Paris, 1988).
- B. A. Dubrovin, *Geometry of Hamiltonian Evolutionary Systems* (Bibliopolis, Naples, 1991).
- Faddeev and V. E. Korepin, "Quantum Theory of Solitons," *Physics Reports* **42C**, 1 (1978).
- S. P. Novikov, S. V. Manakov, L. P. Pitaevskii, and V. E. Zakharov, *Theory of Solitons: The Inverse Scattering Method* (Consultants Bureau, New York, 1984).

Appendix A

Lagrange: A Short Biography

Lagrange is considered one of the greatest mathematicians of the modern age and it is impossible, in a few pages, to quote his enormous contribution to mathematics and physics. Thus, we shall limit ourselves to a short biographic note.

Giuseppe Luigi Lagrangia was born in Torino on January 25, 1736, and died in Paris on April 10, 1813. At the age of 19 he already was Professor of Mathematics at Artillery's School in Torino and soon after associate founder of the Academy of Sciences of Torino. The first fruit of Lagrange's works here was his letter, written when he was still only 19, to Euler, in which he solved the isoperimetrical problem, which for more than half a century, had been a subject of discussion. "To effect the solution he enunciated the principles of the calculus of variations. Euler recognized the generality of the method adopted, and its superiority to that used by himself; and with rare courtesy he withheld a paper he had previously written, which covered some of the same ground, in order that the young Italian might have time to complete his work, and claimed the undisputed invention of the new calculus".[46]

Most of Lagrange's early writings are to be found in the five volumes of Transactions of Turin Academy, usually known as the *Miscellanea Taurinensia.*

The first volume contains a memoir on the theory of sound propagation. In this he indicates a mistake made by Newton, obtains the general differential equation for the motion and integrates it for motion in a straight line. In this volume is to be found a complete solutions of the problem of a string vibrating transversely. In particular, the article points out a lack of generality in the solutions previously given by Taylor, D'Alembert, and Euler; arrives at the conclusion that the form of the curve at any time t is given by $y = a \sin mx \sin nt$, and concludes with a masterly discussion of *echoes*, *beats*, and *compound sounds.* In this volume, other articles concern *recurring series*, *probabilities*, and *calculus of variations.*

The second volume includes remarks on the theory and notation of the calculus of variations, already discussed in the first volume, the derivation of *Least Action Principle* as an illustration of the method, and solutions of various dynamical problems.

The third volume, besides the solutions of additional dynamical problems by means of the calculus of variations, and some articles on the integral calculus, includes the general differential equations of motion for three bodies moving under their mutual attractions.

In a word, in 1761 Lagrange stood without a rival as the foremost mathematician living. In his paper in 1764, on the libration of the moon, he explains, with the aid of the *Principle of the Virtual Work*, why the moon always turns to the earth the same face. Here there was already, in germ, the future generalized equations of the motion.

"In 1766 Euler left Berlin, and Frederick the Great immediately wrote expressing the wish that 'the greatest King in Europe' to have 'the greatest mathematician in Europe' resident at his court. Lagrange accepted the offer and spent the next twenty years in Prussia, where he produces, not only the long series of memoirs published in the Berlin and Torino transactions, but his monumental work, the *Mécanique Analytique*".[46]

Indeed, during these 20 years, Lagrange contributed one memoir per month, on the average, to the Academies of Berlin, Torino, and Paris. All his memoirs are of high scientific level. Moreover, some of them are actually treatises. Among the ones sent to Paris it is worth to mention the memoir on the Jovian system (1766), the essay on the three body problem (1772), the article on the secular equation of the moon (1773), and the treatise on cometary perturbation (1778). For all these memoirs, the *Acadèmie the France*, who had proposed the subjects, awarded a prize to Lagrange.

In 1787, after the death of Frederick, Lagrange "who had found the climate of Berlin trying, gladly accepted the offer of Louis XVI to migrate to Paris. He received similar invitations from Spain and Naples".[46] The decree of October 1793, which ordered all foreigners to leave France, specially exempted him by name. He was offered the presidency of the commission for the reform of weights and measures and the different revolutionary governments loaded him with honors and distinctions. In 1795, Lagrange was appointed to a mathematical chair at the newly-established *École Normale*, which enjoyed only a brief existence of four months, and in 1797, he was made professor at the *École Polytechnique*.

In appearance Lagrange was of medium height, and slightly formed, with pale blue eyes and a colorless complexion. In character, he was nervous and timid, he detested controversy, and to avoid it willingly allowed others to take the credit for what he had himself done. Indeed, no inconsiderable part of the discoveries of his great contemporary, Laplace, consists of the application of the Lagrangian formulae to the facts of nature. Even the introduction of the *momenta* and of the *Poisson bracket* occur in the writings of Lagrange as well as *the theory of reduction of the problem of* n *bodies.*

Appendix B
Concerning the Lie Derivative

Let us observe that from

$$\sum_{h=1}^{n} \frac{\partial x_0^h}{\partial x^i} \frac{\partial x^k}{\partial x_0^h} = \delta_i^k ,$$

it follows that

$$\frac{d}{dt} \sum_{h=1}^{n} \frac{\partial x_0^h}{\partial x^i} \frac{\partial x^k}{\partial x_0^h} = 0 ,$$

or equivalently,

$$\sum_{h=1}^{n} \left(\frac{d}{dt} \frac{\partial x_0^h}{\partial x^i} \right) \frac{\partial x^k}{\partial x_0^h} + \sum_{h=1}^{n} \frac{\partial x_0^h}{\partial x^i} \left(\frac{d}{dt} \frac{\partial x^k}{\partial x_0^h} \right) = 0 .$$

Then, we may write

$$\begin{aligned} \sum_{h=1}^{n} \left(\frac{d}{dt} \frac{\partial x_0^h}{\partial x^i} \right) \frac{\partial x^k}{\partial x_0^h} &= - \sum_{h=1}^{n} \frac{\partial x_0^h}{\partial x^i} \left(\frac{d}{dt} \frac{\partial x^k}{\partial x_0^h} \right) \\ &= - \sum_{h=1}^{n} \frac{\partial x_0^h}{\partial x^i} \frac{\partial}{\partial x_0^h} \left(\frac{dx^k}{dt} \right) \end{aligned}$$

$$= -\sum_{h=1}^{n} \frac{\partial x_0^h}{\partial x^i} \frac{\partial X^k}{\partial x_0^h}$$

$$= -\sum_{h=1}^{n} \frac{\partial x_0^h}{\partial x^i} \frac{\partial X^k}{\partial x_0^h}$$

$$= -\frac{\partial X^k}{\partial x^i}\,.$$

In this way, multiplying by $\partial x_0^j/\partial x^k$ and summing over k, we obtain

$$\sum_k \frac{\partial x_0^j}{\partial x^k} \sum_{h=1}^{n} \left(\frac{d}{dt}\frac{\partial x_0^h}{\partial x^i}\right) \frac{\partial x^k}{\partial x_0^h} = -\sum_k \frac{\partial x_0^j}{\partial x^k}\frac{\partial X^k}{\partial x^i}\,,$$

$$\sum_{h=1}^{n} \left(\frac{d}{dt}\frac{\partial x_0^h}{\partial x^i}\right) \left(\sum_k \frac{\partial x_0^j}{\partial x^k}\frac{\partial x^k}{\partial x_0^h}\right) = -\sum_{k=1}^{n} \frac{\partial x_0^j}{\partial x^k}\frac{\partial X^k}{\partial x^i}\,,$$

and then

$$\frac{d}{dt}\frac{\partial x_0^j}{\partial x^i} = -\sum_{k=1}^{n} \frac{\partial x_0^j}{\partial x^k}\frac{\partial X^k}{\partial x^i}\,.$$

Therefore,

$$\sum_{j=1}^{n} \left(\frac{d}{dt}\frac{\partial x_0^j}{\partial x^i}\right) \frac{\partial}{\partial x_0^j} = -\sum_{k=1}^{n} \frac{\partial X^k}{\partial x^i} \sum_{j=1}^{n} \frac{\partial x_0^j}{\partial x^k}\frac{\partial}{\partial x_0^j} = -\sum_{k=1}^{n} \frac{\partial X^k}{\partial x^i}\frac{\partial}{\partial x^k}\,.$$

Appendix C

Concerning the Kepler Action Variables

The two closed curves of integration in the integrals

$$\begin{cases} J_\vartheta = \dfrac{1}{2\pi} \oint d\vartheta \sqrt{\pi_\vartheta^2 - \dfrac{\pi_\varphi^2}{\sin^2 \vartheta}}\,, \\ J_r = \dfrac{1}{2\pi} \oint dr \sqrt{2mE + \dfrac{2mk}{r} - \dfrac{\pi_\vartheta^2}{r^2}}\,, \end{cases}$$

are fixed by requiring the vanishing of the corresponding velocities or, better, of the corresponding momenta p_ϑ and p_r expressed, of course, in terms of variables π_ϑ and π_φ. In this way, the integration limits are fixed by

$$\begin{cases} p_\vartheta^2 \equiv \pi_\vartheta^2 - \dfrac{\pi_\varphi^2}{\sin^2 \vartheta} = 0\,, \\ p_r^2 \equiv 2mE + \dfrac{2mk}{r} - \dfrac{\pi_\vartheta^2}{r^2} = 0\,. \end{cases}$$

Therefore, the "ϑ" integration must be performed between the limits ϑ_1 and ϑ_2 given by the solutions of

$$\sin^2 \vartheta = \frac{\pi_\varphi^2}{\pi_\vartheta^2} = \cos^2 \alpha\,,$$

where Eq. (4.25) has been used. Since ϑ itself always lies between 0 and π, where $\sin \vartheta > 0$, we have $\sin \vartheta_1 = \sin \vartheta_2 = \cos \alpha$. Thus, the integration goes from $\vartheta_1 = \pi/2 - \alpha$ to $\pi/2$ to $\vartheta_2 = \pi/2 + \alpha$ and again back to ϑ_1; the $\sin\vartheta$ goes from $\cos\alpha$ to 1, then to $\cos\alpha$. In this way, we obtain

$$J_\vartheta = \frac{4\pi_\vartheta}{2\pi}\int_{\pi/2-\alpha}^{\pi/2}\frac{1}{\sin\vartheta}\sqrt{\sin^2\alpha - \cos^2\vartheta}d\vartheta$$

$$= \frac{2\pi_\vartheta}{\pi}\sin^2\alpha\int_0^{\pi/2}\frac{\cos^2\tau}{1-\sin^2\alpha\sin^2\tau}d\tau\,,$$

with τ defined by

$$\cos\vartheta = \sin\alpha\sin\tau\,.$$

Therefore, with $x \equiv \tan\tau$, we have

$$J_\vartheta = \frac{2\pi_\vartheta}{\pi}\int_0^{+\infty}\left[\frac{dx}{1+x^2} - \cos^2\alpha\frac{dx}{1+x^2\cos^2\alpha}\right]$$

$$= \frac{2\pi_\vartheta}{\pi}\left(\frac{\pi}{2} - \frac{\pi}{2}\cos\alpha\right)$$

$$= \pi_\vartheta(1-\cos\alpha)\,,$$

and then, by using again Eq. (4.25),

$$J_\vartheta = \pi_\vartheta - \pi_\varphi\,.$$

The "r" integration requires the application of *the method of residues.* The roots r_1 and r_2 of the equation

$$2mE + \frac{2mk}{r} - \frac{\pi_\vartheta^2}{r^2} = 0$$

are positive if $E < 0$ and correspond to the radii of turning points.

In the complex z plane, the function

$$f(z) = \sqrt{2mE + \frac{2mk}{z} - \frac{\pi_\vartheta^2}{z^2}}$$

has two branch points at

$$z_\pm = -\frac{k}{2E}\left[1 \pm \sqrt{1 + \frac{2\pi_\vartheta^2 E}{mk^2}}\right]$$

and a simple pole at $z = 0$, so that

$$J_r = i(\mathcal{R}(z = 0) + \mathcal{R}(z = +\infty)) .$$

Since $\mathcal{R}(z = 0) = \sqrt{-\pi_\vartheta^2}$ and $\mathcal{R}(z = +\infty) = mk/\sqrt{2mE}$, we finally obtain

$$J_r = -\pi_\vartheta + \frac{mk}{\sqrt{-2mE}} .$$

Appendix D

Concerning the Reduced Phase Space

In order to prove the relation

$$(Ad_g\xi)_{\mathcal{M}}(p) = (\Phi_g)_{*\Phi_{g^{-1}}(p)}(\xi_{\mathcal{M}}(\Phi_{g^{-1}}(p))),$$

we start by observing that

$$\begin{aligned}(Ad_g\xi)_{\mathcal{M}}(p) &= \frac{d}{dt}\Phi(e^{tAd_g\xi}, p)|_{t=0} \\ &= \frac{d}{dt}\Phi(ge^{t\xi}\cdot g^{-1}, p)|_{t=0} \\ &= \frac{d}{dt}\Phi(ge^{t\xi}, \Phi_{g^{-1}}(p))|_{t=0}\,.\end{aligned}$$

Therefore, since Φ is an action

$$\begin{aligned}(Ad_g\xi)_{\mathcal{M}}(p) &= \frac{d}{dt}\Phi(ge^{t\xi}, \Phi_{g^{-1}}(p)|_{t=0} \\ &= \frac{d}{dt}\Phi_g\circ\Phi(e^{t\xi}, \Phi_{g^{-1}}(p))|_{t=0} \\ &= (\Phi_g)_{*\Phi_{g^{-1}}(p)}(\xi_{\mathcal{M}}(\Phi_{g^{-1}}(p))),\end{aligned}$$

so that we have

$$(Ad_{g^{-1}}\xi)_{\mathcal{M}}(p) = (\Phi_{g^{-1}})_{*\Phi_g(p)}(\xi_{\mathcal{M}}(\Phi_g(p)))\,.$$

Appendix E

On the Canonical Differential 1-Form

Let $\mathcal{M}$ be a differentiable manifold and $T^*\mathcal{M}$ its cotangent bundle. The map

$$\tau : T^*\mathcal{M} \to \mathcal{M},$$

which associates with every differential 1-form on $T_q\mathcal{M}$ the point $q \in \mathcal{M}$, is a surjective differentiable map.

Let $V_{\alpha_q} \in T_{\alpha_q}(T^*\mathcal{M})$ be a tangent vector on the cotangent bundle at the point $\alpha_q \in T_q^*\mathcal{M}$; the derivative

$$\tau_* : T(T^*\mathcal{M}) \to T\mathcal{M}$$

of the natural projection τ maps the vector V_{α_q} to the vector $\tau_{*\alpha_q}(V_{\alpha_q})$, which is tangent to $\mathcal{M}$ at the point q.

The map

$$\theta : \alpha_q \in T^*\mathcal{M} \to \theta(\alpha_q) = \theta_{\alpha_q} \in T^*_{\alpha_q}(T^*\mathcal{M}),$$

defined by

$$\theta_{\alpha_q}(V_{\alpha_q}) = \alpha_q(\tau_{*\alpha_q}(V_{\alpha_q})), \tag{E.1}$$

is called the *canonical* 1-*form* on the cotangent bundle $T^*\mathcal{M}$.

If the manifold $\mathcal{M}$ is supposed to be a Lie group G, the diagram

$$\begin{array}{ccc} T^*G & \xrightarrow{\Phi^*_{g^{-1}}} & T^*G \\ \tau\downarrow & & \downarrow\tau \\ G & \xrightarrow[\Phi_g]{} & G \end{array} \tag{E.2}$$

is a commutative diagram (here $\Phi_g \equiv L_g$ and, for every $g \in G$, $\Phi^*_{g^{-1}}$ is the symplectic diffeomorphism of the induced action of G on the cotangent bundle T^*G).

Indeed,

$$\tau(\Phi^*_{g^{-1}}(\alpha_h)) = \tau(\alpha_{gh}) = gh\,, \quad \forall\, \alpha_h \in T^*G$$

and

$$\Phi_g(\tau(\alpha_h)) = \Phi_g(h) = gh\,, \quad \forall\, \alpha_h \in T^*G\,.$$

The diagram

$$\begin{array}{ccc} T^*G & \xrightarrow{\tau} & G \\ \xi_{T^*G}\downarrow & & \downarrow\xi_G \\ T(T^*G) & \xrightarrow[\tau_*]{} & TG \end{array}$$

where ξ_G and ξ_{T^*G} are given by

$$\xi_G : g \in G \to \xi_G(g) = \frac{d}{dt}\Phi_{e^{t\xi}}(g)|_{t=0} \in T_gG \quad \forall\, \xi \in \mathcal{G}$$

and

$$\xi_{T^*G} : \alpha_g \in T^*G \to \xi_{T^*G}(\alpha_g) = \frac{d}{dt}\Phi^*_{e^{-t\xi}}(\alpha_g)|_{t=0} \in T_{\alpha_g}(T^*G)\,,$$

is a commutative diagram too.

Of course, by using the commutative diagram (E.2), we have

$$\begin{aligned} \tau_{*\alpha_h}(\xi_{T^*G}(\alpha_h)) &= \frac{d}{dt}\tau(\Phi^*_{e^{-t\xi}}(\alpha_h))|_{t=0} \\ &= \frac{d}{dt}\Phi_{e^{t\xi}}(\tau(\alpha_h))|_{t=0} \end{aligned}$$

$$= \frac{d}{dt}\Phi_{e^{t\xi}}(h)|_{t=0}$$

$$= \xi_G(h)$$

$$= \xi_G(\tau(\alpha_h))\,.$$

From Eq. (10.24), we have

$$J_\xi(\alpha_g) = (i_{\xi_{T^*G}}\theta)(\alpha_g) = \theta_{\alpha_g}(\xi_{T^*G}(\alpha_g)) = \alpha_g(\tau_{*\alpha_g}(\xi_{T^*G}(\alpha_g))) = \alpha_g(\xi_G(g))\,.$$

Let us go back to the general case.

It is not difficult to prove that

$$\beta^*\theta = \theta$$

for every differential 1-form on $T^*\mathcal{M}$,

$$\beta : q \in \mathcal{M} \to \beta(q) = \beta_q \in T_q^*\mathcal{M}\,. \tag{E.3}$$

We start by observing that the derivative of Eq. (E.3) defines a map

$$\beta_* : T\mathcal{M} \to T(T^*\mathcal{M})\,,$$

so that, if

$$V_q = \frac{d}{dt}q(t)|_{t=0}\,,$$

i.e. $V_q \in T_q\mathcal{M}$ with an integral curve $q(t)$, where $q(0) = q$, then

$$(\beta^*\theta)_q(V_q) = \theta_{\beta_q}(\beta_{*q}(V_q)) = \beta_q(\tau_{*\beta_q}(\beta_{*q}(V_q)))\,.$$

On the other hand,

$$\tau_{*\beta_q}(\beta_{*q}(V_q)) = \frac{d}{dt}\tau(\beta(q(t)))|_{t=0} = \frac{d}{dt}q(t)|_{t=0} = V_q\,,$$

so that

$$(\beta^*\theta)_q(V_q) = \beta_q(V_q)\,, \quad \forall V_q \in T_q\mathcal{M}\,.$$

If $\omega = -d\theta$ is the canonical symplectic form on $T^*\mathcal{M}$, we have

$$\beta^*\omega = -\beta^*d\theta = -d\beta^*\theta = -d\beta$$

for every differential 1-form β on $T^*\mathcal{M}$.

Appendix F

Concerning Rigid Body Equations

Let us observe that given the map

$$Ad^*_\bullet\alpha : g \in G \to (Ad^*_\bullet\alpha)(g) = Ad^*_g\alpha \in \mathcal{G}^*\,, \quad \alpha \in \mathcal{G}^*\,,$$

its derivative, at identity, gives the map

$$(Ad^*_\bullet\alpha)_{*e}(\xi) = \frac{d}{dt}Ad^*_{e^{t\xi}}\alpha|_{t=0} = ad^*_\xi\alpha\,, \quad \forall\, \xi \in \mathcal{G}\,.$$

Similarly, the derivative of the map

$$Ad^*_\bullet Ad^*_g\alpha : h \in G \to Ad^*_h Ad^*_g\alpha \in \mathcal{G}^*$$

is given by

$$(Ad^*_\bullet Ad^*_g\alpha)_{*e}(\xi) = \frac{d}{dt}Ad^*_{e^{t\xi}} Ad^*_g\alpha|_{t=0} = ad^*_\xi(Ad^*_g\alpha)\,.$$

On the other hand, we also have

$$\begin{aligned}(Ad^*_\bullet Ad^*_g\alpha)_{*e}(\xi) &= \frac{d}{dt}Ad^*_{e^{t\xi}} Ad^*_g\alpha|_{t=0} = \frac{d}{dt}Ad^*_{ge^{t\xi}}\alpha|_{t=0} \\ &= \frac{d}{dt}Ad^*_{L_g e^{t\xi}}\alpha|_{t=0} = (Ad^*_\bullet\alpha)_{*g}((L_g)_{*e}(\xi))\,,\end{aligned}$$

so that

$$(Ad^*_\bullet\alpha)_{*g}((L_g)_{*e}(\xi)) = ad^*_\xi(Ad^*_g\alpha)\,.$$

If $g = g(t)$ is a curve in G and $\alpha(t) = Ad^*_{g(t)^{-1}}\alpha$ is a curve in $\mathcal{G}^*$, the above relation is equivalent to Eq. (10.49).

Appendix G

The Gelfand–Levitan–Marchenko Equation

Let us consider the stationary Schrödinger equation (with $\hbar = 1, m = 1/2$), on the real line $\Re$,

$$\frac{d^2}{dx^2}\varphi + (k^2 - U(x))\varphi = 0\,, \tag{G.1}$$

where the potential $U(x)$ is assumed to be a *fast-decreasing* function at $\pm\infty$:

$$\lim_{x\to\pm\infty} U(x) = 0\,.$$

If $\phi(x,k)$ is a solution of Eq. (G.1) with the following asymptotic behavior:

$$\phi(x,k) \underset{x\to\infty}{\sim} \exp[ikx] \tag{G.2}$$

then, by parity, $\phi(x,-k)$ is a solution of Eq. (G.1) whose asymptotic behavior is given by

$$\phi(x,-k) \underset{x\to\infty}{\sim} \exp[-ikx]\,.$$

Moreover, it is not difficult to prove[14,16] that the solution $\phi(x,k)$ can be expressed in the following form:

$$\phi(x,k) = \exp[ikx] + \int_x^\infty A(x,y)\exp[iky]dy\,. \tag{G.3}$$

To this end let us observe that

- the following theorem holds:

  Theorem 45 (Titchmarch) *A necessary and sufficient condition for a real function $F(q) \in L_2(-\infty, +\infty)$ be the real limit*

$$F(q) = \lim_{b \to 0} F(q + ib), \quad \forall q \in \Re$$

  of a function $F(z)$ holomorphic in the upper complex plane $(b > 0)$ and satisfying the condition

$$\int_{-\infty}^{+\infty} |F(q + ib)| dq = \bigcirc(\exp[-2\alpha b]),$$

  is that

$$\varphi(t) \equiv \frac{1}{2\pi} \int_{-\infty}^{+\infty} F(q) \exp[-iqt] dq = 0, \quad \forall t < \alpha;$$

- for $b > 0$, $k \neq 0$ and $M(x) \equiv \int_x^\infty |U(y)| dy < \infty$, a constant C exists such that

$$|g(x,k) - 1|^2 < C, \tag{G.4}$$

  where

$$g(x,k) = \phi(x,k) \exp[-ikx].$$

Indeed, by multiplying Eq. (G.1), written for $\varphi \equiv \phi(y,k)$, by $\sin k(y - x)$, we obtain

$$U(y)\phi(y,k) \sin k(y-x) = \frac{d}{dy}\left[\frac{d}{dy}\phi(y,k) \sin k(y-x) - k\phi(y,k) \cos k(y-x)\right],$$

which, integrated between x and ∞ with the boundary condition expressed by Eq. (G.1), gives

$$\int_x^\infty U(y)\phi(y,k) \sin k(y-x) dy = k[\phi(x,k) - e^{ikx}],$$

or

$$\phi(x,k) = e^{ikx} - \int_x^\infty \frac{\sin k(x-y)}{k} U(y)\phi(y,k) dy.$$

The above equation, expressed in terms of $g(x,k) = \phi(x,k)\exp[-ikx]$, reads

$$g(x,k) = e^{ikx} + \int_x^\infty \frac{\exp[-2ik(x-y)] - 1}{2ik} U(y) g(y,k) dy\,.$$

A solution of the above equation is expressed by the expansion

$$g(x,k) = \sum_{n \in \aleph} g_n(x,k)\,,$$

with

$$g_{n+1}(x,k) = \int_x^\infty \frac{\exp[-2ik(x-y)] - 1}{2ik} U(y) g_n(y,k) dy\,, \quad g_0(x,k) = 1\,,$$

whose uniform convergence can be easily checked. Indeed, since $y > x$ and $k = q + ib$ with $b > 0$, we have

$$\left|\frac{\exp[-2ik(x-y)] - 1}{2ik}\right| \le \left|\frac{1}{2ik}\right| + \left|\frac{\exp[2b(x-y)]}{2ik}\right| < \frac{1}{|k|}\,,$$

so that, with $M(x) = \int_x^\infty |U(y)| dy$,

$$|g_1(x,k)| < \frac{M(x)}{|k|}\,.$$

Moreover,

$$|g_{n+1}(x,k)| \le \frac{1}{|k|} \int_x^\infty |U(y)|\, |g_n(y,k)| dy\,,$$

so that from the *induction hypothesis*

$$|g_n(x,k)| < \frac{M^n(x)}{|k|^n}\,,$$

we have

$$|g_{n+1}(x,k)| < \frac{1}{|k|} \int_x^\infty |U(y)| \frac{M^n(x)}{|k|^n} dy < \frac{M^{n+1}(x)}{|k|^{n+1}}\,.$$

Thus,

$$|g(x,k) - 1| < \exp\left[\frac{M(x)}{|k|}\right] - 1\,.$$

It follows that, for $b > 0$, the function

$$h(x,k) = \phi(x,k) - \exp[ikx] = \exp[ikx][g(x,k) - 1]\,, \quad k = q + ib\,,$$

is square integrable with respect $q \in \Re$, since

$$\begin{aligned} |h(x,k)| &= |\exp[ikx][g(x,k) - 1]| \\ &= \exp[-bx]\,|g(x,k) - 1|\,, \end{aligned}$$

so that

$$|h(x,k)|^2 = \exp[-2bx]|g(x,k) - 1|^2 < C\exp[-2bx] \propto \bigcirc(\exp[-2bx])\,.$$

Therefore,

$$\int_{-\infty}^{+\infty} |h(x,k)|^2 dq = \bigcirc(\exp[-2bx])\,,$$

and we can apply Titchmarch' theorem, to write

$$A(x,y) \equiv \frac{1}{2\pi}\int_{-\infty}^{+\infty} h(x,k)\exp[-iqy]dq = 0\,, \quad \forall\, y < x\,,$$

whose inversion, for $y > x$, gives

$$h(x,k) = \int_x^{+\infty} A(x,y)\exp[iky]dy\,, \quad y > x\,.$$

From the expression of $h(x,k)$ and the above equation, we have Eq. (G.3).

Thus, it has been shown that the solution

$$\phi(x,k) = \exp[ikx] + \int_x^{\infty} A(x,y)\exp[iky]dy \qquad \text{(G.5)}$$

is analytic in the complex open upper plane defined by $\Im m k > 0$. Moreover, every solution of (G.1) can be expressed as a linear combination of $\phi(x,k)$ and $\phi(x,-k)$, the last being linearly independent.

Therefore, two solutions $\psi(x,k)$ and $\psi(x,-k)$ of (G.1), having the following asymptotic behavior:

$$\psi(x,k) \underset{x\to-\infty}{\sim} \exp[-ikx]\,,$$

$$\psi(x,-k) \underset{x\to-\infty}{\sim} \exp[ikx]\,,$$

can be expressed as

$$\begin{aligned}\psi(x,k) &= \beta(k)\phi(x,k) + \alpha(k)\phi(x,-k)\,,\\ \psi(x,-k) &= \beta(-k)\phi(x,-k) + \alpha(-k)\phi(x,k)\,.\end{aligned} \tag{G.6}$$

The inverse relations are given by

$$\begin{aligned}\phi(x,k) &= \bar{\beta}(k)\psi(x,k) + \bar{\alpha}(k)\psi(x,-k)\,,\\ \phi(x,-k) &= \bar{\beta}(-k)\psi(x,-k) + \bar{\alpha}(-k)\psi(x,k)\,,\end{aligned} \tag{G.7}$$

with

$$\begin{aligned}\bar{\beta}(k)\beta(k) + \bar{\alpha}(k)\alpha(-k) = 1 \quad & \bar{\beta}(k)\alpha(k) + \bar{\alpha}(k)\beta(-k) = 0\,,\\ \bar{\alpha}(k)\beta(k) + \bar{\beta}(-k)\alpha(k) = 0 \quad & \bar{\beta}(k)\beta(k) + \bar{\alpha}(-k)\alpha(k) = 1\,.\end{aligned}$$

The coefficients $\alpha(k)$, $\beta(k)$, $\bar{\alpha}(k)$ and $\bar{\beta}(k)$ can be easily expressed in terms of the functions $\phi(x,k)$ and $\psi(x,k)$.

Indeed, by using Eq. (G.6), the Wronskian W of $\phi(x,k)$ and $\psi(x,k)$,

$$W[\phi(x,k),\psi(x,k)] \equiv \phi(x,k)\frac{d}{dx}\psi(x,k) - \psi(x,k)\frac{d}{dx}\phi(x,k)\,,$$

is related to the Wronskian W of $\phi(x,k)$ and $\phi(x,-k)$ by

$$W[\phi(x,k),\psi(x,k)] = \alpha(k)W[\phi(x,k),\phi(x,-k)]\,.$$

Since the Wronskian of two solutions of the Schrödinger equation does not depend on x, we have

$$\begin{aligned}W[\phi(x,k),\phi(x,-k)] &= \lim_{x\to\infty} W[\phi(x,k),\phi(x,-k)]\\ &= W[e^{ikx}, e^{-ikx}]\\ &= -2ik\,,\end{aligned}$$

so that

$$\alpha(k) = -\frac{W[\phi(x,k),\psi(x,k)]}{2ik}\,. \tag{G.8}$$

Alternatively, by using Eqs. (G.7) instead of Eq. (G.6), we obtain

$$W[\phi(x,k),\psi(x,k)] = \bar{\alpha}W[\phi(x,k),\phi(x,-k)]\,,$$

so that

$$\bar{\alpha}(k) = \alpha(k) = -\frac{W[\phi(x,k), \psi(x,k)]}{2ik}\,.$$

Similarly, from

$$W[\phi(x,-k), \psi(x,k)]$$
$$= \beta(k)W[\phi(x,-k), \phi(x,k)] = \bar{\beta}(-k)W[\phi(x,-k), \phi(x,k)]\,,$$

we also obtain

$$\bar{\beta}(-k) = -\beta(k) = -\frac{W[\phi(x,-k), \psi(x,k)]}{2ik}\,,$$

so that

$$|\alpha(k)|^2 = 1 + |\beta(k)|^2\,. \tag{G.9}$$

From the first of Eq. (G.6), we can introduce the *scattering function* defined by

$$\frac{\psi(x,k)}{\alpha(k)} = \phi(x,-k) + \frac{\beta(k)}{\alpha(k)}\phi(x,k)\,, \tag{G.10}$$

which well describes the following physical process:

A wave $\phi(x,-k) \sim \exp[-ikx]$, *coming from* $+\infty$ *on the obstacle represented by the potential* $U(x)$, *is partially reflected at* $+\infty$ *as* $(\beta(k)/\alpha(k))\phi(x,k) \sim (\beta(k)/\alpha(k))\exp[-ikx]$, *and partially transmitted at* $-\infty$ *as* $(\psi(x,k)/\alpha(k)) \sim (1/\alpha(k))\exp[-ikx]$.

The ratios

$$R(k) = \frac{\beta(k)}{\alpha(k)}\,, \quad T(k) = \frac{1}{\alpha(k)}$$

are called the *reflection coefficient* and the *transmission coefficient*, respectively. They, owing to Eq. (G.9), satisfy the relation

$$|R(k)|^2 + |T(k)|^2 = 1\,.$$

Moreover, $R(k)$, $\alpha(k)$ and $\beta(k)$ are analytic in the complex open upper plane $\Im m k > 0$.

By multiplying both sides of Eq. (G.10) by $\exp[iky]$ and integrating over k from $-\infty$ to $+\infty$, we write

$$\int_{-\infty}^{+\infty} \frac{\psi(x,k)}{\alpha(k)} e^{iky} dk = \int_{-\infty}^{+\infty} \phi(x,-k) e^{iky} dk + \int_{-\infty}^{+\infty} R(k)\phi(x,k) e^{iky} dk \,. \tag{G.11}$$

Let us evaluate the two sides, henceforth denoted by D and B, separately.

By using Eq. (G.5), that is

$$\phi(x, \pm k) = \exp[\pm ikx] + \int_{x}^{\infty} A(x,y) \exp[\pm iky] dy \,,$$

we have[†]

$$\begin{aligned}\int_{-\infty}^{+\infty} \phi(x,-k) e^{iky} dk &= \int_{-\infty}^{+\infty} e^{ik(y-x)} dk \\ &\quad + \int_{x}^{\infty} dz \left(A(x,z) \int_{-\infty}^{+\infty} e^{ik(y-z)} dk \right) \\ &= \int_{x}^{\infty} dz \left(A(x,z) \int_{-\infty}^{+\infty} e^{ik(y-z)} dk \right) \\ &= 2\pi A(x,y) \,.\end{aligned}$$

Therefore,

$$\begin{aligned} B &= 2\pi A(x,y) + \int_{-\infty}^{+\infty} R(k) e^{ik(x+y)} dk \\ &\quad + \int_{x}^{\infty} dz \left(A(x,z) \int_{-\infty}^{+\infty} R(k) e^{ik(z+y)} dk \right) \\ &= 2\pi \left\{ A(x,y) + F_c(x+y) + \int_{x}^{\infty} A(x,z) F_c(z+y) dz \right\} ,\end{aligned}$$

where

$$F_c(x) = \frac{1}{2\pi} \int_{-\infty}^{+\infty} R(k) e^{ikx} dk$$

is the Fourier transform of the reflection coefficient.

[†]We observe that $A(x,y)$ is defined only for $y > x$ and that in this case we have $\delta(x \pm y) \equiv (1/2\pi) \int_{-\infty}^{+\infty} dk \exp[ik(x \pm y)] = 0$.

In order for evaluate the integral on the left-hand side of Eq. (G.11),

$$D = \int_{-\infty}^{+\infty} \frac{\psi(x,k)}{\alpha(k)} e^{iky} dk\,,$$

let us observe that $\alpha(k)$ is an analytic function in the complex open upper plane $\Im m k > 0$, where it has simple zeroes corresponding to bound states. Indeed, if $\alpha(k)$ vanishes at the point k_0, then

$$W[\phi(x,k_0), \psi(x,k_0)] = 0\,,$$

so that $\phi(x,k_0)$ and $\psi(x,k_0)$ are linearly dependent. Therefore, we obtain

$$\psi(x,k_0) = \beta(k_0)\phi(x,k_0)\,. \tag{G.12}$$

On the other hand, the solution $\phi(x,k)$ decreases exponentially for $x \to \infty$ as well as $\psi(x,k)$ when $x \to -\infty$, since $\Im m k > 0$, so that we can conclude that $\phi(x,k_0)$ and $\psi(x,k_0)$ are the wave functions of a bound states if $\Im m k_0 > 0$.

Since, k_0^2 is real, k_0 will be purely imaginary $k_0 = i\chi_0$.

In order to show that k_0 is a simple zero, let us introduce, for the sake of simplicity, the notation

$$\phi = \phi(x,k)\,, \quad \psi = \psi(x,k)\,, \quad \dot{\alpha}(k) = \frac{d\alpha(k)}{dk}\,, \quad \dot{\phi} = \frac{d\phi}{dk}\,, \quad \dot{\psi} = \frac{d\psi}{dk}\,,$$

and let us consider the Schrödinger equations for ϕ and ψ:

$$\phi'' + k^2\phi = U\phi\,, \quad \psi'' + k^2\psi = U\psi\,.$$

By taking the derivative, with respect to k, of the second equation, we have

$$\dot{\psi}'' + k^2\dot{\psi} = U\dot{\psi} - 2k\psi\,.$$

The difference between the first Schrödinger equation multiplied by $\dot{\psi}$ and the above equation multiplied by ϕ, gives

$$\phi''\dot{\psi} - \dot{\psi}''\phi = 2k\phi\psi\,,$$

or equivalently,

$$-\frac{d}{dx} W[\phi, \dot{\psi}] = 2k\phi\psi\,,$$

so that

$$-W[\phi,\dot{\psi}]|_{-l}^{x} = 2k\int_{-l}^{x}\phi\psi dx\,, \tag{G.13}$$

where l is an arbitrary parameter.

With the same procedure we also have

$$-W[\dot{\phi},\psi]|_{x}^{l} = 2k\int_{x}^{l}\phi\psi dx\,. \tag{G.14}$$

On the other hand, by using Eq. (G.8), we can write

$$\frac{d}{dk}(2ik\alpha) = 2i\alpha(k) + 2ik\dot{\alpha}(k) = -W[\dot{\phi},\psi] - W[\phi,\dot{\psi}]\,. \tag{G.15}$$

Let us now observe that

- the above equation does not depend on x;
- for $k = k_0$, both ϕ and ψ vanish;
- both $W[\dot{\phi},\psi]$ and $W[\phi,\dot{\psi}]$ vanish at $x = \pm l \to \pm\infty$.

Therefore, in such limits, by adding Eqs. (G.13) and (G.14), Eq. (G.15) gives

$$2ik_0\dot{\alpha}(k_0) = 2k_0\int_{-\infty}^{+\infty}\phi(x,k_0)\psi(x,k_0)dx = 2k_0\beta(k_0)\int_{-\infty}^{+\infty}\phi^2(x,k_0)dx \neq 0\,, \tag{G.16}$$

where Eq. (G.12) has been used.

The above equation shows that k_0 is a simple zero of $\alpha(k)$.

Let us continue the calculation of the integral D, by applying the *residues method* according to which

$$\int_{\partial\mathcal{D}} f(z)dz = 2\pi i\sum_{j=1}^{n}\mathcal{R}(z_j)\,,$$

for every domain $\partial\mathcal{D}$ completely belonging to the field of the analyticity of the function ϕ and containing a finite number of singular isolated points z_k.

For our purpose, let us choose the half-circle, with an infinite radius, contained in the upper plane, $\Im m k > 0$.

For an infinite radius of the half-circle, the factors of the type $\exp[ik(y \pm x)]$, with $y > x$, give rise to a vanishing contribution from the integral along the boundary. Therefore, the only contribution to the integral comes from the integral along the real k axis; that is, from D. Thus,

$$D = 2\pi i \sum_{j=1}^{n} \mathcal{R}(i\chi_j) ,$$

where the χ's correspond to the bound states; i.e. $\alpha(i\chi_j) = 0$.

On the other hand,

$$\begin{aligned}
\mathcal{R}(i\chi_j) &= \lim_{k\to\chi_j} (k - i\chi_j)\frac{\psi(x,k)}{\alpha(k)} e^{iky} \\
&= e^{-\chi_j y}\psi(x,i\chi_j) \lim_{k\to\chi_j} \frac{(k - i\chi_j)}{\alpha(k)} \\
&= e^{-\chi_j y}\psi(x,i\chi_j) \lim_{k\to\chi_j} \frac{(k - i\chi_j)}{\alpha(k) - \alpha(i\chi_j)} \\
&= e^{-\chi_j y}\psi(x,i\chi_j) \lim_{k\to\chi_j} \frac{(k - i\chi_j)}{\alpha(k) - \alpha(i\chi_j)} \\
&= e^{-\chi_j y}\psi(x,i\chi_j) \left(\left.\frac{d\alpha(k)}{dk}\right|_{k-i\chi_j} \right)^{-1} \\
&= ie^{-\chi_j y}\psi(x,i\chi_j) \frac{1}{\beta(i\chi_j)\int_{-\infty}^{+\infty} \phi^2(x,i\chi_j)dx} \\
&= ie^{-\chi_j y}\psi(x,i\chi_j) \frac{c_j^2}{\beta(i\chi_j)} ,
\end{aligned}$$

where Eq. (G.16) has been used and where the c's are the normalization constants of the ϕ's, specified below.

The waves functions of bound states $\zeta(x,i\chi_j)$ are given by

$$\zeta(x,i\chi_j) = c_j\phi(x,i\chi_j) = c_j\frac{\psi(x,i\chi_j)}{\beta(i\chi_j)} ,$$

where Eq. (G.12) has been used. The c's are the normalization constants of the ϕ's defined by

$$1 = \int |\zeta(x,i\chi_j)|^2 dx = c_j^2 \int |\phi(x,i\chi_j)|^2 dx ,$$

and appear in the asymptotic behavior

$$\zeta(x, k_j) \underset{x\to\infty}{\sim} c_j \exp[-k_j x]$$

of the ζ's.

Thus, we have shown that

$$\mathcal{R}(i\chi_j) = ic_j^2 e^{-\chi_j y}\phi(x, i\chi_j)\,.$$

Therefore, by using for ϕ the expression given by Eq. (G.5), we finally have

$$D = -2\pi \sum_j c_j^2 \left[e^{-\chi_j(x+y)} + \int_x^{\infty} A(x, z)e^{-\chi_j(z+y)}dz \right].$$

By setting

$$F_b(x) = \sum_j c_j^2 e^{-\chi_j x}$$

and

$$F(x) \equiv F_b(x) + F_b(x) = F_c(x) = \frac{1}{2\pi}\int_{-\infty}^{+\infty} R(k)e^{ikx}dk + \sum_j c_j^2 e^{-\chi_j x}\,,$$

we can write Eq. (G.11); i.e. $B = D$, in the following form:

$$A(x, y) + F(x + y) + \int_x^{\infty} A(x, z)F(z + y)dz = 0\,.$$

The above equation, which is an integro-differential equation of Volterra type, is known as the *Marchenko equation* for $x \in \Re$ and as the *Gel'fand–Levitan equation* for $x \in \Re^+ \cup \{0\}$.

The *Gel'fand–Levitan–Marchenko* equation allows us to recover the interaction potential U once the scattering data $S \equiv \{\chi_j, c_j, R(k)\}$ are given, since the following relation can be proven‡:

$$U(x) = -2\frac{d}{dx}A(x,x)\,.$$

In order to prove that the above relation holds, let us compute, from Eq. (G.5), the derivative of $\phi(x,k)$ with respect to x.

We have

$$\frac{d}{dx}\phi(x,k) = ike^{ikx} + \int_x^\infty A_x(x,y)e^{iky}dy - A(x,y)|_{y=x}e^{ikx}\,,$$

so that

$$\begin{aligned}\frac{d^2}{dx^2}\phi(x,k) &= -k^2e^{ikx} + \int_x^\infty A_{xx}(x,y)e^{iky}dy - A_x(x,y)|_{y=x}e^{ikx}\\ &\quad - [A_x(x,y)|_{y=x} + A_y(x,y)|_{y=x}]e^{ikx} - ikA(x,y)|_{y=x}e^{ikx}\,.\end{aligned}$$

On the other hand, performing the integral in Eq. (G.5) by part twice, with the assumption

$$\lim_{x\to\infty} A(x,y) = 0\,,$$

we obtain

$$\begin{aligned}\phi(x,k) &= e^{ikx} + \frac{1}{ik}A(x,y)e^{iky}|_x^\infty - \frac{1}{ik}\int_x^\infty A_y(x,y)e^{iky}dy\\ &= e^{ikx} - \frac{1}{ik}A(x,x)e^{ikx} - \frac{1}{ik}\int_x^\infty A_y(x,y)e^{iky}dy\\ &= e^{ikx} - \frac{1}{ik}A(x,x)e^{ikx} - \frac{1}{(ik)^2}A_y(x,y)e^{iky}|_x^\infty\end{aligned}$$

‡It is worth recalling that $A(x,y)$ is defined only for $y > x$. Thus,

$$U(x) = \lim_{y\to x}\left(\frac{\partial A(x,y)}{\partial x} + \frac{\partial A(x,y)}{\partial y}\right)$$

Of course, $A(x,y)$ can be defined by continuity at $y = x$, and if it is differentiable at the point, the potential will be given by

$$U(x) = -2\frac{dA(x,x)}{dx}\,.$$

$$+\frac{1}{(ik)^2}\int_x^\infty A_{yy}(x,y)e^{iky}dy$$

$$= e^{ikx} - \frac{1}{ik}A(x,x)e^{ikx} + \frac{1}{k^2}A_y(x,y)|_{y=x}e^{ikx}$$

$$- \frac{1}{k^2}\int_x^\infty A_{yy}(x,y)e^{iky}dy\,.$$

By replacing the previous expressions of ϕ and ϕ'' on the left-hand side of Schrödinger's equation, we obtain

$$-2A_x(x,x)e^{ikx} + \int_x^\infty [A_{xx}(x,y) - A_{yy}(x,y)]e^{iky}dy = U\phi(x,k)\,,$$

and, by using once again Eq. (G.5) on the right-hand side, also

$$-2A_x(x,x)e^{ikx} + \int_x^\infty [A_{xx}(x,y) - A_{yy}(x,y)]e^{iky}dy$$

$$= U\left[e^{ikx} + \int_x^\infty A(x,y)e^{iky}dy\right]\,,$$

which finally gives

$$U(x) = -2\frac{d}{dx}A(x,x)\,,$$

$$A_{xx}(x,y) - A_{yy}(x,y) - U(x)A(x,y) = 0\,.$$

Bibliography

Books

1. R. Abraham and J. E. Marsden, *Foundations of Mechanics* (Benjamin/Cummings, Reading, Mass., 1978).
2. V. I. Arnold, *Mathematical Methods of Classical Mechanics* (Mir, Moscow 1976).
3. V. I. Arnold, V. V. Kozlov, and A. I. Neishtadt, *Mathematical Aspects of Classical and Celestial Mechanics*, Dynamical Systems III (Springer-Verlag, Berlin, 1988).
4. D. V. Alekseevskij, A. M. Vinogradov, and V. V. Lychagin, "Basic ideas and concepts of differential geometry," *Encyclopaedia of Mathematical Sciences*, Vol. 28, ed. R. V. Gamkrelidz, (Springer-Verlag, Berlin, 1992).
5. F. A. Berezin, *Introduction to Superanalysis*, ed. Kirillov, (D. Reidel, Dordrecht, 1987).
6. L. Brillouin, *Wave Propagation in Periodic Structures* (Dover, 1953).
7. I. Choquet-Bruhat, C. Dewitt-Morette, and Dillard-Bleik, *Analysis, Manifold and Physics* (North-Holland P. C., 1977).
8. F. Calogero and A. De Gasperis, "Spectral transform and solitons 1," *Studies in Mathematics and its Applications*, Vol. 13 (Amsterdam, 1982).
9. C. Cercignani, *Spazio, tempo e movimento* (Zanichelli, Bologna).
10. R. Chernoff and J. Marsden, *Lect. Notes in Math* (Springer-Verlag, Berlin, 1974), p. 425.
11. C. Chevalley, *Theory of Lie Groups* (Princeton University Press, Princeton, 1946).

12. A. Connes, *Non-Commutative Differential Geometry*, Pub. I.H.E.S. **62**, 257 (Paris, 1986).
13. R. Cushman and L. Bates, *Classical Integrable Systems* (1993).
14. V. de Alfaro and T. Regge, *Potential Scattering* (North-Holland, Amsterdam, 1965).
15. B. De Witt, *Supermanifolds* (Cambridge University Press, London, 1984).
16. A. di Giacomo, *Lezioni di Fisica Teorica* (Edizioni ETS, Pisa, 1992).
17. P. A. M. Dirac, *Lectures in Quantum Mechanics*, Belfer Graduate School of Sciences (Yeshiva University, New York, 1964).
18. P. A. M. Dirac, *Lectures on Quantum Field Theory*, Belfer Graduate School of Science (Yeshiva University, New York, 1966).
19. B. Dubrovin, S. Novikov, A. Fomenko, Géométric Contemporaine, MIR (Moscow 1979, 1982).
20. J. J. Duistermaat, "Global action-angle coordinate," *Comm. Pure Appl. Math.*, Vol. XXXIII (1980), p. 687.
21. G. Eilemberger, *Solitons* (Springer-Verlag, Berlin, 1981).
22. A. Fasano and S. Marmi, *Meccanica Analitica* (Boringhieri, Torino, 1994).
23. L. D. Faddeev and L. A. Takhtadjan, *Hamiltonian Method in the Theory of Solitons* (Springer-Verlag, Berlin, 1986).
24. G. Gallavotti, *Meccanica Elementare* (Boringhieri, Torino, 1986).
25. F. Gantmacher, *Lectures on Analytical Mechanics* (Mir, Moscow, 75).
26. E. Goursat, *Lecons sur l' integration des equations aux derivees partielles du premier ordre* (Hermann, Paris, 1921).
27. A. Hanson, T. Regge, and C. Teitelboim, *Constrained Hamiltonian Systems* (Accademia Nazionale dei Lincei, Roma, 1976).
28. A. Kirillov, *Eléments de la théorie des représentations* (Mir, Moscow, 1974).
29. A. Kolmogorov and S. Fomine, *Éléments de la théorie des fonctions et de l' analyse fonctionelle* (Mir, Moscow, 1977).
30. S. Kobaiashi and K. Nomizu, *Foundations of Differential Geometry, I, II* (Interscience Publishers, New York, 1963).
31. I. S. Krasil'shchik, V. V. Lichagin, and A. M. Vinogradov, *Geometry of Jet Spaces and Nonlinear Partial Differential Equations* (Gordon and Breach, 1986).
32. B. Kupershmidt, *Elements of Superintegrable systems* (D. Riedel, Dordrecht, 1987).
33. C. Lanczos, *The Variational Principles of Mechanics* (Dover, 1986).
34. G. Landi, "An introduction to noncommutative spaces and their geometries," *Lectures Notes in Physics, New Series m: Monographs* (Springer-Verlag, Berlin, 1997).
35. S. Lang, *Linear Algebra* (Addison-Wesley, Reading, Mass., 1966).
36. J. W. Leech, *Classical Mechanics*, Science Paperbacks (Chapman and Hall, London, 1965).
37. T. Levi-Civita and U. Amaldi, *Lezioni di Meccanica Razionale* (Zanichelli, Bologna, 1929).

38. S. Lie, *Theorie der transformationsgruppen*, Zweiter Abschnitt, unter mitwirkung von. Prof. Dr. Frederich Engel (Teubner, Leipzig, 1890).
39. L. Landau and E. Lifchitz, *Mechanics* (Mir, Moscow, 1966).
40. S. D. Lindenbaum, *Analytical Dynamics* (World Scientific, 1994).
41. E. Mach, "Die Mechanik," in *ihrer Entwickelung historisch-kritisch dargestellt* (1883).
42. G. Marmo, E. J. Saletan, A. Simoni, and B. Vitale, *Dynamical Systems* (John Wiley, 1985).
43. J. Moser, "Various aspects of integrable Hamiltonian systems," in *Progr. Math.*, Vol. 8 (Birkhäuser, Boston, 1980).
44. E. Nelson, *Tensor Analysis* (Princeton University Press, Princeton, 1967).
45. A. M. Perelomov, *Integrable Systems of Classical Mechanics and Lie Algebras* (Birkhäuser Verlag Basel, 1990).
46. M. Rasetti, *Modern Methods in Equilibrium Statistical Mechanics* (World Scientific, 1986).
47. W. W. Rouse Ball, *A Short Account of the History of Mathematics* (Dover, New York, 1960).
48. E. J. Saletan and A. H. Cromer, *Theoretical Mechanics* (John Wiley, New York, 1971).
49. J. A. Schouten, *Ricci Calculus*, 2nd edition (Springer, Berlin, 1954).
50. R. Schmid, *Infinite Dimensional Hamiltonian Systems* (Bibliopolis, Naples, 1987).
51. B. Schutz, *Geometrical Methods of Mathematical Physics* (Cambridge University Press, 1980).
52. M. Spivak, *A Comprensive Introduction to Differential Geometry* (Publish or Perish, Boston, 1970).
53. J. M. Souriau, *Structure des systèmes dynamiques* (Dunod, Paris, 1970).
54. L. Tonelli, *Fondamenti di Calcolo delle Variazioni* (Zanichelli, Bologna, 1921).
55. I. Vaisman, *Lectures on the Geometry of Poisson Manifolds* (Birkhäuser Verlag Basel, 1990).
56. F. W. Warner, *Foundations of Differentiable Manifolds and Lie Groups* (Springer-Verlag, Berlin, 1983).
57. H. Weyl, *Space, Time and Matter*, 4th edition (Dover, New York, 1952).
58. E. T. Whittaker, *A Treatise on the Analytical Dynamics of Particles and Rigid Bodies* (Cambridge University Press, London, 1904, 4th edition, 1959).
59. A. Wintner, *The Analytical Foundations of Celestial Mechanics* (Princeton University Press, Princeton, 1941).

Articles

60. M. J. Ablowitz, D. J. Kaup, A. C. Newell, and H. Segur, *Phys. Rev. Lett.* **30**, 1262 (1973); *Stud. Appl. Math.* **53**, 249 (1974).
61. M. Adams, T. Ratiu, and R. Schmid, *Math. Ann.* **273**, 529 (1986); **275** (1986).

62. M. Adler, "On the trace functional for formal pseudodifferential operators and the symplectic structure for Korteweg–de Vries type equations," *Inv. Math.* **50**, 219 (1979).
63. V. I. Arnold, *Ann. Inst. Fourier Grenoble* **16**(1), 319 (1966).
64. O. Babelon and C. M. Viallet, *SISSA* **54**, EP (1989).
65. C. Bartocci and U. Bruzzo, "Some differential-geometric approach to supermanifolds," *J. Geom. and Phys.* **4**, 391 (1987).
66. S. Benenti, *Lect. Notes in Math.* No. 836, 512 (1980).
67. A. M. Bloch, *Phys. Lett.* **A116**, 353 (1986).
68. A. Bogoyavlenskij, *Comm. Math. Phys.* **180**, 529 (1996).
69. F. Cantrijn and A. Ibort, *Introduction to Poisson Supermanifolds*, preprint, 1989.
70. G. Cassinelli, E. De Vito, and A. Levrero, *The Dynamical Evolution of Quantum Systems in the Adiabatic Approximation*, preprint (Università di Genova, 1996).
71. D. V. Choodnovsky and G. V. Choodnovsky, *Il Nuovo Cimento* **40**, 339F (1977).
72. R. Cirelli and L. Pizzocchero, "On the integrability of quantum mechanics as an infinite dimensional Hamiltonian system," *Nonlinearity* **3** (1990).
73. J. D. Cole, *Q. Appl. Math.* **9**, 225 (1951).
74. M. Crampin, G. Marmo, and C. Rubano, *Phys. Lett.* **A37**, 88 (1983).
75. A. Das and S. Roy, *J. Math. Phys.* **31**, 2145 (1990).
76. A. Das, When-Jui Huang, and S. Roy, *J. Math. Phys.* **32**, 2733 (1991).
77. S. De Filippo, G. Landi, G. Marmo, and G. Vilasi, "Tensor field defining a tangent bundle structure," *Ann. Inst. H. Poinc.* **53**, 205 (1989); "An algebraic description of the electron-monopole dynamics," *Phys. Lett.* **B220**, 576 (1989); G. Landi and G. Marmo, "Algebraic Lagrangian formalism," *SISSA*, Trieste (SISSA 90/90/FM, 1990).
78. S. De Filippo, G. Landi, G. Marmo, and G. Vilasi, "An algebraic characterization of complete integrability for Hamiltonian systems," in *Differential Geometric Methods in Theoretical Physics*, Rapallo (June, 1990), *Lect. Notes in Physics* 375 (Springer-Verlag, 1991).
79. S. De Filippo, G. Marmo, M. Salerno, and G. Vilasi, *On the Phase Manifold Geometry of Integrable Nonlinear Field Theory*, preprint IFUSA (Salerno, 1982), unpublished.
80. S. De Filippo, G. Marmo, M. Salerno, and G. Vilasi, "Phase manifold geometry of Burgers hierarchy," *Lett. Nuovo Cimento* **37**(3), 105 (1983).
81. S. De Filippo, G. Marmo, M. Salerno, and G. Vilasi, "A new characterization of complete integrable systems," *Il N. Cimento* **B83**, 97 (1984).
82. S. De Filippo, G. Marmo, and G. Vilasi, "A geometrical setting for the Lax representation," *Phys. Lett.* **B117**(6), 418 (1982).
83. S. De Filippo and G. Vilasi, "Geometrical methods for infinite dimensional dynamical systems," Proceedings of the Second World Conference on Mathematics (Las Palmas, Spain, 1982), p. 236.
84. S. De Filippo, M. Salerno, and G. Vilasi, *Lett. Math. Phys.* **9**, 85 (1985).

85. L. A. Dikii, *Funkt. Anal. Pril.* **6**(4), 83 (1972).
86. P. A. M. Dirac, "Generalized Hamiltonian dynamics," *Canad. J. Math.* **2**, 129 (1950).
87. P. Di Stasio and G. Vilasi, *Lett. Math. Phys.* **11**, 299 (1986).
88. I. Ya Dorfman, *Deformation of Hamiltonian Structures and Integrable Systems* (1983).
89. V. G. Drinfeld and V. V. Sokolov, *Lie Algebras and Korteweg–de Vries Type Equations*, VINITI, *Contemporary Problems of Mathematics. Recent Developments* (Moscow, 1984) (*in Russian*), pp. 81–180.
90. B. A. Dubrovin, V. B. Matveev, and S. P. Novikov, *Russ. Math. Survey* **31**(1), 59 (1976).
91. A. Einstein, "Die Grundlage der allgemeinen Relativitätstheorie," *Annalen der Physik* **49**, 769 (1916).
92. L. Faddeev, "Symplectic structures and quantization of the Einstein gravitation theory," *Actes du Congrés Internat. Math.* **3**, 35 (1970).
93. L. D. Faddeev and V. E. Zakharov, *Funct. Anal. Appl.* **5**, 18 (1971).
94. L. D. Faddeev and L. A. Takhtajan, *Theor. Math. Phys.* **21**, 160 (1974).
95. A. S. Fokas and B. Fuchsteiner, "On the structure of symplectic operators and Hereditary symmetries," *Lett. Nuovo Cimento* **28**(8), 299 (1980); A. S. Fokas and B. Fuchsteiner, *Nonlinear Anal. TMA* **5**, 423 (1981); A. S. Fokas and R. L. Andersen, "On the use of isospectral eigenvalue problem for obtaining hereditary symmetries for Hamiltonian systems," *J. Math. Phys.* **23**(6), 1066 (1982).
96. A. P. Fordy and P. P. Kulish, "Nonlinear Schrödinger equation and simple Lie algebras," *Comm. Math. Phys.* **89**(4), 423 (1983).
97. A. Frolicher and A. Nijenhuis, *Indag. Math.* **23**, 338 (1956).
98. A. Frölicher and A. Nijenhuis, "Theory of vector valued differential forms I–II," *Nederl. Acad. Westinch. Proc.* **A59**, 338 (1965).
99. B. Fuchsteiner, "The Lie algebra structure of the NEE admitting infinite dimensional abelian symmetry group," *Prog. Theor. Phys.* **65**(3), 861 (1981).
100. B. Fuchsteiner, "The Lie algebra structure of the degenerate Hamiltonian and bi-Hamiltonian system," *Prog. Theor. Phys.* **68**(4), 1082 (1982).
101. B. Florko and A. Yanovski, *On Magri's theorem for complete integrability*, Communication JINR (Joint Institute for Nuclear Research), Dubna, E5-83-831 (1983).
102. C. S. Gardner, S. M. Greene, M. D. Kruskal, and R. M. Miura, *Phys. Rev. Lett.* **19**, 1095 (1967).
103. I. M. Gel'fand and B. M. Levitan, "On the determination of a differential equation from its spectral function," *Isv. Akad. Nauk SSSR Ser. Mat.* **15** (1951); *Amer. Math. Soc. Transl.* **1**, 253 (1955).
104. I. M. Gelfand and I. Ya. Dorfman, "The Schouten bracket and Hamiltonian operators," *Funct. Anal. Appl.* **14**, 71 (1980).
105. V. S. Gerdjikov and P. P. Kulish, "The generating operator for $n \times n$ linear system," *Physica D* **D3**(3), 549 (1981).

106. V. S. Gerdjikov and A. B. Yanovski, "Gauge-covariant formulation of the generating operator" (1. The Zakharov–Shabat system; 2. Systems on homogeneous spaces), *Phys. Lett.* **A103**(5), 232 (1984); **A110**(2), 53 (1985); *Comm. Math. Phys.* **103**, 549 (1986).
107. R. Giachetti, "Hamiltonian systems with symmetry: An introduction," *La rivista del Nuovo Cimento* **4**(12) (1981).
108. J. Grabowski, G. Landi, and G. Marmo, "Generalized reduction procedure: Symplectic and Poisson formalism," *Forthscritte der Physik* **42**(5) (1994).
109. J. Gotay, J. Nester, and G. Hinds, "Presymplectic manifolds and the Dirac–Bergman theory of constraints," *J. Math. Phys.* **19**, 2388 (1978).
110. F. Guerra and L. Morato, *Phys. Rev.* **D27**, 1774 (1983); F. Guerra and R. Marra, *Phys. Rev.* **D28**, 1916 (1983); **D29**, 1647 (1984).
111. J. Haantjes, "On X_m-forming sets of eigenvectors," *Nederl. Acad. Westinch. Proc.* **A58**(2), 158 (1955).
112. W. R. Hamilton, *Trans. R. Irish Acad.* **15**, 69 (1828).
113. W. R. Hamilton, *Phil. Trans. Roy. Soc. Part I* 247 (1834); *Phil. Trans. Roy. Soc. Part II* 95 (1835). In *The Mathematical Papers of Sir W. R. Hamilton* (Cambridge University Press, 1940).
114. J. Hietarinta, *Quantum Integrability and Classical Integrability* (Turku University, 1984).
115. E. Hopf, *Comm. Pure Appl. Math.* **3**, 201 (1950).
116. C. G. J. Jacobi, *Crelle's J.* **17**, 97 (1837); *Liouville's J.* **3**, 60, 161 (1837).
117. R. Jackiw and V. E. Zakharov, *Phys. Rev.* **I**, 133 (1979).
118. D. J. Kaup, *J. Math. Anal. Appl.* **54**, 849 (1976).
119. B. Kay and H. E. Moses, *Il Nuovo Cimento* **3**, 276 (1956).
120. B. G. Konopelchenko and V. G. Dubrovski, "General Nth order differential spectral problem: general structure of the integrable equations, nonuniqueness of the recursion operator and gauge invariance," *Ann. Phys.* **156**(2), 256 (1984).
121. D. J. Korteweg and G. de Vries, *Philos. Mag. Ser. 5* **39**, 422 (1895).
122. Y. Kosmann-Schwarzbach, "Geométrie des systemes bihomiltoniennes", Seminaire de Mathemetiques Superieurs aur les systemes dynamiques nonlineaires, Pub IRMA Lille Vol. 2, I (1986).
123. B. Kostant, "Graded manifolds, graded Lie theory and prequantization," *Differential Geometric Methods in Mathematical Physics*, LNM 570 (Springer-Verlag, Berlin, 1977), pp. 177–306.
124. P. P. Kulish and A. G. Reymann, "Hierarchy of symplectic forms for the Schrödinger equation and for the Dirac equation on the line," Zap. LOMI (Notes of the Leningrad Branch of the Mathematical Institute) **77** (1978) (*in Russian*), pp. 134–147.
125. P. P. Kulish and E. K. Sklyanin, "About the Yang–Baxter equation solutions," Zap. LOMI (Notes of the Leningrad Branch of the Mathematical Institute) **95** (1980) (*in Russian*), pp. 129–160.
126. B. A. Kupershmidt, *Phys. Lett.* **A109**, 136 (1985).

127. G. Landi, "An algebraic setting for gauge theories," Doctor Philosophiae Thesis, SISSA (Trieste, 1988); G. Landi and G. Marmo, "Lie algebra extensions and Abelian monopoles," *Phys. Lett.* **B195**, 429 (1987); Extensions of Lie superalgebras and supersymmetric Abelian gauge fields, *Phys. Lett.* **B193**, 61 (1987); Graded Chern–Simons terms, *Phys. Lett.* **B192**, 61 (1987); Algebraic reduction of the 't Hooft–Polyakov monopole to the Dirac monopole, *Phys. Lett.* **B201**, 101 (1988); Einstein algebras and the algebraic Kaluza–Klein monopole, *Phys. Lett.* **B210**, 68 (1988); Algebraic instantons, *Phys. Lett.* **B215**, 338 (1988); Algebraic differential calculus for gauge theories, *Nuclear Physics B* (Proc. Suppl.) **A18**, 171 (1990).
128. G. Landi, G. Marmo, and G. Vilasi, "An algebraic approach to integrability," *Journal of Group Theory in Physics* **2**, 1 (1994).
129. G. Landi and G. Marmo, "Algebraic Lagrangian Formalism," *SISSA*, Trieste (SISSA 90/90/FM, 1990).
130. G. Landi, G. Marmo, G. Sparano, and G. Vilasi, "A general reduction procedure for dynamical systems," *Mod. Phys. Lett.* **A6**(37), 3445 (1991).
131. G. Landi, G. Marmo, and G. Vilasi, "Remarks on the complete integrability of dynamical systems with fermionic variables," *J. Phys.* **A25**, 4413 (1992).
132. G. Landi, G. Marmo, and G. Vilasi, "Recursion operators: Meaning and existence for completely integrable systems," *J. Math. Phys.* **35**(2), 808 (1994).
133. P. D. Lax, *Comm. Pure Math.* **21**, 467 (1968); **28**, 141 (1975); *SIAM Rev.* **18**, 351 (1975).
134. D. L. Lebedev and J. I. Manin, "Gel'fand-Dikii, Hamiltonian operator and the adjoint representation of the Volterra group," *Funct. An. Appl.* **13**(4), 40 (1979).
135. D. A. Leǐtes, "New Lie superalgebras and mechanics," *Soviet Math. Dokl.* **18**, 1277 (1977).
136. S. Lie, "Theorie der partiallen gleichungen erster Ordnung," *Ann. Math.* Bd **IX**, 1875.
137. J. Liouville, *Acta Math.* **20**, 239 (1897).
138. L. Lusanna, *Phys. Rep.* **185**, 1 (1990); *Riv. Nuovo Cimento* **14**(3), 1 (1991); *Int. J. Mod. Phys.* **A8**, 4193 (1993).
139. F. Magri, "A simple model of the integrable Hamiltonian equations," *J. Math. Phys.* **19**, 1156 (1978).
140. F. Magri, "A geometrical approach to the nonlinear solvable equations," in *Lecture Notes in Physics* (Springer-Verlag, New York, 1980), p. 120.
141. F. Magri and C. Morosi, "A geometrical characterization of integrable Hamiltonian systems through the theory of Poisson–Nijenhuis manifolds," Quaderni del Dipartimento di Matematica (Università di Milano, 1984).
142. F. Magri, C. Morosi, and O. Ragnisco, "Reduction techniques for infinite-dimensional Hamiltonian systems. Some ideas and applications," *Comm. Math. Phys.* **99**, 115 (1985).
143. S. V. Manakov, *Funkt. Anal. Pril.* **10**, 93 (1976).

144. Y. Manin and A. O. Radul, *Comm. Math. Phys.* **98**, 65 (1985).
145. V. A. Marchenko, "On the reconstruction of the potential energy from phases of the scattered waves," *Dokl. Akad. Nauk. SSSR* **104**, 695 (1955).
146. G. Marmo, "A geometrical characterization of completely integrable systems," *Proceedings on Geometry and Physics* (Florence, 1982), p. 257.
147. G. Marmo, "Nijenhuis operators in classical dynamics," *Proceedings of Seminar on Group Theoretical Methods in Physics* (USSR Academy of Sciences, Yurmala, Latvian SSR, 1985).
148. G. Marmo, N. Mukunda, and J. Samuel, "Dynamics and symmetry for constrained systems: a geometrical analysis," *La rivista de Il Nuovo Cimento* **6**, 2 (1983).
149. G. Marmo and C. Rubano, *Il N. Cimento* **B78**, 70 (1983).
150. G. Marmo and E. J. Saletan, *Il Nuovo Cimento* **B40**, 67 (1977).
151. G. Marmo, G. Morandi, and G. Mukunda, "A geometrical approach to the Hamilton–Jacobi form of dynamics and its generalizations," *La rivista del Nuovo Cimento*, No. 8 (1990).
152. G. Marmo and G. Vilasi, "When do recursion operators generate new conservation laws?" *Phys. Lett.* **B277**, 137 (1992).
153. G. Marmo and G. Vilasi, "Symplectic structures and quantum mechanics," *Mod. Phys. Lett.* **B10**(12), 545 (1996).
154. G. Marmo, A. M. Vinogradov, and G. Vilasi, "The local structure of n-Jacobi and n-Poisson manifold," *J. Geom. Phys.* **25**(1/2) (1997).
155. J. E. Marsden and T. Ratiu, "Reduction of Poisson manifolds," *Lett. Math. Phys.* **11**, 161 (1986).
156. A. S. Mischenko, *Funkt. Anal. Pril.* **4**(3), 73 (1970).
157. R. M. Miura, C. S. Gardner, and M. D. Kruskal, *J. Math. Phys.* **9**, 1204 (1968).
158. R. M. Miura, *SIAM Rev.* **3**, 412 (1976); C. Cercignani, *Riv. Nuovo Cimento* **7**, 429 (1977); M. J. Ablowitz, *Lect. Appl. Math.* **58**, 17 (1978); L. D. Faddeev and V. E. Korepin, *Phys. Rep.* **C42**, 1 (1978); V. E. Zakharov and S. V. Manakov, *Phys. Rev.* **1**, 133 (1979).
159. G. Morandi, C. Ferrario, G. Lo Vecchio, G. Marmo, and C. Rubano, "The inverse problem in the calculus of variations and the geometry of the tangent bundle," *Phys. Rep.* **188**, 147 (1990).
160. N. N. Nehorošev, "Action-angle variables and their generalizations," *Trans. Moscow Math. Soc.* **26**, 180.
161. A. Nijenhuis, *Indag. Math.* **13**, 200 (1951); **15**, 233 (1953); **16**, 17 (1954).
162. A. Nijenhuis, *Indag. Math.* **49**, 2 (1987).
163. M. A. Olshanetsky and A. M. Perelomov, *Phys. Rep.* **71**, 313 (1981).
164. J. P. Olver, *J. Math. Phys.* **18**, 1212 (1977).
165. R. S. Palais, "The symmetries of solitons," *Bulletin of the American Mathematical Society*, **34**(4), 339 (1997).
166. H. Poincaré, *Acta Math.* **13**, 1 (1899).
167. T. Ratiu, *Indiana Univ. Math. J.* **29**(4), 609 (1980).

168. H. J. Reshetihin and L. D. Faddeev, "Hamiltonian structures of the integrable field theory models," *Theoretical and Mathematical Physics* **56**(3), 323 (1983).
169. A. Rogers, "A global theory of supermanifolds," *J. Math. Phys.* **21**, 1352 (1980); "Graded manifolds, supermanifolds and infinite-dimensional Grassmann algebras," *Comm. Math. Phys.* **105**, 375 (1986).
170. A. G. Reiman and M. A. Semenov Tyan-Shanskii, *Invent. Math.* **54**, 81 (1979). A. G. Reymann and M. A. Semenov-Tian Shanski, "Families of Hamiltonian structures, hierarchies of Hamiltonian functions and reductions for first order matrix-valued differential operators," *Funct. Anal. Appl.* **14**(2), 77 (1980).
171. M. Rothstein, "The axioms of supermanifolds and a new structure arising from them," *Trans. AMS* **297**, 159 (1986).
172. A. Scott, F. Chu, and D. Mc Laughlin, *Proc. IEEE* **61**, 1443 (1973).
173. M. A. Semenov-Tian-Shanski, "What is the classical r-matrix," *Func. Anal. Appl.* **17**(4), 17 (1973).
174. V. A. Soroka, *Lett. Math. Phys.* **17**, 201 (1989).
175. R. Sorkin, *J. Phys.* **A10**, 717 (1977).
176. A. P. Stone, "Higher order conservation laws," *J. Diff. Geometry* **3**, 447 (1969).
177. A. P. Stone, "Some remarks on the Nijenhuis tensor," *Canad. J. Math.* **25**(5), 903 (1973).
178. F. Strocchi, *Rev. Mod. Phys.*, Vol. 4 (1966).
179. K. Sundermeyer, "Constrained dynamics," *Lecture Notes in Physics*, Vol. 169 (1982).
180. W. W. Symes, *Physica* **D1**, 339 (1980).
181. G. Vilasi, "On the Hamiltonian structure of KdV and sine–Gordon theories," *Phys. Lett.* **B94**(2), 195 (1980).
182. G. Vilasi, "Invariant endomorphism and conserved functionals of the Liouville equation," *Phys. Lett.* **B174203** (1986).
183. G. Vilasi, "Recursion operator and Γ-scheme for Kepler Dynamics," *Conference Proceedings*, Vol. 48, *National Workshop on Nonlinear Dynamics*, Società Italiana di Fisica (Bologna, 1995).
184. A. M. Vinogradov, "Local symmetries and conservation laws," *Acta Aplicandae Mathematicae*, No. 2 (1984), pp. 21–78.
185. A. M. Vinogroolov and B. A. Kupershmit, "The Structures of Hamiltonian Mechanics", *Russian Math. Surveys* **32:4**, 177–243 (1977).
186. D. V. Volkov, A. I. Pashnev, V. A. Soroka, and V. I. Tkach, *JETP Lett.* **44**, 70 (1986).
187. E. Witten, *Nucl. Phys.* **B188**, 513 (1981).
188. V. E. Zakharov and B. G. Konopelchenko, "On the theory of recursion operator," *Commun. Math. Phys.* **94**, 483 (1984).
189. V. E. Zakharov and S. V. Manakov, "About the complete integrability of nonlinear Schrödinger equation," *Theoretical and Mathematical Physics* **19**(3), 332 (1974); *Sov. Scient. Rev.* **1**, 133 (1979).

190. V. E. Zakharov and A. B. Shabat, *Sov. Phys. JEPT* **34**, 62 (1972); F. Lund and T. Regge, *Phys. Rev.* **114**, 1524 (1976); A. S. Bugadov and L. A. Takhtajan, *Sov. Phys. Dokl.* **22**, 428 (1977); M. Chaichain and P. Kulish, *Phys. Lett.* **B78**, 413 (1978); V. E. Zakharov and A. V. Minkailov, *Sov. Phys. JETP* **47**, 1017 (1978); V. E. Zakharov and A. B. Shabat, *Funct. Anal. Appl.* **13**, 166 (1979); R. D'Auria, T. Regge, and S. Sciuto, *Phys. Lett.* **B89**, 363 (1980).
191. V. E. Zakharov and L. A. Takhtadjan, "Equivalence between nonlinear Schrödinger equation and Heisenberg ferromagnet equation," *Theoretical and Mathematical Physics* **38**(1), 26 (1979).

Index

www.ingramcontent.com/pod-product-compliance
Ingram Content Group UK Ltd.
Pitfield, Milton Keynes, MK11 3LW, UK
UKHW020403150926
8624IPUK00004B/319

9 789810 233082